Fearing Food: Risk, Health and Environment

Fearing Food: Risk, Health and Environment

edited by Julian Morris and Roger Bate

OXFORD AUCKLAND BOSTON JOHANNESBURG MELBOURNE NEW DELHI

Butterworth-Heinemann
Linacre House, Jordan Hill, Oxford OX2 8DP
225 Wildwood Avenue, Woburn, MA 01801-2041
A division of Reed Educational and Professional Publishing Ltd

A member of the Reed Elsevier plc group

First published 1999

British Library Cataloguing in Publication Data
A catalogue record for this book is available from the British Library

Library of Congress Cataloguing in Publication Data
A catalogue record for this book is available from the Library of Congress

ISBN 0 7506 4222 X

Typeset by Avocet Typeset, Brill, Aylesbury, Bucks
Printed and bound in Great Britain by Biddles Ltd, Guildford and King's Lynn

Contents

Executive summary

Environmentalists have for a long time attacked the use of pesticides, fertilisers and other aspects of intensive farming. Yet, as Dennis Avery shows, intensive farming has enabled growth in food production at a rate greater than population growth, thereby ensuring that people are better fed than ever before. In addition, as Bruce Ames and Lois Gold demonstrate, synthetic pesticides enable the production of larger quantities of fresh fruit and vegetables, which means that people are better protected against cancer. Moreover, the synthetic pesticides themselves are often less toxic than natural pesticides. Overall, synthetic pesticides present a net gain in health terms.

Alarmists are inaccurately claiming that antibiotic resistance in animals is spreading to humans and they demand bans on animal antibiotic use. Roger Bate explains that using antibiotics in young animals keeps meat prices low and does not contribute to antibiotic resistance in humans.

Jean-Louis L'hirondel argues that nitrate fertilisers, far from being a threat to human health, are in fact beneficial. Nevertheless, overuse of fertiliser can be harmful to the environment – causing eutrophication of lakes and streams. However, command-and-control measures are not an efficient way of controlling this 'non-point source' pollution; as Bruce Yandle shows, marketable permits are much better suited to the job.

Michael Wilson and his co-authors suggest that many environmental problems associated with agriculture can be reduced by using genetically modified organics (GMOs), which have the potential to improve yields and quality whilst simultaneously reducing associated inputs (including fertiliser, herbicide and pesticide).

Indur Goklany argues that new technologies offer the possibility of producing more food on the same amount of land, thereby reducing the amount of land that will be converted from wildlife habitat.

Barrie Craven and Christine Johnson argue that instances of food poisoning in the UK may have been exacerbated by over-cautious government regulation and he critically assesses the creation of the UK Food Standards Agency: will it be yet another addition to an already bloated bureaucracy?

David Warburton and his co-authors demonstrate that dietary guidelines from around the world are inconsistent. This is shown to be a result of the fact that they are typically not justified by a balanced assessment of the scientific evidence. They are therefore often non-productive and in some cases are even positively harmful.

Linda Whetstone shows that the original aims of the Common Agricultural Policy are not being fulfilled. Subsidies to farming, intended to ensure food and job security, have resulted in over-production and overspending. Inadvertently, the CAP has caused environmental degradation and undermined farming itself.

Lynn Scarlett argues that packaging enhances the shelf life of food and reduces wastage during transport. Importing food allows society to take advantage of different environmental and socio-economic conditions that exist in different countries, benefiting people in both importing and exporting nations.

Peter Bowbrick exposes five famine fallacies. Although political unrest undoubtedly exacerbates food shortages, speculation and over-consumption by certain groups do not cause famines. It is primarily a decline of available food, which policy-makers need to address by food imports, that is the cause of famine.

Rod Hunter provides a critical assessment of the European regulations affecting the research, development and distribution of genetically modified organisms.

Julian Morris discusses the role of information and legal duties on the provision of food that is produced in an environmentally sound manner and is safe for consumption. He argues that in general private law provides better protection of the environment and of consumers than do statutory regulations.

Biographies

Bruce N. Ames
Dr Ames is Professor of Biochemistry and Molecular Biology and Director of the National Institute of Environmental Health Sciences Center, University of California, Berkeley. He is a member of the National Academy of Sciences and was on their Commission on Life Sciences. He was a member of the National Cancer Advisory Board of the National Cancer Institute (1976–82). His many awards include: the General Motors Cancer Research Foundation Prize (1983), the Tyler Prize for environmental achievement (1985), the Gold Medal Award of the American Institute of Chemists (1991), the Lovelace Institutes Award for Excellence in Environmental Health Research (1995), the Honda Foundation Prize for Ecotoxicology (1996) and the Japan Prize (1997). His 380 publications have resulted in his being the 23rd most-cited scientist (in all fields) (1973–84).

Dennis Avery
Dennis T. Avery is a senior fellow of Hudson Institute and Director of Hudson's Center for Global Food Issues. He is an internationally recognised expert on agriculture – especially on the ways in which technology and national farm policies interact to affect farm output.

Before joining Hudson, Mr Avery served for nearly a decade (1980-88) as the senior agricultural analyst for the US Department of State, where he was responsible for assessing the foreign-policy implications of food and farming developments worldwide.

At Hudson, Mr Avery continues to monitor developments in world food production, farm product demand, the safety and security of food supplies, and the sustainability of world agriculture. Mr Avery has written several books and has contributed articles to *Science* and other learned journals. He is a frequent contributor to the Hudson Institute's American Journal and a recent article in that journal stimulated nationwide debate over the safety of organic food.

Mr Avery grew up on a Michigan dairy farm and studied agricultural economics at Michigan State University and the University of Wisconsin. He holds awards for outstanding performance from three different government agencies and was awarded the National Intelligence Medal of Achievement in 1983.

Roger Bate

Roger Bate founded the Environment Unit at the Institute of Economic Affairs in 1993 and co-founded the European Science and Environment Forum with Dr John Emsley and Professor Frits Böttcher in 1994. Mr Bate is the editor of 'What Risk?' (Butterworth Heinneman, 1997), a collection of papers that critically assess the way risk is regulated in society. He has also written several scholarly papers and numerous shorter scientific articles, for newspapers and magazines, including the Wall Street Journal, the *Financial Times*, *Accountancy* and *LM Magazine*. His most recent book is a co-edited collection of papers 'Environmental Health: Third World Problems, First World Preoccupations' (Butterworth Heinemann, 1999).

Peter Bowbrick

Dr Bowbrick took his first degree in economics and postgraduate degrees at Oxford University (Dip. Ag. Econ.), Nottingham University and Henley. He has published widely, mostly on food marketing issues. He has been research assistant at Cambridge University, visiting academic then research associate at Oxford University, research fellow at Henley Management College, and senior research officer in the Irish food research institute.

His forte is putting theory into practice in the real world. He has worked in more than 25 countries in Africa, Asia, the Caribbean, Eastern Europe and Western Europe, and indeed has spent most of his life in the Third World. He has worked on projects for Food and Agricultural Organisation, the World Bank, the Asian Development Bank, UN Development Programme, the three branches of the EU and on bilateral projects including US Agency for International Development and GTZ (German Technical Assistance). His experience includes the administration of food and agricultural policy as a civil servant, as well as normal consultancy work.

Barrie Craven

Dr Craven is reader in economics at the University of Northumbria. He is a graduate of the Universities of Hull and Newcastle-upon-Tyne. He has published in the field of monetary economics in the *Journal of Monetary Economics*, the *Manchester School* and the *European Journal of Finance*. More recently, he has researched public policy and health care issues. Current research is focused on the resource strategies associated with AIDS where results are published in several journals including the *Journal of Public Policy* and *Financial Accountability and Management*. He has taught at Curtin University, Western Australia and at California Polytechnic, USA.

Indur Goklany

Dr Goklany obtained his PhD in Electrical Engineering from Michigan

State University and has over twenty-five years experience addressing both science and policy aspects of environmental and natural resource issues in state and federal government and in the private sector. At EPA's Chicago region office, he developed particulate matter and sulphur dioxide control strategies, atmospheric modelling and regulations for the US's industrial heartland. Subsequently, as Chief, Technical Assessment Division, National Commission on Air Quality, he switched his attention to investigating national impacts of pollution and its control. After a period as a Washington-based consultant on energy and environmental issues, he helped develop EPA's first ever new source emission trade (bubble), for which he received an EPA bronze medal. Working with EPA's Regulatory Reform Staff, he was also responsible for helping EPA adopt the emissions trading policy statement in the mid-1980s.

At the Department of the Interior, where he is Manager, Science and Engineering, Office of Policy Analysis, he served on various national and international panels and groups dealing with global climate change and acid rain before focusing on issues related to population, biodiversity and natural resources. He has published extensively in various peer-reviewed journals on air pollution, climate change, biodiversity and the role of technology in creating as well as solving environmental problems. Being Indian-born and bred, he is keenly aware that one cannot solve Third World problems with First World sensibilities.

Dr Goklany's contribution to this book reflect his personal views rather than those of any organisation he has associated with, past or present.

Lois S. Gold
Dr Lois Swirsky Gold is Director of the Carcinogenic Potency Project at the National Institute of Environmental Health Sciences Center, University of California, Berkeley and a Senior Scientist at the Berkeley National Laboratory, Berkeley. She has published 80 papers on analyses of animal cancer tests and implications for cancer prevention, interspecies extrapolations and regulatory policy. The Carcinogenic Potency Database (CPDB), published as a CRC handbook, analyses results of 5000 chronic, long-term cancer tests on 1300 chemical. Dr Gold has served on the Panel of Expert Reviewers for the National Toxicology Program, the Boards of the Harvard Risk Management Group.

Jean-Louis L'hirondel
Dr L'hirondel has been a practising specialist in rheumatology at the Regional and Hospital Centre at Caen, France since 1974 and a former intern of the Hospitals of Paris. He has written widely on nitrates and their toxicity to man, co-authoring the paper, 'Man and Nitrates: the toxic myth' with his father, Dr Jean L'hirondel, who was Professor of Clinical Paediatrics at Caen from 1962 to 1982. Since his father's death

in 1995, Dr Jean-Louis L'hirondel has continued his professional interest in nitrates.

Rod Hunter

Rod Hunter is a partner in the Brussels office of the international law firm Hunton and Williams. His practice encompasses advice on regulatory, trade and transactional matters. His transactional practice has involved environmental advice to clients on matters involving industrial facilities throughout Europe and the United States.

Mr Hunter also serves as Director of Regulatory Studies at the Centre for the New Europe (CNE), a Brussels-based think tank. In 1998, he was awarded a Japan Society Public Policy Fellowship to study how Japan develops and applies regulations.

Before joining Hunton & Williams, Mr Hunter served during 1988 as associate to the Chief Justice of Australia, Sir Anthony Mason. Prior to that, he clerked for a year for US Circuit Judge Boyce Martin, Jr., of the Sixth Circuit Court of Appeals. He served during 1982 and 1983 in Washington, D.C., as an aide to US Senator John W. Warner, Jr. While in law school, Mr Hunter was a member of the *Virginia Law Review's* Managing Board (Notes Editor).

Mr Hunter's articles have appeared in, amongst others, the *Wall Street Journal*, the *National Law Journal*, the *International Environment Reporter*, and the *European Environmental Law Review*. He is also co-author of the *European Community Deskbook* (Environmental Law Institute 1992; 2nd edition forthcoming 1998).

Christine Johnson

Christine Johnson is a medical editor and a freelance science journalist from Los Angeles. Her primary interest is the prognostic value of medical diagnostic tests and their use in measuring and assessing health risks. Her work has been translated into four languages and has appeared in *Christopher Street, Spin, Continuum* and *Medicina Holistica*.

Julian Morris

Julian Morris graduated from Edinburgh University with an MA in economics and holds Masters degrees from University College London (in environment and resource economics) and Cambridge University (in land economics). Following a period as an econometric forecaster at Commerzbank in Frankfurt (where he purportedly predicted the unpredictable), Julian became a Research Fellow at the Institute of Economic Affairs in 1993.

Julian is the author of several papers and books on environmental issues, ranging from land degradation and climate change to ecolabels and environmental management. His articles and book reviews have appeared in *The Sunday Times*, the *Wall Street Journal Europe*, *Nature*, *Economic Affairs* and various other journals.

In July 1998 Julian became Co-Director of the IEA Environment Unit, where he continues to think the unthinkable. The views expressed by Julian in his chapter are his alone and not of the IEA (which has no corporate view), its Advisors, Directors, or Trustees.

Lynn Scarlett
Lynn Scarlett is executive director of the *Reason Public Policy Institute* and vice president for research at the Reason Foundation, a Los Angeles-based public policy research organisation. Her own research focuses primarily on environmental issues, with a particular emphasis on solid and hazardous waste, recycling, urban air quality, environmental risk issues, and market-based environmental policy.

Scarlett is the author of numerous articles on environmental policy, privatisation, and local economic development published in both academic and general-audience publications. Her general-audience articles have appeared in the *Wall Street Journal*, the *Los Angeles Times*, *Reader's Digest* and numerous other newspapers and magazines. She has appeared frequently on national television and radio talkshows to discuss environmental policy.

Since 1993, Ms Scarlett has served as Chair of the 'How Clean Is Clean' Working Group of the Washington, D.C.-based National Environmental Policy Institute, founded by former Congressman Don Ritter (R-PA). She is also a member of the Enterprise for the Environment Task Force, a year-long project on environmental policy reform chaired by former EPA Administrator William Ruckelhaus.

In 1994, Ms Scarlett was appointed by Governor Pete Wilson to chair The California Interim Inspection & Maintenance Review Committee, charged with making recommendations on revising the state's vehicle smog check program. She was reappointed in 1996 to chair the permanent committee. She is currently part of a special consulting team to California's Department of Conservation charged with evaluating the state's container recycling laws.

David M. Warburton
Professor Warburton is Director of the Human Psychopharmacology Group at the University of Reading – one of the premier centres in the world for the study of drugs and human performance. He is the founder of ARISE (Associates for Research into the Science of Enjoyment) and he researches the Psychobiology of Pleasure from Alcohol, Chocolate, Coffee and Nicotine. He has a BSc from the University of London and an PhD (with Highest Distinction) at Indiana University which he attended on an English Speaking Union Scholarship and a Fulbright Grant. He is also a Fellow of the British Psychological Society and Chartered Psychologist.

Linda Whetstone
Linda Whetstone has an economics degree from London University. She wrote two Research Monographs for the Institute of Economic Affairs on milk marketing and the market for animal semen and contributed to the Omega Publications of the Adam Smith Institute, as well as writing articles and taking part in discussions on radio and television. Her particular interest is agricultural subsidies and she has had a lifelong involvement with farming. She is Trustee of the IEA and a Director of Atlas UK. The views she expresses in her chapter are hers alone and in no way reflect those of the IEA (which has no corporate view) or of Atlas (which similarly has no corporate view).

Michael M. Wilson
Professor Wilson is currently Science Director of Horticultural Research International, having previously been Acting Director of the Scottish Crop Research Institute. He is recognised both nationally and internationally for his fundamental discoveries on the molecular mechanisms responsible for the release and activation of viral RNA during the early events of plant cell infection; he first introduced the concept of co-translational disassembly of virus particles. Subsequent findings concerned the identification and exploration, for both fundamental plant virus research and strategic biotechnological purposes, of a translational enhancer sequence and an RNA packaging sequence from a plant virus. Both have generated wide commercial interest and aided basic mechanistic studies on resistance to viruses in crops genetically engineered to express a viral coat protein. For these innovative discoveries with generic utility, he was awarded the 1984 Herpes Vaccine Research Trust Prize by the Society for General Microbiology, a unique honour for a plant molecular virologist.

During 3 years in the USA as a Professor of Plant Pathology and founder member of the New Jersey Advanced Technology Center for Agricultural Molecular Biology, his group created the first genetically transformed, herbicide tolerant, recreational turfgrass for the US Golf Association. More conventional, fundamental molecular studies on a number of important plant viruses have been described in over 90 refereed papers, as well as several invited reviews and chapters in high quality publications and an equal number of Abstracts. Proceedings and Invited Symposium Presentations. He has co-edited two major books on molecular plant virology and virus genetic engineering, and was invited to guest-edit an important collection of 'Seminars' on the early events of plant and animal virus infection cells. He has held several Editorial posts on international journals of virology, and is (co-) inventor on five suites of biotechnology patents.

From 1995 to 1997, Professor Wilson was Senior Editor of the *Journal of General Virology*. From 1988 to 1995, he was Associate Editor and from 1990 to 1993 Senior Editor of *Molecular Plant-Microbe Interactions*.

From 1993 until 1998, Professor Wilson was a Committee Member of the Church of Scotland, Society, Religion and Technology Project entitled, 'Ethics of Genetic Engineering of Non-Human Life'. He holds honorary professorships at Dundee University and Zhejiang Academy of Agricultural Sciences, China.

Bruce Yandle

Bruce Yandle is Alumni Professor of Economics and Legal Studies at Clemson University and is a member of the Board of Advisors of Political Economic Research Center in Montana. His writings on environmental policy include many journal articles, special reports and monographs. He is author of *The Political Limits of Environmental Regulation* (Quorum 1989). Yandle's current research focuses on the rise of markets for environmental goods. Yandle received his PhD in economics from Georgia State University and served as Executive Director of the Federal Trade Commission from 1982 to 1984.

Introduction

Julian Morris

Environmental and consumer organisations have for a long time attacked the use of pesticides, fertilisers and other aspects of intensive farming on the grounds that these technologies are bad for the environment and bad for our health. However, their arguments have typically been informed by a poor or partial understanding of the scientific evidence regarding the impact of these technologies. In addition, their arguments rarely if ever take account of the offsetting benefits to health and the environment that accrue from the use of these technologies. As a result, consumers have become unnecessarily worried about the impacts of new technologies – the most recent example of which is 'genetic modification' – and government officials have been cajoled into implementing unnecessary and often counterproductive regulations.

The not so silent spring

The current fear of modern agricultural and food technologies, such as pesticides, preservatives and genetic modification, can be traced in large part to the publication in 1962 of Rachel Carson's *Silent Spring*. Although several of the chapters in Carson's book had appeared earlier in the *New Yorker* magazine and there had also been a scare in 1959 over the use of the herbicide aminotriazole on cranberries (ACSH, 1998, p. 6; Wildavsky, 1995, p. 11), it was the widespread coverage of *Silent Spring* and its publication outside the US that really began this fear of food (Whelan, 1993, p. 96)., Carson, an eloquent writer whose previous books on marine biology (her academic speciality) had sold well, asserted that the widespread use of DDT (dichlorodiphenyltrichloroethane) and other pesticides, as well as herbicides and fungicides, was resulting in the death of many birds, and that before too long might even lead to their extinction. The evidence that Carson provided was mainly circumstantial: numbers of certain birds, especially predators such as eagles and peregrine falcons, appeared to have declined over the period since synthetic pesticides had been introduced. Carson's explanation for this phenomenon was that earthworms living in fields sprayed with DDT would contain relatively high levels of the pesticide, and that birds eating them would concentrate it further. She argued that at such concentrations DDT was interfering in the production of eggshells, causing them to be thinner and thereby increasing the chances that the eggs would be destroyed during brooding.

Whilst some subsequent research indicates that DDT may have caused thinning of the eggshells of some birds (Wildavsky, 1995, pp. 71-72), the apparent correlation between DDT use and die-off of certain bird species was largely spurious. Bird populations had either been declining for other reasons since before the introduction of DDT or began to decline long after DDT's introduction. Analysis of records kept by the Audubon Society (a US organisation similar to the Royal Society for the Protection of Birds) shows that, during the period of most intensive use of DDT populations of at least 26 different kinds of bird were increasing. Perhaps most telling is the fact that the population of bald eagles in the US increased from about 200 in 1941, six years before the widespread use of DDT, to nearly 900 in 1960 (Lieberman and Kwon, 1998, p.8). The population of migrating ospreys observed at Hawk Mountain, Pennsylvania, likewise increased from 191 in 1946 to 630 in 1972 (ibid.). In Britain, the decline in peregrine falcons, which Carson had specifically blamed on DDT, ceased in 1966, when DDT was still in widespread use (ibid.), and a study carried out by the British government's Advisory Committee on Toxic Chemicals concluded that "There is no close correlation between the decline in population of predatory birds, particularly the peregrine falcon and the sparrow hawk, and the use of DDT" (ACTC, 1969).

The obsessive focus of early environmental activists on the apparent correlation between DDT use and bird die-off led to restrictions on the use of DDT in Britain, America and elsewhere. The immediate and direct consequence of this was an increase in the use of alternative pesticides, including other organochlorine compounds, such as aldrin and dieldrin, as well as organophosphate (OP) compounds, such as malathion and parathion. This led, in Arizona and California, to a decline in the number of honeybees, which are more sensitive to these alternatives than they are to DDT, and consequently to a decline in the productivity of crops that require honeybees for pollination (Wildavsky, 1995, p. 74). A 1969 Environmental Protection Agency-imposed ban on the use of DDT to control the tussock moth led to severe infestations of the magnificent Douglas fir trees in northwestern USA. Foresters attempted to use alternative pesticides, but those that were available at the time simply did not do the job. Between 1969 and 1974, when the US EPA permitted a limited application of DDT to control the moths, thousands of Douglas fir trees had already died as a result of moth infestations (ibid.). The other organochlorine and OP pesticides are also far more toxic to mammals than is DDT and it is probable that many people have suffered injury or died as a result of the restriction on use of DDT (see below).

For pro-environment, read against modern technology

In his introduction to the British edition of Carson's book, Julian Huxley wrote, "The present campaign for mass chemical control, besides being fostered by the profit motive, is another example of our exaggeratedly technological and quantitative approach." I shall refrain from commenting on the pejorative description of the profit motive and the utterly spurious militaristic implications of talking about a campaign for mass chemical control because these are merely incidental. The real issue is Huxley's assumption that we should use less technology and rely less on quantitative analysis, an assumption that is simultaneously astounding and typical of the approach taken by environmentalists then and since. And, like future generations of environmentalists, Huxley is essentially hypocritical, for he is just as technocratic and quantitative as the farmers he berates. He wants to achieve the 'optimum conservation of resources' and sees this as being more likely if 'the ecological approach' is taken than if synthetic chemicals are used. But how are we to know whether or not we have achieved the optimum conservation of resources if we are not using some quantitative measure? Moreover, is 'the ecological approach' not itself a technology?

Such irrational attacks on modern agricultural technologies are neither informative nor helpful. Indeed by making assumptions about the cause of what is perhaps a serious problem without carefully considering alternative explanations, it is possible that regulations may be put in place that will have the opposite effect to that intended.

A danger to mankind?

Whilst Carson's primary focus was the silent spring that would result from the extinction of bird species, she also claimed that similar processes would soon begin to affect humans. "For the first time in history", she wrote, "every human being is subjected to contact with dangerous chemicals from the moment of conception until death" (Carson, 1963, p.13). By 'dangerous' Carson evidently meant synthetic – produced by man. Human beings have of course been subjected to contact with dangerous natural chemicals since they first walked the earth: trace amounts of chemicals such as hydrogen (which can be highly explosive), carbon monoxide (which can be poisonous), and (in rather larger quantities) dihydrogen monoxide (which can lead to asphyxiation) have been ever present. What Carson failed to add was that these chemicals, although in principle dangerous, are mostly, like their natural counterparts, entirely harmless to humans at the concentrations present in the atmosphere and waterways. She made the then common assumption that there is no safe dose for any carcinogen. This has subsequently been shown to be erroneous: the human immune system is able effectively to dispense with low doses of most carcinogens (Wilson, 1997). To paraphrase Paracelsus: the dose makes the poison.

The ironic truth is that people in developed countries such as the US and UK had been subject to truly dangerous man-made chemicals, such as soot and hydrochloric acid in the atmosphere and lead in the water supply, for many years, but ambient concentrations of these substances had been falling or were beginning to fall in most parts of Britain and the US by the time Carson came to write her tome (US EPA, 1992). Nevertheless, there were and are still many places in the world where air and water pollution cause problems. But in the developed world things have become considerably better over the past half-century (Haywood and Jones, 1999). In part this is no doubt due to the imposition of environmental regulations, although much has occurred simply because it is more efficient to produce less waste and pollution (Morris, 1998).

Many environmental regulations are the result of lobbying by environmental organisations. This is true even for the regulations that were put in place during the 19th century (Ashby and Anderson, 1981). Because of their origin, these regulations have often not been the most efficient means of dealing with a particular problem (Buchanan and Tullock, 1975). Nevertheless, it is probable that most of the environmental regulations that were imposed until the 1960s provided benefits that exceeded their costs. However, since *Silent Spring* there has been an increasing tendency for environmental organisations to lobby for regulations limiting substances that may not be doing any harm, or the harmful effects of which are more than offset by the benefits they provide. For example, as Bruce Ames and Lois Swirsky Gold point out in Chapter 2, pesticides can actually be beneficial for health, in that they can increase crop yields and thereby reduce the cost of fresh fruit and vegetables. Because fruit and vegetables contain essential vitamins and micro-nutrients, they help boost the body's immune system, enabling it to ward of disease (Ames, 1998). So even though there may be some potential negative effects resulting from eating pesticide residues in food, these are more than compensated for by the benefits in terms of greater availability and lower cost of these immunity-boosting fresh fruit and vegetables. These benefits accrue especially to the poorest in society who would otherwise not be able to buy as much fresh fruit and vegetables.

Moreover, there may be other trade-offs that should be considered before drastic action is recommended. For example, in spite of Rachel Carson's criticism of organophosphate (OP) pesticides – some of which, as she points out, are deadly not only to insects but also to man – the major response to Carson's book and subsequent lobbying by 'consumer' activists such as Ralph Nader, was to restrict and ultimately to ban the use of DDT in all developed and now many developing countries. This led to an increase in the use of OPs, such as parathion and malathion, and almost certainly therefore indirectly caused death and serious injury to many human beings. (OPs prevent the breakdown of the neurotransmitter acetylcholine by destroying the cholinesterase which performs this function, so the central nervous system becomes flooded with acetylcholine). In sufficient quantities, OPs can cause severe damage to the central nervous system, leading to permanent injury and even death. OPs may also be responsible for prion-related diseases such as BSE and nvCJD. Carson (1963, p. 24) herself notes several horrific cases of parathion poisoning leading to death and serious injury. Of course, Carson might counter that all organochlorine and OP pesticides should be banned. But this would also have had adverse consequences. Few alternatives to these chemicals were available at the time DDT was banned and these alternatives tended to be less effective and more expensive (Wildavsky, 1995, pp. 72-80). Dennis Avery (Chapter 1) points out that the use of synthetic pesticides, fertilisers and new crop types has meant that productivity has risen faster than the growth in population. Indur Goklany (Chapter 13) shows that this has resulted in a significant decrease in the number of people who are malnourished and the number of people living in areas with reported famine dropped from an average of 700 million in 1950-56 to 35 million in 1992. If the least expensive pesticides had been banned, how many more people might have died?

Another consequence of the subsequent widespread ban on use of DDT has been the reduction in application of DDT in malaria areas. DDT proved to be the most effective means of killing the malaria mosquito and has saved many millions of lives. In Sri Lanka alone, DDT reduced the number of cases of malaria from over two million in 1948 to 17 cases in 1963 when DDT use was abandoned; it has now returned to over two million cases per year (DeGregori 1998; Kemm, 1999:5). Whilst it is true that DDT resistance was beginning to emerge at the time of the ban, its premature withdrawal almost certainly led to the death of several million people. (In any case, resistance to DDT may be only temporary. Roberts et al. (1997) compared the incidence of malarial in various South American countries in 1993 and 1995 and found that those countries that have recently discontinued their use of DDT have experienced increases in the incidence of malaria. But Ecuador, which increased its use of DDT, experienced a significant (approximately 60 per cent) decline in malaria).

The question is whether these effects were a price worth paying for reducing the use of a chemical that was having no observable effect on man and a questionable impact on birdlife?

The organochlorine war continues

After the DDT fiasco, things on the organochlorine front quietened down for a while. Then, in 1969, a rodent test indicated that one of the most popular organochlorine herbicides, 2,4,5-trichlorophenoxyacetic acid (2,4,5-T), caused birth defects in mice. The putative cause of this effect was the presence in 2,4,5-T of dioxin (the generic name given to a particular group of organochlorines) and in particular the highly toxic 2,3,7,8-tetrachlorodibenzo-*p*-dioxin, or TCDD, which is known to have teratogenic (causes birth defects) and carcinogenic effects. The EPA banned 2,4,5-T for use on domestic food crops. In a subsequent review by the National Academy of Science, it transpired that in the test in which the birth defects had been observed, the 2,4,5-T had been contaminated with much higher levels of TCDD than were present in production batches. At the levels present in the form of residue on food, 2,4,5-T was almost certainly not carcinogenic. The NAS recommended overturning the ban. However, following criticism from the consumer activist Ralph Nader and the science journalist Nicholas Wade, the head of the EPA decided to keep the ban (Lieberman and Kwon, 1998, p. 28)

These examples illustrate how environmental and consumer groups have selectively utilised the results of relatively high-dose animal experiments to justify their claims that organochlorines cause disease in man at the low doses which we might consume them in our food and at which we might be exposed to them in the environment. The evidence for such impacts, as attested above, is poor at best and relies upon an assumption that there is a linear relationship between dose and response. Whilst it may have been scientifically acceptable in the 1960s to assert that there is no safe dose for carcinogens, this is not so today: we now have empirical evidence to the contrary (Wilson, 1997). Radioactive substances such as radon are carcinogenic at high doses but seem to be beneficial at low doses (Cohen, 1995). Similarly, very high doses of Vitamin A are carcinogenic (Jukes, 1974, p.14), whilst low doses are clearly beneficial. However, in spite of our improved understanding of the science, environmentalists have continued with their campaign against organochlorines, focusing on ever more minuscule levels of exposure to ever less hazardous substances.

A current favourite of environmentalists is the supposed adverse health effects of polyvinyl chloride (PVC) food packaging. A link was putatively established by a study in Sweden in the 1980s, which found that food packaged in PVC became contaminated with dioxins. However, a subsequent study by British government scientists found that even when the amount of dioxins present in samples of foods that had been stored in PVC wrap is deliberately overestimated (it was assumed, unrealistically, that all the dioxin present in the wrap migrated to the food) and even when the likely effect of these dioxins at low doses is deliberately overestimated (by assuming a 'tolerable daily intake' well below the level at which any effect on health is likely), PVC food wrap was found to be 'unlikely to pose a risk to health' (MAFF, 1995a).

Another current favourite of environmentalists is the supposed adverse health effects of phthalate plasticisers. Phthalates are present in some food packaging manufactured from paper and board and in some inks used in food packaging (MAFF, 1995b), however tests indicate that most of the phthalate present in food arises from general environmental contamination, not from the packaging. Phthalates have been found in rodent tests to mimic the effects of oestrogens and to have negative effects on male reproductive organs. However, the likely human exposure to phthalates from food is more than 100 times below that at which mild effects were observed in rodents (MAFF, 1996). Our current knowledge indicates that phthalates from food products are not causing any harm. Moreover, we are constantly exposed to far higher levels of oestrogen-mimicking substances – both natural, in the form for example of soy, and artificial, in the form of the birth control pill, the residues of which contaminate water supplies (Tolman, 1999).

Similar concerns have been raised with regard to the presence of polychlorinated biphenyls (PCBs), which are also known to be rodent carcinogens. Tests (based on a small sample) indicate that PCBs are present in about a quarter of all packaged food. However, they are present in such low concentrations that they are extremely unlikely to have any impact on humans eating them (MAFF, 1999). Moreover, the fact that PCBs are present in much higher concentrations in packaging strongly suggests that they do not migrate from packaging into food (ibid.).

The POPs protocol

The attack on organochlorines has been expanded recently to an attack on organohalogen compounds in general (these are compounds made by combining naturally occurring carbon-based substances, such as ethane, with chemicals from the halogen group, which includes bromine, chlorine and fluorine).

During the coming year, a UN treaty will probably be agreed that will set global phase-out dates for 12 organohalogen substances, including DDT, Dieldrin, 2,4,5-T, Aldrin and Lindane, which because of their slow rate of degradation and putatively negative effect on the environment have become known pejoratively as 'persistent organic pollutants' (POPs). The ban is being promoted by western governments and supported by environmental organisations. However, those in developing countries are not so keen.

In response to concerns raised by officials from developing country governments, it is possible that DDT will be given a long phase-out (currently projected for 2007). However Africa's farmers may face a more immediate problem. DDT and most of the other pesticide POPs have not been sprayed on crops for several years, but because of their persistence they remain in tiny quantities in the soil, and hence seep into new crops. Western concerns about these residues may lead to stricter regulations on the importing of food from Africa. Such legislation would cause immense harm to Africans without providing any benefit to western consumers. In fact western consumers will suffer because food prices will rise. However, it is likely to be welcomed by western farmers who would otherwise face competition from cheaper imports.

Constituencies supporting such legislation tend to be influential and well funded, whilst delegates from less developed countries are often wary of speaking out at international meetings – especially when to do so might jeopardise the provision of 'aid' money. In addition, public health bodies such as the World Health Organisation (WHO) have only observer status at the POPs meetings and their pleas against the DDT ban are falling on deaf ears.

Protecting the ozone layer or merely increasing the cost of carrots?

Other organohalogens are being restricted on different grounds. Environmentalists claim that methyl bromide, a widely used crop fumigant, is damaging the earth's ozone layer. Under the auspices of the Montreal Protocol to the Vienna Convention on Ozone Depleting Substances, developed nations have agreed to stop producing methyl bromide by 2005 and developing nations must follow suit by 2015. This leaves American and European users with two serious problems: first, how to get along without methyl bromide and, second, how to compete with other nations, such as Mexico, that will have a methyl bromide advantage for ten more years.

Although the deadline for the ban is fast approaching, there are no viable alternatives to methyl bromide for many of its most important uses, which range from preparing the soil before planting, to protecting food against pests during storage, processing and shipment. According to the US Department of Agriculture, "the likelihood of developing new, effective fumigant alternatives to methyl bromide appears very remote" (quoted in Lieberman, 1998). In addition, several imports and exports to the US are required by law to be treated with methyl bromide, so international trade may be disrupted. For example, the Japanese government insists that American-grown fruit be treated with methyl bromide prior to shipment. Given the arduous regulatory gauntlet any substitute must pass through, it is highly unlikely than an adequate solution will emerge before the ban takes effect.

The US Department of Agriculture has predicted that the costs of the ban to the US agricultural sector will be over $1 billion annually. However, the ultimate costs of the ban will be paid by consumers in

Europe and America, who will suffer higher prices for strawberries, tomatoes, peppers, carrots and several other methyl bromide-dependent fruits and vegetables, as well as the possibility of reduced availability and increased food safety problems.

Meanwhile, the environmental pretext for banning methyl bromide lacks merit. Methyl bromide became the focus of attention only after chlorofluorocarbons (used mainly as refrigerants) were banned on similar grounds. Ironically, recent research shows that most methyl bromide is created naturally, so the ban will have little effect on overall quantities in the environment. Evidence also indicates that most methyl bromide molecules harmlessly break down long before they can possibly affect the ozone layer (Lieberman, 1998).

Fear of food additives

Whilst concern about the impact of agriculture and food production technologies on health and the environment is a relatively recent phenomenon, concern about the impact of food additives on man has a very much longer – and more scientifically sound – history.

The first piece of British legislation concerning food additives was the Assize of Bread of 1266, which was intended to prevent bread from being adulterated with impurities such as stone – which was used to make it heavier and hence increase the profits of bakers. Whilst the addition of stone to bread was probably not harmful to health, some later adulterations certainly were. During the 1850s 15 people died after consuming sweets manufactured using a mould constructed of arsenic, which had been used in mistake for plaster of paris (Coultate, 1998). In the same era, at a public banquet in Nottingham, three guests died after eating a green blancmange coloured with copper arsenite (ibid.). Whilst such acute poisonings were probably relatively rare, the use of nutritionally dubious additives such as copper sulphate in beer and *bole armenium* in sauces was common, with apparently negative effects upon the health of much of the British population (ibid.). However, their use gradually died out as their ill effects became known and responsible manufacturers ceased using them. For example, Crosse and Blackwell gave up 'coppering' pickles – at some initial cost, because the public had become used to bright green pickles and sales dropped until the rather more drab brown variety became accepted – once it became apparent that this process was harmful to health.

Over the course of the past century, the use of harmful food additives has gradually declined, however over the same period concern over the use of additives seems to have increased.

Cyclamates and saccharin – the perils of dieting

Fear of food additives began in earnest in October 1969, after Jacqueline Verrett, an official at the US Food and Drug Administration appeared on prime time US national television news declaring that cyclamates (salts of cyclamic acid), then a popular sugar substitute, caused malformations in chick embryos. Earlier studies had found that injecting cyclamate into the bladders of mice could lead to bladder cancer, but these had been criticised for the unrealistic method of introduction of the cyclamate. Likewise, senior FDA officials criticised Verrett for her premature announcement of a scientific study that was far from a replication of the conditions pertaining to human beings – the cyclamate had been injected directly into the embryos, which is not a conventional procedure for consuming soda. Nevertheless, shortly after the announcement by Verrett, Abott Laboratories, the manufacturer of cyclamate, released a study showing that 8 out of 240 rats fed a mixture of ten parts cyclamate to one part saccharin (the mixture typically used in food products) developed bladder tumours (Lieberman and Kwon, 1998, p.12). Never mind that the rats had consumed in human terms the equivalent of 350 cans of diet soda per day, now even the company producing the chemical felt obliged to make public scientific studies that were of little relevance in terms of assessing the human impact of cyclamate. The message was simple: animals had become ill as a result of consuming cyclamate, so cyclamate is obviously a threat to human health and should be banned. And so it was; first as an additive and then even as a prescription drug (ibid.).

Cyclamates were to a large extent replaced in US food products by the more bitter saccharin. However, in 1972 a study showed that saccharin too induces bladder cancer in rats – albeit that in the tests saccharin represented up to 5 per cent of their entire food intake. Given that saccharin is 300 times sweeter than sugar, to obtain equivalent levels of sweetness from sugar would have meant increasing the dietary intake of the rats 15-fold and giving them only sugar – hardly a recipe for a long and healthy life! Nevertheless, after a second study in 1977 showed similar results, the FDA moved to ban saccharin (Lieberman and Kwon, 1998, p. 18).

Faced with the prospect of the only remaining source of non-sugar sweetener being removed from the market, diabetics and dieters in the US were outraged and lobbied Congress to overturn the ban. Congress complied by imposing a moratorium on the ban (ibid.).

Subsequent studies have shown that at the lower doses more likely to be ingested by humans neither cyclamates nor saccharin induces cancer in rodents, and epidemiological studies show that at the doses at which saccharin is normally consumed it does not induce bladder cancer in humans. Both products are now available in the EC and

Canada but cyclamate remains banned in the US (Lieberman and Kwon, 1998, pp 12 and 19).

Nitrites

A year after the proposed saccharin ban, the FDA was faced with calls to ban nitrites, a common preservative used in bacon and other cured red meat to prevent botulism. In animal tests, nitrites had been implicated as a cause of lymphatic cancer. However, this time the FDA decided to weigh up the relative risk presented by nitrites with that presented by botulism and chose not to impose a ban. It turns out that 'smoking', the traditional alternative to direct application of nitrites, actually results in significantly higher levels of nitrites, so a ban would probably have been counterproductive (Lieberman and Kwon, 1998, p. 14). In addition, for most people nitrites in cured meat are likely to represent only a small proportion of total ingested nitrites. Moreover, our bodies naturally convert nitrates into nitrites, and most of us consume relatively large amounts of nitrates since these occur in many 'natural' food products, including spinach, lettuce, celery and radishes.

The environmental and health benefits of modern farming techniques

The tendency of environmentalists and consumer activists to focus only on the tiny possible negative effects of modern agricultural and food production technologies and to ignore the benefits is probably causing a great deal of harm both to humans and to the environment. As noted above, intensive farming practices, including the use of pesticides, fertilisers and new crop types, have enabled growth in food production at a rate greater than population growth, thereby ensuring that people are better fed than ever before. This has occurred in spite of many unnecessary regulatory interventions. How much faster might food production have grown and how many fewer people might have died of famine had these interventions not taken place, we will never know. Nor can we easily calculate how many fewer people would have died from malaria and how many fewer people would have suffered from OP poisoning if DDT use had been permitted to continue. (I am not saying that no restrictions on pesticide and herbicide are justifiable; merely that the likely negative effects of some of the restrictions that were imposed were greater than the likely negative effects of less restricted use.)

Of course, there are still some concerns raised by the use of fertiliser, herbicide and pesticide. Fertiliser run-off can cause eutrophication of streams, leaving them starved of oxygen and less able to support life. Herbicides often kill more than merely the weeds that they are designed to attack, and this can have a knock-on effect for other parts of the ecosystem, leading to local reductions in wildlife. As emphasised above, these effects are generally less severe than environmental activists make out, and newer technologies have meant that they are becoming less severe still (Hayward and Jones, 1999). For instance, new methods for applying fertiliser and for modelling demand for it means that farmers save money because they are able to use less of the stuff. However, in much of the world farmers still have perverse incentives to use too much fertiliser because their activities are subsidised in one way or another. The experience of New Zealand and elsewhere suggests that when subsidies are removed, use of fertiliser declines as farmers cease attempting to use more marginal (and hence more nutrient-hungry) lands (Avery, 1997; OECD, 1998; and see also Chapter 8). The US is due soon to remove its agricultural subsidies; Australia, Canada, Brazil, Hungary and Thailand are doing likewise, so the pressure on the EU to follow suit will be great (Avery, 1997).

If getting rid of the perverse subsidies is not sufficient, then some regulation of the use of fertilisers may still be necessary. If that is the case, then rather than imposing uniform restrictions on use, Bruce Yandle suggests that acceptable ambient concentrations of nitrates and phosphates should be set and farmers allocated discharge permits, which they can then trade with one another (see Chapter 12). Such a system of tradeable permits would reduce levels of nitrate and phosphate at lower cost than command and control measures. This is a win-win situation for the environment: pollution is reduced and fewer resources are consumed in the process.

Michael Wilson (Chapter 5) suggests that many environmental problems associated with agriculture can be reduced by using genetically modified organisms (GMOs), which have the potential to improve yields and quality whilst simultaneously reducing associated inputs (including fertiliser, herbicide and pesticide). Yet environmental organisations have campaigned against the importation and production of GMO food crops on the grounds that they may be harmful to health or that their use might result in damage to the environment. Michael Wilson dismisses these claims, arguing that there is no evidence of such effects, pointing out that the process by which GMOs are created ensures that the resultant organism has more predictable characteristics than organisms created using conventional cross breeding techniques. Once again, the environmental organisations seem to be crying wolf.

Virtual risks

In this brief review, it has been argued that the early claims relating to health and environmental risks posed by pesticides, herbicides and food additives, were based on plausible scientific theories backed up by semi-plausible scientific data. The objections that have been levelled at such claims are essentially empirical: nice as the theories are, they do not in fact describe the real world. Using more comprehensive data it has been possible to show that many of the hypothesised risks are not worthy of concern. The more recent scares, however, are based upon scientific theories that are only semi-plausible and are backed up by almost no data at all. Furthermore, in the case of BSE there appears to be an alternative explanation – organophosphate poisoning leading to an autoimmune response – that is equally if not more plausible than the one that has been the focus of the scare (Axelrad, 1998). As John Adams (1997) points out, the risks posed by BSE-infected beef and GM foods are best described as 'virtual'. The greatest threat posed by such virtual risks is that our attempts to avoid them will result in an increase in exposure to real risks and a consequent increase in mortality and morbidity. An example of this would be an increase in consumption of 'organic' and 'natural' foods in preference for foods manufactured using pesticides, fertiliser, GM technology, preservative and PVC packaging. Dennis Avery recently analysed data collected by the US Center for Disease Control on the incidence of poisoning with E-Coli O157:H7. Interestingly, Avery found that people eating natural and organic foods were eight times more likely to contract this new and potentially deadly bacterium (Avery, 1997). The reason for this is that some 'natural' and 'organic' foods are produced using manure as a fertiliser and if manure is not treated appropriately (as it often is not) then it can contain these harmful bacteria, which can then spread onto the food being produced.

Lynn Scarlett (Chapter 6) argues that packaging and transport can help to reduce the incidence of food poisoning and simultaneously reduce waste. Modern packaging enables food to be stored for longer without degrading. This reduces waste at all stages of the chain from producer to consumer. In addition, packaging has enabled the production of a vast array of ready made food products, from arrabbiata sauce to chicken tikka. Because much food poisoning occurs in the home, as a result of unhygienic conditions in the kitchen, these packaged foods should in principle reduce the incidence of food poisoning. In addition to preventing contamination, packaging helps to prevent degradation of food during transport. This means that goods can be shipped from afar, by plane or boat, ensuring that we have fresh fruit and vegetables all year round – which, as was noted above, is beneficial for our health.

Another virtual risk being promoted by environmentalists and consumer activists is the possibility that antibiotic resistance in animals might spread to humans. This claim has led to successful demands for severe restrictions and bans on the use of antibiotics for farm animals. However, Roger Bate (Chapter 4) explains that using antibiotics in young animals keeps meat prices low and does not materially contribute to antibiotic resistance in humans. Again, the claims are shown to be scientifically unjustified and the benefits clearly outweigh the minuscule risks.

Similarly, Jean-Louis L'hirondel (Chapter 3) argues that nitrate fertilisers, far from being a threat to human health (as the scaremongers would have us believe), are in fact beneficial.

Food safety regulation: public v private

The behaviour of Crosse and Blackwell (see above) illustrates the importance of reputation in ensuring the production of high quality goods. Once the company realised that the use of coppering might have negative health effects, it chose to discontinue the procedure in order to ensure that its good name as a purveyor of quality produce was not sullied. The continued presence of the Crosse and Blackwell brand today, over 100 years later, attests to the success of this corporate philosophy. This may be contrasted with the experience of the Carbolic Smoke Ball Company, whose eponymous 'Smoke Balls' were supposed to prevent influenza and which offered £100 to anyone who succumbed to the 'flu whilst using them in the specified manner. A Mr Carlill duly bought a ball and used it in the specified manner but nevertheless came down with the 'flu. Carbolic refused to pay, so Carlill successfully sued for breach of contract (*Carlill v Carbolic Smoke Ball Co* [1892] 2 QB 484; affd [1893] 1 QB 256). Such behaviour is not good for one's reputation; unsurprisingly Carbolic is no longer around.

Over time food companies with good reputations should outcompete those with less good reputations. This is especially true in an environment in which incomes are gradually rising, because the demand for high quality food is positively associated with income.

In addition to reputation, legal protection of agreements made between buyers and sellers is desirable. This has two significant effects: first it provides a means of redress for people who are injured by products that do not conform to the claims made by the manufacturer (as in *Carlill v Carbolic*). Second, it sends a signal to the market, creating or reinforcing the reputation of the manufacturer.

In many cases, the precise terms of a contract between buyer and seller may not be expressly specified. When one purchases a sandwich, for example, it should not be necessary for the seller to state that the sandwich is fit for human consumption – that should be implicit given the nature of the transaction. The law protects such implicit terms as though they are express terms. Thus, when one buys an item of food from a retailer specialising in the sale of food for human consumption, one is generally entitled to compensation if one is harmed by poisons in the food – although, as seems perfectly reasonable, it is necessary for the injured party to prove that the food did in fact contain the poison when it was sold (*Daniels and Daniels v R. White & Sons Ltd and Tabbard* [1938] All ER 258).

Furthermore, it may be desirable for the law to protect people who are injured when they consume a product even though they are not contractually bound to the producer. Perhaps the most famous instance of this was the case of *Donoghue v Stevenson* [1932] AC 562, in which the plaintiff drank from a bottle of ginger beer bought by her friend and apparently became ill because the bottle contained a decomposing snail. The defendant, the producer of the ginger beer, was found to have been negligent. However, it should still be necessary for the injured party to establish that the product left the manufacturer substantially in the state in which it was consumed by the injured party.

Thus, private law combined with the manufacturer's desire to protect and enhance his reputation, ensure to a very great extent that the food sold to consumers is fit for consumption. It could be said that these mechanisms provide for an effective system of private regulation of food quality.

The position of public regulation in ensuring the quality of food is more dubious. As noted above, there has been a tendency for regulation of food technologies to be driven by the lobbying of environmentalists and consumer activists and not to be based on sound science. As a result, it is possible for the manufacturer of a perfectly sound technology to acquire a bad reputation because it produces or uses a technology that falls foul of such a regulation.

Because of this possible adverse consequence of regulation, companies are by and large very careful to ensure that they abide by the letter of the regulatory restrictions that are imposed upon them. However, as the burden of regulation becomes ever more extreme, this is becoming more and more difficult. In his review of the regulations pertaining to genetically modified organisms, Rod Hunter (Chapter 11) notes that, contrary to the claims made by environmental organisations, GMOs are very heavily regulated – probably excessively so. Barrie Craven and Christine Johnson (Chapter 9) argue that the same is true of food safety in general. Indeed, they argue that regulatory intervention may actually increase the incidence of food poisoning.

Environmental regulation: private v public

Ownership of the banks of a stream or river confers stringent protection from interference by polluters. As Lord Macnaghten stated in *Young & Co v Bankier Distillery Co* [1893] AC 698:

> A riparian owner is entitled to have the water of the stream, on the bank of which his property lies, flow down as it has been accustomed to flow down to his property, subject to the ordinary use of the flowing water by upper proprietors, and to such further use, if any, on their part in connection with the property as may be reasonable in the circumstances. Every riparian owner is thus entitled to the water of his stream, in its natural flow, without sensible diminution or increase, and without sensible alteration in its character or quality.

Thus, if a farmer pollutes a stream with fertiliser run-off, affected riparian owners should be entitled to compensation and may even seek an injunction against the polluting farmer.

Similar rules apply in principle to airborne pollutants, including certain crops – if those crops cross-breed in a way that negatively affects other farmers (See Morris, 1999, for a review). The rules even apply to noxious smells. One of the oldest cases of nuisance involved a pig farm, which caused a stench that was unbearable for the neighbouring property owner. The judge ruled that the operation of the pig farm should cease (*Aldred's Case* (1610) 9 Co. 57b.).

However, where many people are affected by pollution it may not be in any single individual's interest to bear the burden of taking an action. One way around this problem is for an interest group to indemnify an injured party. For example, in 1948 John Eastwood KC formed the Anglers Co-operative Association (the ACA, now the Anglers Conservation Association), with the express intention of using the common law to protect the quality of water in rivers and streams by indemnifying the owners of riparian rights against the costs of taking actions against polluters. To date, the ACA has taken many thousands of actions against polluters, and has thereby effectively protected many rivers in England and Wales.

However, if the local population desires a higher level of environmental protection, the only realistic solution is public regulation. One problem with such regulations, however, is that they may be construed by judges to imply that farmers have permission to pollute up to the limits specified, so derogating the rights of owners and possessors of property, who might otherwise have obtained an injunction against and/or compensation from the polluter (Morris, 1999). Another problem with conventional regulations is that they tend to be technologically restrictive. As Bruce Yandle points out in Chapter 12, a system of tradable permits is likely to be a more flexible and efficient means of achieving the same objective.

Conclusions

Modern agriculture and food technologies, including synthetic pesticides, fertiliser, genetic modification, plastic packaging, and air freight provide benefits both to human health and to the environment. Whilst some level of regulation of food safety may have been desirable in the past (for example because of the excessive cost of taking a case to court and the difficulty of proving harm), the evidence suggests that current regulatory intervention in these technologies tends to be overburdensome – imposing costs greater than any benefits that pertain. Moreover, the existence of a highly bureaucratised system of regulation of food technologies provides incentives for environmental and consumer activists to continue making scientifically unjustified claims about the effects of food technologies, perpetuating a vicious regulatory circle.

In Britain and the US, common law rules of contract and tort already exist that by-and-large provide consumers with an adequate means of obtaining redress when harm does arise. Most developed countries have similar laws. In addition, companies have strong incentives not to fall foul of such private legal actions – first, because of the direct cost in terms of compensation, and second in terms of the impact on their reputation. As the population becomes wealthier and the importance of this reputation effect increases, the incentive to avoid inflicting harm on consumers increases commensurately.

The time has come to think about deregulating the food and agriculture industries, to remove subsidies from agriculture and to begin implementing a more efficient system for controlling run-off from farms.

References

Accum, F. (1820). Treatise on Adulterations of Food and Culinary Poisons.

ACTC (1969): Review of Organochlorine Pesticides in Britain, Advisory Committee on Toxic Chemicals, Department of Education and Science, London: HMSO.

Adams, J. (1997). 'Cars, Cholera, Cows and Contaminated Land: virtual risk and the management of uncertainty,' in Bate (1997), pp. 285-314.

Ames, B. N. (1998). 'Micronutrients Prevent Cancer and Delay Aging,' *Toxicology Letters*.

Ashby, E. and Anderson, M. (1981). *The Politics of Clean Air*, Oxford.

Avery, D. T. (1997). The coming collapse of western Europe's farm policies, *Outlook*, **1**(9), pp. 2–25.

Axelrad, J. (1998). An autoimmune response causes transmissible spongiform encephalopathies, *Medical Hypotheses*, **50**, pp. 259–264.

Bate, R. ed. (1997). *What Risk?* Oxford: Butterworth-Heinemann.

Buchanan, J. M. and Tullock, G. (1975). Polluters profits and political response, American Economic Review.

Carson, R. (1963). *Silent Spring*, London: Hamish Hamilton.

Cohen, B. L. (1995). Tests of the Linear-No Threshold Theory of Radiation Carcinogenesis for Radon Decay Products, *Health Physics*, **68**, pp. 157–174.

Coultate, T. (1998). There is Death in the Pot, mimeo, South Bank University.

De Gregori, T.R. (1998). Counter to Conventional Wisdom: In defense of DDT and against chemophobia, Rockwell Lecture, University of Houston.

EPA (1992). National Air Pollution Emission Estimates, 1940-1990, Washington DC: US Environmental Protection Agency.

Hayward, S. and Jones, L. (1999). Environmental Indicators in North America and the United Kingdom, IEA Environment Working Paper No. 3, London: Institute of Economic Affairs.

Jukes, T. H. (1974). Insecticides in Health, Agriculture and the Environment, *Naturwissenschaften* **61**, cited by Wildavsky (1995), p. 60.

Kemm, K (1999). Malaria and the DDT Story, in Mooney and Bate, *Environmental Health: Third World Problems, First World Preoccupations*. Oxford: Butterworth-Heinemann (1999).

Lieberman, A. J. and Kwon, S. C. (1998): Facts versus Fears – A Review of the Greatest

Unfounded Health Scares of Recent Times, New York: American Council on Science and Health, www.acsh.org.

Lieberman, B (1998). Doomsday déjà vu: Ozone Depletion's Lesson for Global Warming, Working Paper, Cambridge: European Science and Environment Forum

MAFF (1995a): Dioxins in PVC Food Packaging, Joint Food Safety and Standards Group Food Surveillance Information Sheet No. 59, London: Ministry of Agriculture Fisheries and Food, April,
www.maff.gov.uk/food/infsheet/1995/no59/59dioxin.htm.

MAFF (1995b): Phthalates in Paper and Board Packaging, Joint Food Safety and Standards Group Food Surveillance Information Sheet No. 60, London: Ministry of Agriculture Fisheries and Food, May,
www.maff.gov.uk/food/infsheet/1995/no60/60phthal.htm.

MAFF (1996): Phthalates in Food, Joint Food Safety and Standards Group Food Surveillance Information Sheet No. 82, London: Ministry of Agriculture Fisheries and Food, March, www.maff.gov.uk/food/infsheet/1996/no82/82phthal.htm.

MAFF (1999): Survey of Retail Paper and Board Food Packaging Materials for Polychlorinated Biphenyls (PCBs), Joint Food Safety and Standards Group Food Surveillance Information Sheet No. 174, London: Ministry of Agriculture Fisheries and Food, April,
www.maff.gov.uk/food/infsheet/1999/no174/174pcb.htm.

Mooney, L. and Bate, R. (eds.) (1999). Environmental Health: *Third World Problems, First World Preoccupations*, Oxford: Butterworth-Heinemann.

Morris, J. (1998). Business and the Environment, Paper Presented to the Prince of Wales Trust Seminar on Business and the Environment, March.

Morris J. (1999): Private Law and Environmental Protection in England, in *Private Law and the Environment in Europe*, ed. B. Pozzo, Cambridge: Cambridge University Press.

OECD (1998). Agricultural Policies in OECD Countries, Measurement of Support and Background Information, Paris: OECD.

Roberts, D. J, Laughlin, L. L., Hseih, P., and Legters, L. J. (1997). 'DDT, Global Strategies, and a Malaria Control Crisis in South America,' Emerging Infectious Diseases, Vol. 3(3).

Tolman, J (1999). 'Nature's Hormone Factory,' in Mooney and Bate, *Environmental Health: Third World Problems, First World Preoccupations*. Oxford: Butterworth-Heinemann (1999).

Whelan, E. (1993): *Toxic Terror*, 2nd ed., Buffalo, NY: Prometheus Books.

Wildavsky, A. (1995). *But is it True? A Citizen's Guide to Environmental Health and Safety Issues*, Cambridge, MA: Harvard University Press.

Wilson, J. D. (1997). Thresholds for Carcinogens: a review of the relevant science and its implications for regulatory policy, in Bate *What Risk?*, Butterworth-Heinemann (1997), pp. 3-36.

Section 1:
But is it true?

1 The fallacy of the organic Utopia

Dennis Avery

Summary

Farming that shuns modern methods and technology cannot provide food for any more than a niche market. It is a fine paradox that organic farming, which claims to be 'kinder to the earth' fails to fulfil key environmental criteria, for instance by introducing alien species to control pests and using the bare-earth technique, with its risk of erosion. If all agriculture were organic, biodiversity and wildlands would certainly be lost to the plough. Popular fears of chemicals and technology are distracting attention from the positive effects of high-yield agriculture while a faith in going back to nature blinds its proponents to some fundamental disadvantages.

Introduction

Organic farming has been put forward as one of the major pillars of a new, more-sustainable human society that would be 'kinder to the earth'. Unfortunately, organic farming cannot deliver on that promise. In fact, organic farming is an imminent danger to the world's wildlife and a hazard to the health of its own consumers.

Organic farming today supplies less than 1 per cent of food for affluent countries and a declining proportion of the food in the Third World. This is good. A thoughtful world would not allow organic farming to be more than a niche source of fruits and vegetables for chemophobes who might otherwise endanger their health by refusing to eat fresh produce.

However, as a global food production system, organic farming would be an environmental disaster:

- Organic farming would force us to plough up an additional 5 to 10 million square miles of wildlands to make up for its lower yields and the global shortage of organic nitrogen. We would trade wildlands for huge tracts of clover and alfalfa (Avery 1997).

- Expanding agriculture onto poorer-quality land would inflict severe losses on wildlife species. The world is already farming most of its good-quality land. A major expansion would clear large tracts of poorer-quality land, which typically have more biodiversity (Huston 1993). For example, there are more wildlife species in a few square miles of tropical rain forest than in the whole of North America.
- Organic farming uses biological controls instead of modern pesticides. Unfortunately, biological controls are far more likely to wipe out wild species than farm chemicals. Pesticides have not caused a single known species extinction in 50 years of broad use. Farm pesticides are applied only on fields, with very limited spillover into wildlands where most of the wild species live. Bio-controls move out on their own into the ecosystem. We already know that introduced species are one of the major threats to the world's wild species. One bio-control insect released by the US government (the European flowerhead weevil) now threatens to eradicate two of America's native thistle species and the pictured-wing fly that depends on one of them.
- Global organic farming would incur staggering levels of soil erosion as millions more square miles of more-fragile land would have to be farmed to provide food. Equally unfortunate, organic farmers refuse to use the only truly sustainable tillage systems available to most of the world (conservation tillage and no-till farming) because those systems depend on chemical weed killers.

Worldwide organic-only farming would also be dangerous to public health:

- The world's food supply would become far more erratic and uncertain under organic farming. Low organic yields would force the use of far more marginal land, prone to drought, pests and early frosts. The fields would be far more susceptible to periodic plagues of pests because biological pest controls are the weakest and most erratic in the farmers' defence arsenal.
- Food costs would be prohibitively high for the poor in the low-production years. The world would also have to store more food reserves – unprotected by storage pesticides – and thus suffer much higher losses to cereal beetles, weevils, moulds and other threats to stored food.
- Eating five fruits and vegetables per day is humanity's strongest weapon against cancer; it cuts total cancer risk in half, no matter how the produce was grown (see Ames' Chapter). Only about one-tenth of First World consumers eat enough produce for full protection, even though the stores are full of cheap and attractive fruits and vegetables. Organic produce is often twice as expensive and typ-

ically less attractive due to pest damage. How many consumers would eat enough fruits and vegetables if the stores had only organic suppliers?

Saving room for wildlife with high-yield farming

High-yield farming gained fame as a way to feed the world's expanding human population and lower food costs for the poor – but today its biggest advantage is that it feeds more people from less land.

The crop yield increases achieved since World War II have saved at least 15 million square miles of wildlife habitat from being ploughed for low-yield crops. We are feeding more than twice as many people today and feeding them better diets – on the same amount of cropland used in 1950. Africa is the exception to this happy state of affairs, because Africa is not yet using high-yield farming (Avery 1997).

Additional land has been saved by modern meat production, with its highly bred genetics, effective veterinary medicines, scientifically balanced feed rations and confinement housing systems. (Without 'hog hotels' the world might need more than a million square miles of land in 2050 just to house its breeding hogs outdoors. The US would need all the cropland in the State of Pennsylvania just to raise its chickens on free range.)

Modern food processing also saves wildlands because it allows us to grow crops where the yields are highest (often two or three times as high). Processing prevents post-harvest losses as we transport the food wherever the people choose to live.

Why we will have to triple farm output by 2050

The human population is rapidly stabilising. Births per woman in the Third World have plummeted from 6.5 in 1960 to 3.1 today. As stability is 2.1 births per woman, poor countries have come three-fourths of the way to stability in one generation. First World families average 1.7 births per woman reflecting an inverse population growth curve. Affluence, urbanisation and food security have encouraged this decline in birth rates.

These human population trends project to a peak of 8.5 to 9 billion humans, reached about the year 2035. That peak will be followed by a slow, modest decline in human numbers. However, the price of population stability is that we will have to feed virtually all these people high-quality diets, including lots of meat, milk, fruit, vegetables and provide the cotton and natural fibre wardrobes that affluent people will demand for themselves and their children.

There is no vegetarian trend in the world and it is probably too late to count on vegetarianism to save the wildlands. 'Vegetarian conservation' would require at least 50 per cent of the world population to become vegan, foregoing all livestock products. The First World average is currently less than 0.5 per cent vegan.

Instead of becoming vegetarians, the world is in the midst of the biggest surge in meat and milk demand ever seen. China's meat consumption has doubled in the past seven years. India has doubled its milk consumption since 1980. Indonesia is clearing tropical forest to grow chicken feed and cattle pasture.

In addition to feeding the people, the world will have to feed its pets. The US has 113 million cats and dogs living amongst its 270 million people. Pet stores in Hong Kong and Shanghai are now beginning to get beyond the traditional caged birds and crickets, selling a few kittens and the occasional small puppy. With a one-child population policy how many companion pets will an affluent Chinese population of 1.3 billion have in 2050? And woe unto whoever stands between a pet-owner and that animal's preferred diet.

All told, experts predict that the world must be prepared to produce three times as much food from its farms in 2050 as today (McCalla 1994). Hopefully, we will be able to harness enough selective breeding, chemistry, engineering and biotechnology to again triple the yields on the land we are already farming.

It is impossible to believe that we can triple the yields on organic farms or that humans will voluntarily give up high-protein diets.

Organic yields are too low

A British farm manager who handles 50,000 acres of organic and mainstream crops summed up the objective data from a dozen First World countries. He said, 'I'm lucky to get half as much yield from my organic acres.'

The *American Journal of Alternative Agriculture* recently published a best-case organic yield achievement. The Rodale Institute, a famous organic research centre, acquired a field near its headquarters and shifted it from chemically supported farming to organic production. The field has good soils and lies in a well-watered region. Rodale spent eight years building up the organic content of the field's soils, adding manure imported from nearby feedlots and testing the most productive crop rotations. Then, under ideal conditions, the newly organic field produced yields that were only 21 per cent below those of neighbouring mainstream farms – using 40 per cent more labour.

Researchers at Texas A&M concluded that US field crop production would drop by 24-57 per cent without pesticide protection (Knutson et al. 1990, Taylor et al. 1993). They concluded fruit and vegetable yields would drop by 50 to 100 per cent without pesticides and that many production areas would have to quit producing altogether.

The shortage of organic plant nutrients

Despite the substantial penalty of low yields, organic farming's biggest environmental shortcoming is the global shortage of 'natural' nitrogen fertiliser. Experts at the US Department of Agriculture have calculated that the available animal manure and sustainable biomass resources in America provide only about one-third of the plant nutrients needed to support current US food production. And, most of the world has far less pasture and manure per capita than the US. Globally, we may have less than 20 per cent of the organic plant nutrients needed to sustain current food output – let alone tripling farm output for the 21st century.

Organic proponents talk of millions of tons of compostable materials, such as urban sewage sludge, being 'wasted'. But the nitrogen in all of America's sewage sludge equals only about 2 per cent of the nitrogen being applied through commercial fertilisers. Nor could we ignore the problems of heavy metals and live pathogens in the sewage sludge.

There are indeed 'millions of tons' of compost materials, such as tree chippings and grass clippings, that could be gathered. However, a recent New Jersey recycling report suggests it would cost perhaps $100 per ton just to gather the wastes – with additional costs for hauling and distributing them on farmland.

Moreover, the 'millions of tons' of compost must be compared to the 2 billion tons of manure, corncobs and other farm waste already being spread annually on American farms. They make compost look like just an expensive way to add 1 or 2 per cent to the American agricultural biomass.

The only realistic way for the world to get millions of additional tons of organic nitrogen is to plough millions of square miles of forest to expand the world's green manure crops. Growing six million square miles of clover and alfalfa would take a land area equal to all of Europe and half the US away from wildlife.

The safety of modern pesticides

It is a continuing mystery how organic believers can be so casual about the pathogens in manure and sewage sludge – which cause thousands of First World deaths per year – and so incredibly fearful of pesticide residues which have never caused a documented consumer death.

The World Health Organisation estimates that pesticides cause about 200,000 human deaths per year. However, more than 90 per cent of these deaths are suicides. For example, pesticides are a relatively cheap and painless aid to suicide in India, while chewing the leaves of a local tree produces an agonising death. Another 7 per cent of the pesticide deaths are accidental household poisonings: children getting into rat poison, or an adult drinking something from an unlabelled bottle. The remaining few percentage points of the 200,000 deaths are farm workers who apply products carelessly, or return to a field too soon after the spraying of one of the harsher pesticides, such as methyl parathion.

The US National Research Council has published *Carcinogens and Anti-Carcinogens in the Human Diet* (1996), a conclusive 417-page report which says pesticide residues are no significant health risk to consumers. By implication, this report also says the much-weaker traces of pesticides found in drinking water are even less of a threat. The report was careful to emphasise that consumers should not let any fear of pesticides interfere with eating lots of fruits and vegetables.

The National Cancer Institute of Canada's Panel on Pesticides and Cancer came to the same conclusion in 1997: 'The Panel concluded that it was not aware of any definitive evidence to suggest that synthetic pesticides contribute significantly to overall cancer mortality. The Panel also concluded that it did not believe that any increased intake of pesticide residues associated with increased intake of fruits and vegetables poses any increased risk of cancer.'

In other words, we are still looking for the first victim of pesticide residues, 50 years after we began using them broadly and after billions of dollars in medical research spent trying to find such a victim.

> All substances are poisons. There is none which is not a poison. The right dose differentiates a poison from a remedy (Paracelsus 1492-1541).

In contrast, the US Centres for Disease Control estimate that 250 Americans are being killed every year by the virulent new strain of E. coli 0157. In 1996, 8 per cent of these fatalities were linked to organic or 'natural' foods even though such foods made up only about 1 per cent of the nation's food sales volume. Thousands of E. coli O157 victims survived, but many incurred permanent damage to their livers and kidneys.

The CDC estimates that salmonella killed another 250 Americans, from an estimated 2 to 4 million attacks, about one-quarter of them from a more recent and virulent strain labelled S. *typhimurium*.

Consumer Reports in 1998 bought free-range and mainstream chickens in US stores and compared their bacterial risks. They found nearly three times as much salmonella on the free-range chickens, along with higher levels of the dangerous bacteria listeria and campylobacter.

The higher bacterial risks of organic and natural foods are logical, since animal manure is one of the major sources of plant nutrients for organic food crops. Organic farmers compost most of the manure, with a guideline of 'two months at 130-140 degrees Fahrenheit.' Unfortunately, preliminary studies by Dr Dean Cliver at the University of California/Davis suggest this is neither hot enough nor long enough to kill E. coli O157: H7. Nor do organic farmers typically use thermometers to check their composting.

The US Food and Drug Administration says that it typically finds higher levels of dangerous natural toxins on organic and natural foods. Again, it's logical. Organic fields suffer higher levels of rodent and pest damage and the pest attacks create openings for fungi to attack the grains and fruits. It is the fungi that produce the toxins – including alflatoxin, a potent liver toxin and one of the most carcinogenic substances ever tested.

Science supporting endocrine disruption collapses

In recent years, the public's fear of cancer has waned somewhat. The age-adjusted mortality rate from non-smoking cancers has been declining since 1970 (Cole and Rodu 1996) in the Western World. More people now understand that lots of people get cancer primarily because they are living so long and we have eliminated so many other causes of death. Also, the rapidly increasing understanding that our own genes often code for cancer has been highlighted by the news media.

Recently, the organic and environmental movements attempted to rekindle the fear of pesticides by accusing them of disrupting the endocrine systems in humans and wildlife species. The activists presented two important sets of data: 1) a Finnish study that claimed sperm counts in human males have declined by half in the modern pesticide era; and, 2) a Tulane University study which found 1600 times more endocrine disruption when two pesticides were tested together.

However, the claim that male sperm counts have declined was quickly discredited. The Finnish authors assumed male sperm counts were the same worldwide; in reality, they widely vary. When the study was corrected for these differences, there was no evidence of decline (Le Fanu 1999).

The Tulane finding of high endocrine disruption rates through synergy between pesticides has been withdrawn. No other laboratory was able to repeat Tulane's results – and neither was Tulane. Meanwhile, toxicologists point out that we probably get millions of times more endocrine impact from the foods we eat (such as peas) than from pesticide residues.

The world is left with an overwhelming body of evidence that says chemically-supported farming is safer for consumers – and for the environment – than organic farming.

No nutritional advantage from organic foods

For decades, organic believers have claimed organic foods had higher nutritional value. This defied logic, since virtually all crop plants need exactly the same menu of plant nutrients. Without them, the plants will not grow. With them, the plants will produce their normal grains and fruits, with their normal nutritional content.

Tufts University in Massachusetts hosted an international conference in 1997 on 'Agricultural Production and Nutrition.' For the most part, the following studies were done by alternative agriculturists trying to demonstrate that their products are more nourishing. Researchers from dozens of countries reported they could find no nutritional differences between organic and chemically-supported foods (AJAA 1997).

- Colorado State compared the vitamin contents of carrots and broccoli and found no differences between conventional and organic produce. Colorado State also grew potatoes under four different systems, analysed nine minerals and seven vitamins – and found no clear differences.
- A Lithuanian study compared three farming systems on carrots, potatoes and cabbage. 'Traditional chemical analyses of nutritional value did not show strong, consistent effects ... on produce quality.'
- A Norwegian study found *conventional* carrots had more beta-carotene and more carotinoids (both important anti-oxidants). The organic carrots had more aluminium.
- An American study found more soluble iron in conventional spinach, but modestly more available iron in the organically grown spinach.

The varieties of vegetables grown seem to make a bigger difference in nutritional levels than the farming system and even the impact of the varieties was not large.

Modern farming is the most sustainable

People are rightly concerned about the sustainability of agriculture. Nothing is more important to the long-term sustainability of human society. It is clear, however, that modern high-yield farming is the most sustainable practised by humanity in the 10,000 years since we left the forests and created fields.

Soil erosion

Soil erosion has always been a major problem. We have only to visit the Mediterranean Basin to realise that early farming systems allowed too much soil degradation and erosion. Primitive plough-based farming systems went straight up and down the hillsides. Ploughing also allowed water to run off the surface, taking too much soil with it and released carbon and reduced organic matter in the soils, making them even more vulnerable to wind and water.

Asia's wet rice culture was one of the few early farming systems that was truly sustainable, in large part because of the laborious land terracing that prevented erosion.

High-yield farming, which really only got started in the 1930s, has delivered major victories in the war against erosion. Tripling the yields on the best-quality land almost automatically cut erosion per ton of food by two-thirds. It allows us to avoid farming steep hillsides and fragile soils, so can cut erosion per ton by *more* than two thirds. High-yield conservationists of that era also developed some important weapons against erosion, including contour ploughing, strip cropping and cover crops.

Conservation tillage

However, it was not until modern chemistry collided with the oil-price crisis of the 1970s that modern farming gained its biggest weapon against erosion – conservation tillage. Chemists of that era were beginning to develop chemical weed killers at the same time farmers were looking for ways to cut fuel costs after oil prices quadrupled. The result was that farmers put away their deep mouldboard ploughs, began using shallow disc-ploughs and relied on herbicides to control weeds. The conservation tillage revolution was born.

Conservation tillage chops crop residues into the top few inches of the soil, creating billions of tiny dams to prevent erosion. Or, in no-till farming, it uses cover crops to protect the soil all year round, using a herbicide to kill the cover crop just before the new seeds are planted through the killed sod.

Conservation tillage cuts water run-off from the field by up to 90 per cent, traps more of the rainfall for the crop and cuts erosion per acre by another 65 to 90 per cent. Subsoil populations of earthworms and soil microbes thrive with year-round supplies of decomposing vegetation and the freedom from being ploughed.

Conservation tillage is now being used on hundreds of millions of acres of land in North America, South America, Australia and Asia. It has even been tested successfully in Africa, where too few farmers can yet afford herbicides.

Paul Johnson, President Clinton's former Administrator of the Natural Resources Conservation Service in the US Department of Agriculture, said that conservation tillage was allowing farmers, for the first time, to build topsoil even as they produced some of the highest crop yields in the history of the world.

Ironically, organic farmers refuse to use conservation tillage, even though it emulates their own long-cherished emphasis on maintaining soil tilth and encouraging subsoil biota. Organic farmers continue to use 'bare-earth' farming systems featuring mouldboard ploughs, steel cultivator shanks and tractor-drawn rotary hoes. This means leaving themselves far more vulnerable to erosion. It also leaves them vulnerable to wet springs, when mechanical cultivators can't get into the fields and weeds can overwhelm the organic crops. The same year, the next-door farmer who used a pre-emergence herbicide, is producing high yields.

Pesticide sustainability

Critics say pesticides aren't 'sustainable' because pests keep developing resistance. That's true of modern antibiotics as well – but we keep investing in the research to develop new antibiotics, instead of accepting death from pneumonia. In the same way, we must increase our research efforts and develop still more pest control strategies. Crop and livestock pests multiply so rapidly that we've never beaten them with any static defence. Moreover, modern monoculture fields create an ideal environment for pests to attack (but they also produce the highest yields per acre).

Biological pest controls offer much promise, but to date they have been the weakest and most erratic of the farmers' pest weapons. They are often slow to build up in the fields, allowing the pests to get a head start on crop destruction. They can often be thrown off stride by unfavourable weather. As mentioned earlier, they also present potential threats to the local eco-systems and thus must be used with great care.

Integrated pest management tries to combine all our weapons against pests: crop rotation, biological pest controls, baited traps, crop timing and many other techniques besides pesticides. This helps minimise pesticide use and costs and helps maintain the effectiveness of a given pesticide for a longer period. However, even IPM ultimately depends on the synthetic chemicals as the last line of defence against crop loss.

Hopefully, our next generations of pesticides will be even safer than the ones used now. However, Roundup and atrazine, key herbicides used in conservation tillage, are among the safest chemicals ever tested.

Glyphosate, the active ingredient in Roundup, is so safe it can be used around trout and quail with no ill effects. Atrazine, while it bonds weakly to the soil and seasonally turns up in drinking water, is so safe that a person would have to drink 154,000 gallons of water per day containing the maximum allowable level – for 70 years – to get above the no-effect level. The sulfanylurea pesticides are no more toxic than table salt. One tablet, the size of an aspirin, treats a whole acre and they degrade into harmless compounds within weeks.

These types of herbicides are the only category of farm pesticides in which First World usage has been rising in recent decades.

Petroleum

Modern farming is criticised for using petroleum as a fuel. But agriculture accounts for only 2 per cent of US petroleum use. Another 1 per cent is used to transport foods from the farms to the table, but transportation allows us to grow food where yields are often two or three times higher. If farming had to produce its own fuel – as ethanol, or biodiesel or even pasture for horses – it would take more land from nature. The gasoline tractor reportedly released about 30 million acres of US farmland, which had been used for pasture for horses, to grow food for people. When the rest of the economy shifts away from fossil fuels to some new (and perhaps more expensive) energy source, so will farmers.

Chemical fertilisers

Chemical fertilisers replace exactly the nutrients that are taken out of the soil by growing plants. Soil testing allows the farmer to tell exactly what nutrients need to be re-supplied.

Will we run out of fertilisers? Our nitrogen is taken from the air, which is mostly nitrogen; that will never run out. The world has very large deposits of potash. Reasonable estimates of our phosphate ore reserves suggest about 250 years' supply – after which we will have to start recovering phosphate from lower-grade deposits. There is no reason to believe that humanity will run out of plant nutrients.

Too much 'fixed' nitrogen?

Is the world 'fixing' too much nitrogen? Will nitrogen oxides from farming disrupt ecosystems? It is true that our world today is fixing twice as much nitrogen as nature used to and that nitrous oxide is being redeposited on the earth's surface. However, even the worriers have been unable to come up with a credible threat. Nitrogen from chemical fertiliser, animal manure and legume crops is emanating from only about 15 per cent of the earth's surface – and being deposited across the entire surface. It would take truly heroic amounts of fixed nitrogen to alter the ecology of the forests and oceans. To date, the additional nitrogen being deposited is only about 4-6 kg per acre per year – about enough to stimulate forest and fish growth and mildly disadvantage wild legumes.

The Soil and Water Conservation Society of the US, which has often been critical of 'modern' farming, published a report in 1995 titled *Farming for a Better Environment*. The Society concluded that modern farmers were conducting the most sustainable agriculture in history, using high-yield seeds, chemical fertilisers, integrated pest management and conservation tillage.

What are the pesticide impacts on wildlife?

The claims that agricultural pesticides wreak havoc on natural ecosystems are unrealistic given today's highly specific, low-volume, rapidly-degrading pesticides and their precision application. There is relatively little wildlife in the fields themselves and the farmers make dedicated efforts not to waste their expensive pesticides outside the fields.

Some pesticides are, indeed, carried from the fields by heavy rains. However, the traces of pesticide found in ground and surface waters are very low, typically a few parts per billion and well below the safety limits. Less than 1 per cent of the herbicides applied in the US leave the root zone of the crops. The percentage for insecticides is probably even lower, since the insecticides are typically more expensive and targeted more directly at crop plants.

The State of Virginia, in the 1990s, proposed a ban on one granular soil insecticide because a study had found 6 per cent of the eagles in the James River estuary were dying of secondary pesticide poisoning. Doves ate the granules, eagles ate the doves and both died. The manufacturer solved the problem by withdrawing it from sale in areas where it impacted sensitive bird populations.

In Britain, reports indicate that the major impact of today's pesticides on birds is to locally reduce the number of insects on which they can feed.

The major bird problem in South Africa is that raptors, especially vultures, too often feed on poisoned baits put out by native farmers to protect their livestock from hyenas. There is now an aggressive education programme in place to teach the farmers safe ways to protect both the vultures and their livestock.

While efforts to reduce these losses are applauded and must be pursued, the wildlife impacts are trivial in comparison to the huge amounts of wildlife habitat saved from destruction by higher yields.

Can we achieve still-higher yields?

The prospects for raising yields still higher are excellent. The world could perhaps double its current average yields of major crops by extending known technologies to all of its farms and supporting all its farms with roads, input supply systems and competitive pricing.

As examples of potential, India averages 2.5 tons per hectare on its mostly-irrigated wheat and Pakistan 2 tons, while Chile averages 3.5 tons. Thanks to an excellent rice-breeding programme, China averages 6 tons per hectare of rice while India and Bangladesh average only half as much. America averages 7 tons per hectare of corn (mostly rain-fed) while the world corn yield average is 2.7 tons.

Dr Norman Borlaug, with the support of the Sasakawa Global 2000 Foundation, is proving that African farmers can double their yields using improved seeds, along with very modest levels of fertiliser and pesticide. (Borlaug won the Nobel Peace Prize in 1970 for his key plant-breeding role in the Green Revolution.)

Traditional approaches to raising crop yields are still gaining steadily: plant breeding continues to raise world grain yields by 1 to 2 per cent annually. Artificial insemination in dairy cattle is raising milk yields by 2 per cent annually. Integrated pest management is cost-effectively raising yields for some farmers.

However, there is no question that the rate of yield gains on the world's farms has slowed in the 1990s compared to the halcyon days of the Green Revolution. For one thing, the expansion of irrigation has virtually stopped. If the world is to save all its current wildlands, it will need to develop additional yield breakthroughs. This is awkward, because high-yield farming has become politically-incorrect.

- In America, the traditional leader of high-yield farming research, Federal spending has been cut by about 30 per cent in real terms. The research programmes of the land-grant colleges, which did much of the basic research for the yield revolution, have been sharply cut back.
- In Europe, public spending on high-yield research was never a high priority and public disapproval is now so intense that European governments are blocking aid programmes, which would send enough fertiliser and pesticides to save Africa's unique wildlife.
- Third World governments are often strapped for cash and are traditionally wary of spending even modest funds on research programmes which will have no political pay-back for a decade or more. Only China and Brazil can be considered to have first-rate agricultural research programmes.
- The private sector is increasing its investments in such high-tech farm inputs as hybrid seeds and biotechnology. However, it may never invest enough in the basic research to keep the overall yield potential advancing rapidly.
- Greenpeace and the World Wildlife Fund have carried on an intensive European campaign against using biotechnology in food production. This is environmentally unconscionable, since it implies we can use genetic engineering to lengthen lives, but not to provide the additional food needed. Biotechnology is the largest piece of unexplored knowledge the world has for raising farm yields and saving wildlands.

Moreover, recent advances show that research still has major farm productivity gains to offer:

- The International Maize and Wheat Improvement Centre (CIMMYT) in Mexico is completing a major re-breeding of wheat which it says will raise potential world wheat yields by 50 per cent. The new plants have a completely re-designed 'plumbing system,' supporting grain heads with more and larger kernels. Yields of up to 18 tons have already been achieved; any yield over eight tons is considered very high.

- Two researchers from a small Mexican government institute have found a genetically-engineered solution to the problem of aluminium toxicity, which cuts yields by up to 80 per cent on much of the tropical world's arable land. They took a gene for citric acid secretion from a bacterium and inserted it in crop plants. (Some of the successful wild plants on acid-soil savannahs secrete citric acid.) The citric acid prevents the uptake of the aluminium ions in the plants and they are thus free to grow unimpeded. The researchers have succeeded with tobacco, papaya and rice and are now working on other cereal grains.
- Cornell University researchers have genetically inserted genes from wild relatives of our crop plants into high-yield breeding programmes. These wild relative genes, too distant for normal cross breeding, have already produced a 50 per cent yield increase in tomatoes and a one-third gain in rice yields. The wild relative strategy may thus be broadly useful in adding both yield gains and genetic diversity to our crop breeding efforts.
- The latest low-pressure centre-pivot irrigation systems use computerised controls and variable rate nozzles to cut both the water and electricity needed for irrigated crop production by 50 per cent or more. The new systems have at least three times the water efficiency of traditional flood irrigation and, therefore, also sharply reduce the problems of waterlogging and salt build-up. As the value of water rises, such systems will replace 'cheaper' systems that waste much of their water. (Agriculture uses about 70 per cent of the water consumed by human societies, most of it for flood irrigation with efficiencies of perhaps 30 per cent.)

Can we save the small family farm?

Saving the family farm has been the mantra of government farm policy in most of the world's affluent countries for at least 200 years. (Stalin's state farms and Mao's communal farms were important exceptions.)

This is one of the driving ethics behind the urbanites in the organic farming movement. Today's affluent urbanised democracies fanaticise that they can return to the traditional small hand-powered family farm of the last century, which is imagined as a more stable and comfortable alternative to today's large commercial farms. However, agricultural technology has not driven the decline of the small family farm.

The decline in the numbers of small family farms is a direct result of the rising value of off-farm jobs. When high urban wages attract farmers to town, they leave behind land and market demand which must be satisfied with less human labour. The result is the modern, high-tech farm, which substitutes technology for hand hoes and stoop labour.

We cannot 'save' the small family farm unless we are willing to subsidise small farmers through direct income payments to be rural residents; in effect, costumed guides in 'living museums.'

The current trends in affluent societies indicate that we are headed for still-larger and increasingly specialised commercial farms in most countries, with a concurrent increase in the number of 'sundowners' – people with off-farm income, who enjoy working small farms as hobbies. Their numbers are likely to grow rapidly as the electronic age permits more and more people to do their jobs from their homes and to live wherever they choose.

Such hobbyists are likely to fill more of our rural residences and to manage a modest percentage of the farmland in affluent countries. However, they will supply only a small proportion of the farm products.

Carping about wonder wheat

When CIMMYT announced its new 'wonder wheat,' with yields of up to 18 tons per hectare, some critics immediately began worrying aloud that this would require the use of high levels of fertiliser on the fields where it was planted.

However, as it takes approximately 25 kg of nitrogen to produce a ton of wheat wherever and however we grow it, Society has a choice: We can produce 18 tons of wheat by carefully applying 450 kg of nitrogen on one hectare of prime, level farmland; or we can get 18 tons of wheat by clearing 17 hectares of forest and applying 25 kg of nitrogen per hectare to each of 18 hectares. The nitrogen leakage from the 18 hectares is almost certain to be higher than from the single hectare, because the quality of the soils is likely to be poorer, the organic content of the soil lower and the slopes steeper.

The debate about 'wonder wheat' sums up the overall debate about the sort of agriculture the world should have.

In the 21st century, I believe that the debate should be framed by 1) the huge increase in farm output which will be demanded; and 2) the urgent need to save room for wildlife.

If we can triple the yields, again, on the land we are already farming, we can supply the three-fold increase in farm demand without ploughing additional wildlands. If we can use 5 per cent of the current wild forest area for high-yield trees, to supply the projected ten-fold increase in timber and paper demand, we can protect the other 95 per cent of the wild forests from ever being logged, let alone cleared. High-yield conservation thus promises to let us save more than 90 per cent of the current wildlands and more than 95 per cent of the existing wild species.

Organic farming offers no such lofty promise. Rather it pledges to keep more people working on more low-yield farms at a terrifying cost:

huge tracts of wildlands ploughed, far more soil erosion and degradation, perhaps a million wildlife species driven to extinction and higher health risks for consumers.

With high costs, low returns and higher levels of pest damage, it is little wonder that organic food has captured less than 1 per cent of the food market in affluent countries, except where it has been heavily subsidised. American organic purchases in 1996 were $3.5 billion, out of a total grocery market of $430 billion. And, since organic foods tend to carry a price premium, it may be that organics supplied only about 0.5 per cent of the market volume.

In West European countries, where organic food production is subsidised by the government, the organic sector has expanded to as much as 3 per cent of the market. Ironically, however, Western Europe is subsidising organic food precisely because of its low yields. The governments hope that organic subsidies will be less expensive than the export subsidies currently used to dump the surpluses generated by farm price supports that are too high. Both the organic subsidies and the high price supports work against the freer farm trade that is needed to save tropical forests in densely populated Asia.

The critical factors in meeting the 21st century food challenge are:

- More public funding for sustainable higher-yield farming research in both the First and the Third Worlds.
- A regulatory welcome for safe and sustainable new farming systems and farming inputs, instead of applying a 'precautionary principle' which pretends the world already has plenty of food and no pressure to triple yields in the 21st Century.
- Free trade in farm products, so that the world can use its best land and highest-yielding farming systems to save as much wildlife habitat around the globe as possible.

References

American Journal of Alternative Agriculture (1997) International Conference on Agricultural Production and Nutrition, 12, 2: 73-92.

Avery, D. T. (1997) Saving Nature's Legacy Through Better Farming, *Issues in Science and Technology*, National Academy of Sciences and the Centre for study of Science and Society, University of Texas/Dallas; Fall: 59-64.

Cole, P. and Rodu, B. (1996) *Declining Cancer Mortality in the United States*, American Cancer Society, Baltimore, Md, August 10.

Le Fanu, J. (1999). The saga of the falling sperm counts. In: Environmental Health: Third World Problems – First World Preoccupations. Mooney & Bate (eds.) Butterworth-Heinemann, Oxford.

McCalla, A. (1994) *Agriculture and Food Needs to 2025*: Why we should be concerned. Monograph from the Director of Agriculture of the World Bank, Washington, DC, October.

2 Pollution, pesticides and cancer misconceptions

Bruce N. Ames and Lois S. Gold

Summary

The major causes of cancer are:

(a) Smoking, which is responsible for about a third of US cancer (and 90 per cent of lung cancer).
(b) Dietary imbalances, for example, lack of dietary fruits and vegetables: the quarter of the population eating the least fruits and vegetables has double the cancer rate for most types of cancer compared with the quarter eating the most.
(c) Chronic infections: mostly in developing countries.
(d) Hormonal factors influenced by lifestyle.

There is no epidemic of cancer, except for lung cancer from smoking. Cancer mortality rates have declined 15 per cent since 1950 (excluding lung cancer and adjusted for the increased lifespan of the population).

Regulatory policy focused on traces of synthetic chemicals is based on misconceptions about animal cancer tests. Recent research contradicts these ideas:

(a) Rodent carcinogens are not rare. Half of all chemicals tested in standard high dose animal cancer tests, whether occurring naturally or produced synthetically, are 'carcinogens'.
(b) There are high dose effects in these rodent cancer tests that are not relevant to low dose human exposures and which can explain the high proportion of carcinogens.
(c) Though 99.9 per cent of the chemicals humans ingest are natural, the focus of regulatory policy is on synthetic chemicals.

Over 1000 chemicals have been described in coffee: 27 have been tested and 19 are rodent carcinogens.

Plants we eat contain thousands of natural pesticides, which protect plants from insects and other predators: 64 have been tested and 35 are rodent carcinogens.

There is no convincing evidence that synthetic chemical pollutants are important for human cancer. Regulations that try to eliminate minuscule levels of synthetic chemicals are enormously expensive: EPA estimates its regulations cost $140 billion/year. The US spends 100 times more to prevent one hypothetical, highly uncertain, death from a synthetic chemical than it spends to save a life by medical intervention. Attempting to reduce tiny hypothetical risks also has costs; for example, if reducing synthetic pesticides makes fruits and vegetables more expensive, thereby decreasing consumption, then cancer will be increased, particularly for the poor.

Improved health comes from knowledge due to biomedical research, and from lifestyle changes by individuals. Little money is spent on biomedical research or on educating the public about lifestyle hazards, compared with the costs of regulations.

Myths and facts about synthetic chemicals and human cancer

Various misconceptions about the relationship between environmental pollution and human disease, particularly cancer, drive regulatory policy. Here, we highlight nine such misconceptions and briefly present the scientific evidence that undermines each.

Misconception No. 1: cancer rates are soaring

Cancer death rates overall in the US (after adjusting for age and excluding lung cancer due to smoking) have declined 15 per cent since 1950 (Kosary et al. 1995, Doll and Peto 1981). The particular types of cancer deaths that have decreased since 1950 are stomach, cervical, uterine, and rectal. The types that have increased are primarily lung cancer (90 per cent is due to smoking, as are 35 per cent of all cancer deaths in the US), melanoma (probably due to sunburns), and non-Hodgkin's lymphoma. (Cancer incidence rates are also of interest, but they should not be taken in isolation, because trends in the recorded incidence rates are biased by improvements in registration and diagnosis (Doll and Peto 1981; Devesa et al. 1995).)

Cancer is one of the degenerative diseases of old age, increasing exponentially with age in both rodents and humans. External factors, however, can markedly increase cancer rates (e.g. cigarette smoking in humans) or decrease them (e.g. caloric restriction in rodents). Life expectancy has continued to rise since 1950. The increases in cancer deaths are due to the delayed effect of increases in smoking and to increasing life expectancy (Doll and Peto 1981, Devesa et al. 1995).

Misconception No. 2: environmental synthetic chemicals are an important cause of human cancer

Neither epidemiology nor toxicology supports the idea that synthetic industrial chemicals are important for human cancer. Epidemiological studies have identified the factors that are likely to have a major effect on reducing rates of cancer: reduction of smoking, improving diet (e.g. increased consumption of fruits and vegetables), and control of infections (Ames et al. 1995). Although some epidemiologic studies find an association between cancer and low levels of industrial pollutants, the associations are usually weak, the results are usually conflicting, and the studies do not correct for potentially large confounding factors like diet. Moreover, the exposure to synthetic pollutants is tiny and rarely seem plausible as a causal factor when compared with the background of natural chemicals that are rodent carcinogens (Gold et al. 1992). Even assuming that the EPA's worst-case risk estimates for synthetic pollutants are true risks, the proportion of cancer that EPA could prevent by regulation would be tiny (Gough 1990). Occupational exposure to carcinogens can cause cancer, though how much has been a controversial issue: a few percent seems a reasonable estimate (Ames et al. 1995). The main contributor has been asbestos in smokers. Exposures to substances in the workplace can be high in comparison with other chemical exposures in food, air, or water. Past occupational exposures have sometimes been high and therefore comparatively little quantitative extrapolation from high dose rodent tests to high dose occupational exposures may be required for risk assessment. Since occupational cancer is concentrated among small groups exposed at high levels, there is an opportunity to control or eliminate risks once they are identified. We estimate that diet accounts for about one-third of cancer risk (Ames et al. 1995) in agreement with the earlier estimate of Doll and Peto (1981). Other factors are lifestyle influencing hormones, avoidance of intense sun exposure, increased physical activity, and reduced consumption of alcohol.

Since cancer is due, in part, to normal ageing, the influence of the major external risk factors for cancer must be diminished (smoking, unbalanced diet, chronic infection and hormonal factors). Cancer will occur at a later age, and the proportion of cancer caused by normal metabolic processes will increase. Ageing and its degenerative diseases appear to be due in good part to the accumulation of oxidative damage to DNA and other macromolecules (Ames et al. 1993b). By-products of normal metabolism - superoxide, hydrogen peroxide, and hydroxyl radical - are the same oxidative mutagens produced by radiation. Oxidative lesions in DNA accumulate with age, so that by the time a rat is old it has about a million oxidative DNA lesions per (Ames et al. 1993b). Mutations also accumulate with age. DNA is oxidized in normal metabolism because antioxidant defences, though numerous, are not perfect. Antioxidant defences against oxidative damage include vitamins

C and E and carotenoids, most of which come from dietary fruits and vegetables.

Smoking contributes to about 35 per cent of US cancer, about one-quarter of heart disease, and about 400 000 premature deaths per year in the United States (Peto et al. 1994). Tobacco is a known cause of cancer of the lung, bladder, mouth, pharynx, pancreas, stomach, larynx, oesophagus and possibly colon. Tobacco causes even more deaths by diseases other than cancer. Smoke contains a wide variety of mutagens and rodent carcinogens. Smoking is also a severe oxidative stress and causes inflammation in the lung. The oxidants in cigarette smoke - mainly nitrogen oxides - deplete the body's antioxidants. Thus, smokers must ingest two to three times more Vitamin C than non-smokers to achieve the same level in blood, but they rarely do. Inadequate concentration of Vitamin C in plasma is more common among single males, the poor, and smokers (Ames et al. 1993b). Men with inadequate diets or who smoke may damage both their somatic DNA and the DNA of their sperm. When the level of dietary Vitamin C is insufficient to keep seminal fluid Vitamin C at an adequate level, oxidative lesions in sperm DNA are increased 250 per cent (Fraga et al. 1991; Ames et al. 1994; Fraga et al. 1996). Paternal smoking, therefore, may plausibly increase the risk of birth defects and appears to increase childhood cancer in offspring (Fraga et al. 1991; Ames et al. 1994; Ji et al. 1997).

Chronic inflammation from chronic infection results in release of oxidative mutagens from phagocytic cells and is a major contributor to cancer (Ames et al. 1995; Christen et al. 1997) White cells and other phagocytic cells of the immune system combat bacteria, parasites, and virus-infected cells by destroying them with potent, mutagenic oxidizing agents. The oxidants protect humans from immediate death from infection, but they also cause oxidative damage to DNA, mutation, and chronic cell killing with compensatory cell division (Shacter et al. 1988,Yamashina et al. 1986) and thus contribute to the carcinogenic process. Antioxidants appear to inhibit some of the pathology of chronic inflammation. We estimate that chronic infections contribute to about one-third of the world's cancer, mostly in developing countries.

Endogenous reproductive hormones play a large role in cancer, including breast, prostate, ovary and endometrium (Henderson et al. 1991; Feigelson and Henderson 1996), contributing to as much as 20 per cent of all cancer. Many lifestyle factors such as lack of exercise, obesity and reproductive history influence hormone levels and therefore risk.

Genetic factors play a significant role in cancer and interact with lifestyle and other risk factors. Biomedical research is uncovering important genetic variation in humans.

Misconception No. 3: reducing pesticide residues is an effective way to prevent diet-related cancer

On the contrary, fruits and vegetables are of major importance for reducing cancer: if they become more expensive by reducing the use of synthetic pesticides, cancer is likely to increase. People with low incomes eat fewer fruits and vegetables and spend a higher percentage of their income on food.

Dietary fruits and vegetables and cancer prevention. Consumption of adequate fruits and vegetables is associated with a lowered risk of degenerative diseases including cancer, cardiovascular disease, cataracts, and brain dysfunction (Ames et al. 1993b). Over 200 studies in the epidemiological literature have been reviewed that show, with great consistency, an association between a lack of adequate consumption of fruits and vegetables and cancer incidence (Block et al. 1992; Steinmetz and Potter 1991; Hill et al. 1994) (Table 2.1). The quarter of the population with the lowest dietary intake of fruits and vegetables compared with the quarter with the highest intake has roughly twice the cancer rate for most types of cancer (lung, larynx, oral cavity, oesophagus, stomach, colon and rectum, bladder, pancreas, cervix, and ovary). Only 22 per cent of Americans meet the intake recommended by the National Cancer Institute (NCI) and the National Research Council

Table 2.1. Review of epidemiological studies on cancer showing protection by consumption of fruits and vegetables

Cancer site	*Fraction of studies showing significant cancer protection*	*Relative risk (median) low versus high quartile) of consumption*
Epithelial		
Lung	24/25	2.2
Oral	9/9	2.0
Larynx	4/4	2.3
Oesophagus	15/16	2.0
Stomach	17/19	2.5
Pancreas	9/11	2.8
Cervix	7/8	2.0
Bladder	3/5	2.1
Colorectal	20/35	1.9
Miscellaneous	6/8	–
Hormone-dependent		
Breast	8/14	1.3
ovary/endometrium	3/4	1.8
Prostate	4/14	1.3
Total	129/172	

Source: Block et al. (1992)

(Hunter and Willett 1993; Block 1992; Patterson et al. 1990): 5 servings of fruits and vegetables per day. When the public is told about hundreds of minor hypothetical risks, they lose perspective on what is important: half the public does not know that fruits and vegetables protect against cancer (NCI 1996).

Micronutrients in fruits and vegetables are anti-carcinogens, and their antioxidants may account for some of their beneficial effect as discussed in Misconception No. 2. However, the effects of dietary antioxidants are difficult to disentangle by epidemiological studies from other important vitamins and ingredients in fruits and vegetables (Steinmetz and Potter 1991; Hill et al. 1994; Block 1992; Steinmetz and Potter 1996).

Folate deficiency, one of the most common vitamin deficiencies, causes extensive chromosome breaks in human genes (Blount et al. 1997) Approximately 10 per cent of the US population (Senti and Pilch 1985) is deficient at the level causing chromosome breaks. In two small studies of low income (mainly African-American) elderly people (Bailey et al. 1979) and adolescents (Bailey et al. 1982), nearly half were folate deficient to this level. The mechanism is deficient methylation of uracil to thymine, and subsequent incorporation of uracil into human DNA (4 million/cell) (Blount et al. 1997).

During repair of uracil in DNA, transient nicks are formed; two opposing nicks causes a chromosome break. Both high DNA uracil levels and chromosome breaks in humans are reversed by folate administration (Blount et al. 1997). Chromosome breaks could contribute to the increased risk of cancer and cognitive defects associated with folate deficiency in humans (Blount et al. 1997). Folate deficiency also damages human sperm (Wallock et al. 1997), causes neural tube defects in the foetus, and 10 per cent of US heart disease (Blount et al. 1997).

Other micronutrients are likely to play a significant role in the prevention and repair of DNA damage, and thus are important to the maintenance of long-term health. Deficiency of vitamin B12 causes a functional folate deficiency, accumulation of homocysteine (a risk factor for heart disease) (Wickramasinghe and Fida 1994), and misincorporation of uracil into DNA (Wickramasinghe and Fida 1994). Strict vegetarians are at increased risk of developing a Vitamin B12 deficiency (Herbert 1996). Niacin contributes to the repair of DNA strand breaks by maintaining nicotinamide adenine dinucleotide levels for the poly ADP-ribose protective response to DNA damage (Zhang et al. 1993). As a result, dietary insufficiencies of niacin (15 per cent of some populations are deficient (Jacobson 1993)), folate, and antioxidants may act synergistically to adversely affect DNA synthesis and repair. Diets deficient in fruits and vegetables are commonly low in folate, antioxidants, (e.g. Vitamin C) and many other micronutrients, and result in significant amounts of DNA damage and higher cancer rates (Ames et al. 1995; Block et al. 1992; Subar et al. 1989).

Optimizing micronutrient intake can have a major impact on health.

Increasing research in this area and efforts to improve micronutrient intake and balanced diet should be a high priority for public policy.

Misconception No. 4: human exposures to carcinogens and other potential hazards are nearly all exposures to synthetic chemicals

On the contrary, 99.9 per cent of the chemicals humans ingest are natural. The amounts of synthetic pesticide residues in plant foods are insignificant compared with the amount of natural pesticides produced by plants themselves (Ames et al. 1990a; Ames et al. 1990b). Of all dietary pesticides that humans eat, 99.99 per cent are natural: they are chemicals produced by plants to defend themselves against fungi, insects, and other animal predators (Ames et al. 1990a; Ames et al. 1990b). Each plant produces a different array of such chemicals. On average, Americans ingest roughly 5 000 to 10 000 different natural pesticides and their breakdown products. Americans eat about 1 500 mg of natural pesticides per person per day, which is about 10 000 times more than they consume of synthetic pesticide residues.

Even though only a small proportion of natural pesticides has been tested for carcinogenicity, half of those tested (35/64) are rodent carcinogens, and naturally occurring pesticides that are rodent carcinogens are ubiquitous in fruits, vegetables, herbs, and spices (Gold et al. 1997b) (Table 2.2).

Cooking foods produces about 2 000 mg per person per day of burnt material that contains many rodent carcinogens and many mutagens. By contrast, the residues of 200 synthetic chemicals measured by FDA, including the synthetic pesticides thought to be of greatest importance, average only about 0.09 mg per person per day (Ames et al. 1990a; Gold et al. 1997). The known natural rodent carcinogens in a single cup of coffee are about equal in weight to an entire year's worth of carcinogenic synthetic pesticide residues, even though only 3 per cent of the natural chemicals in roasted coffee have been tested for carcinogenicity (Gold et al. 1992) (Table 2.3). This does not mean that coffee is dangerous, but rather that assumptions about high dose animal cancer tests for assessing human risk at low doses need re-examination. No diet can be free of natural chemicals that are rodent carcinogens (Gold et al. 1997a).

Misconception No. 5: cancer risks to humans can be assessed by standard high dose animal cancer tests

Approximately half of all chemicals - whether natural or synthetic - that have been tested in standard animal cancer tests are rodent carcinogens (Ames et al. 1996; Gold et al. 1997b) (Table 2.4). What are the explanations for the high positivity rate? In standard cancer tests rodents are given chronic, near-toxic doses, the maximum tolerated

Table 2.2. Carcinogenicity of natural plant pesticides tested in rodents (fungal toxins are not included)

Carcinogens: N = 35	acetaldehyde methylformylhydrazone, allyl isothiocyanate, arecoline.HCI, benzaldehyde, benzyl acetate, caffeic acid, catechol, clivorine, coumarin, crotonaldehyde, cycasin and methylazoxymethanol acetate, 3,4-dihydrocoumarin, estragole, ethyl acrylate, N2-g-glutamyl-p-hydrazinobenzoic acid, hexanal methylformylhydrazine, p-hydrazinobenzoic acid.HCI, hydroquinone, 1-hydroxyanthraquinone, lasiocarpine, d-limonene, 8-methoxypsoralen, N-methyl-N-formylhydrazine, a-methylbenzyl alcohol, 3-methylbutanal methylformylhydrazone, methylhydrazine, monocrotaline, pentanal methyl-formylhydrazone, petasitenine, quercetin, reserpine, safrole, senkirkine, sesamol, symphytine
Noncarcinogens: N = 29	atropine, benzyl alcohol, biphenyl, d-carvone, deserpidine, disodium glycyrrhizinate, emetine.2HCI, ephedrine sulphate, eucalyptol, eugenol, gallic acid, geranyl acetate, b-N-[g-l(+)-glutamyl]-4-hydroxy-methylphenylhydrazine, glycyrrhetinic acid, glycyrrhizinate, disodium, p-hydrazinobenzoic acid, isosafrole, kaempferol, d-menthol, nicotine, norharman, pilocarpine, piperidine, protocatechuic acid, rotenone, rutin sulfate, sodium benzoate, turmeric oleoresin, vinblastine
These rodent carcinogens occur in:	absinthe, allspice, anise, apple, apricot, banana, basil, beet, broccoli, Brussels sprouts, cabbage, cantaloupe, caraway, cardamom, carrot, cauliflower, celery, cherries, chilli pepper, chocolate milk, cinnamon, cloves, cocoa, coffee, collard greens, comfrey herb tea, coriander, currants, dill, eggplant, endive, fennel, garlic, grapefruit, grapes, guava, honey, honeydew melon, horseradish, kale, lemon, lentils, lettuce, licorice, lime, mace, mango, marjoram, mushrooms, mustard, nutmeg, onion, orange, paprika, parsley, parsnip, peach, pear, peas, black pepper, pineapple, plum, potato, radish, raspberries, rhubarb, rosemary, rutabaga, sage, savory, sesame seeds, soybean, star anise, tarragon, tea, thyme, tomato, turmeric, and turnip.

Source: Gold et al

dose (MTD). Evidence is accumulating that it may be cell division caused by the high dose itself, rather than the chemical per se, that is increasing the cancer rate. High doses can cause chronic wounding of tissues, cell death, and consequent chronic cell division of neighbouring cells, which is a risk factor for cancer (Ames et al. 1996). Each time a cell divides it increases the probability that a mutation will occur, thereby increasing the risk for cancer. At the low levels to which humans are usually exposed, such increased cell division does not occur. Therefore, the very low levels of chemicals to which humans are exposed through water pollution or synthetic pesticide residues are likely to pose no or minimal cancer risks.

Table 2.3. Carcinogenicity in rodents of natural chemicals in roasted coffee

Positive. N=19	acetaldehyde, benzaldehyde, benzene, benzofuran, benzo(a)pyrene, caffeic acid, catechol, 1,2,S,6-dibenzanthracene, ethanol, ethylbenzene, formaldehyde, furan, furfural, hydrogen peroxide, hydroquinone, limonene, styrene, toluene, xylene
Not positive: N=8	acrolein, biphenyl, choline, eugenol, nicotinamide, nicotinic acid, phenol, piperidine
Uncertain:	caffeine
Yet to test:	~ 1000 chemicals

Source: Gold et al.

It seems likely that a high proportion of all chemicals, whether synthetic or natural, might be 'carcinogens' if run through the standard rodent bioassay at the MTD, but this will be primarily due to the effects of high doses for the non-mutagens, and a synergistic effect of cell division at high doses with DNA damage for the mutagens (Butterworth et al. 1995; Ames et al. 1993a; Ames and Gold 1990). Without additional data on the mechanism of carcinogenesis for each chemical, the interpretation of a positive result in

Table 2.4. Proportion of chemicals evaluated as carcinogenic.

Chemicals tested in both rats and mice	330/559	(59%)
Naturally-occurring chemicals	73/127	(57%)
Synthetic chemicals	257/432	(59%)
Chemicals tested in rats and/or mice		
Natural pesticides	35/64	(55%)
Mold toxins	14/23	(61%)
Chemicals in roasted coffee	19/28	(68%)
Innes negative chemicals retested[a]	16/34	(47%)
Drugs in the Physician's Desk Reference	117/241	(49%)

[a] The 1969 study by Innes et al. is frequently cited as evidence that the proportion of carcinogens is low, as only 9% of 119 chemicals tested (primarily pesticides) were positive in cancer tests on mice. However, these tests lacked the power of modern tests. We have found 34 of the Innes negative chemicals that have been retested using modern protocols: 16 were positive again about half.

Source: Gold et al.

a rodent bioassay is highly uncertain. The carcinogenic effects may be limited to the high does tested. The recent report of the National Research Council, *Science and Judgement in Risk Assessment* (NCI 1994) supports these ideas. The EPA's draft document *Working Paper for Considering Draft Revisions to the US EPA Guidelines for Cancer Risk Assessment* (NCI 1994) is a step towards improvement in the use of animal cancer test results.

In regulatory policy, the 'virtually safe dose' (VSD), corresponding to a maximum, hypothetical cancer risk of one in a million, is estimated from bioassay results using a linear model. To the extent that carcinogenicity in rodent bioassays is due to the effects of high doses for the non-mutagens, and a synergistic effect of cell division at high doses with DNA damage for the mutagens, then this model is inappropriate. Moreover, as currently calculated, the VSD can be known without ever conducting a bioassay: for 96 per cent of the NCI/NTP rodent carcinogens, the VSD is within a factor of 10 of the ratio MTD/740,000 (Gaylor and Gold 1995). This is about as precise as the estimate obtained from conducting near-replicate cancer tests of the same chemical (Gaylor and Gold 1995).

Misconception No. 6: synthetic chemicals pose greater carcinogenic hazards than natural chemicals

Gaining a broad perspective about the vast number of chemicals to which humans are exposed can be helpful when setting research and regulatory priorities (Gold et al. 1992; Ames et al. 1990b; Gold et al. 1993; Ames et al. 1987). Rodent bioassays provide little information about mechanisms of carcinogenesis and low dose risk. The assumption that synthetic chemicals are particularly hazardous has led to a bias in testing, such that synthetic chemicals account for 77 per cent of the 559 chemicals tested chronically in both rats and mice (Table 2.4). The natural world of chemicals has never been tested systematically. One reasonable strategy is to use a rough index to compare and rank possible carcinogenic hazards from a wide variety of chemical exposures at levels that humans typically receive, and then to focus on those that rank highest (Gold et al. 1992; Ames et al. 1987; Gold et al. 1994b). Ranking is a critical first step that can help to set priorities for selecting chemicals for chronic bioassay or mechanistic studies, for epidemiological research, and for regulatory policy. Although one cannot say whether the ranked chemical exposures are likely to be of major or minor importance in human cancer, it is not prudent to focus attention on the possible hazards at the bottom of a ranking if, using the same methodology to identify hazard, there are numerous common human exposures with much greater possible hazards. Our analyses are based on the HERP index (Human Exposure/Rodent Potency), which indicates what percentage of the rodent carcinogenic potency (TD_{50} in mg/kg/day) a human receives from a given daily lifetime exposure (mg/kg/day). TD_{50} values in our Carcinogenic Potency Database span a 10-million-fold range across chemicals (Gold et al. 1997c) (Table 2.5).

Overall, our analyses have shown that HERP values for some historically high exposures in the workplace and some pharmaceuticals rank high, and that there is an enormous background of naturally occurring rodent carcinogens in typical portions of common foods that cast doubt on the relative importance of low dose exposures to residues of synthetic chemicals such as pesticides (Gold et al. 1992; Ames et al. 1987; Gold et al. 1994a). A committee of the National Research Council/ National Academy of Sciences recently reached similar conclusions about natural as opposed to synthetic chemicals in the diet, and called for further research on natural chemicals (NRC 1996).

The possible carcinogenic hazards from synthetic pesticides (at average exposures) are minimal compared with the background of nature's pesticides, though neither may be a hazard at the low doses consumed (Table 2.5). Table 2.5 also indicates that many ordinary foods would not pass the regulatory criteria used for synthetic chemicals. For many natural chemicals the HERP values are in the top half of the table, even though natural chemicals are markedly underrepresented because so few have been tested in rodent bioassays. Caution is necessary in drawing conclusions from the occurrence in the diet of natural chemicals that are rodent carcinogens. It is not argued here that these dietary exposures are necessarily of much relevance to human cancer. Our results call for a re-evaluation of the utility of animal cancer tests in protecting the public against minor hypothetical risks.

Misconception No. 7: the toxicology of synthetic chemicals is different from that of natural chemicals

It is often assumed that because natural chemicals are part of human evolutionary history, whereas synthetic chemicals are recent, the mechanisms that have evolved in animals to cope with the toxicity of natural chemicals will fail to protect against synthetic chemicals. This assumption is flawed for several reasons (Ames et al. 1990b; Ames et al. 1996).

(a) Humans have many natural defences that make us well buffered against normal exposures to toxins (Ames, Profet, Gold 1990b), and these are usually general, rather than tailored for each specific chemical. Thus they work against both natural and synthetic chemicals. Examples of general defences include the continuous shedding of cells exposed to toxins - the surface layers of the mouth, oesophagus, stomach, intestine, colon, skin, and lungs are discarded every few days; DNA repair enzymes, which repair DNA damaged from many different sources; and detoxification enzymes of the liver and other organs which generally target classes of toxins rather than individual toxins. That defences are usually general, rather than

Table 2.5. Ranking possible carcinogenic hazards from average US exposures

Possible hazard: HERP (%)	*Average daily US exposure*	*Human dose of rodent carcinogen*	*Potency TD_{50} (mg/kg/day)[a]*	
			Rats	*Mice*
140	EDB: workers (high exposure) (before 1977)	Ethylene dibromide, 150 mg	1.52	(7.45)
17	Clofibrate	Clofibrate, 2 g	169	
14	Phenobarbital, 1 sleeping pill	Phenobarbital, 60 mg	(+)	6.09
6.8	1,3-Butadiene: rubber workers (1978-86)	1,3-Butadiene, 66.0 mg	(261)	13.9
6.1	Tetrachloroethylene: dry cleaners with dry-to-dry units (1980-90)b	Tetrachloroethylene, 433 mg	101	(126)
4.0	Formaldehyde: workers	Formaldehyde, 6.1 mg	2.19	(43.9)
2.1	**Beer, 257 g**	Ethyl alcohol, 13.1 ml	9110	(—)
1.4	Mobile home air (14 hours/day)	Formaldehyde, 2.2 mg	2.19	(43.9)
0.9	Methylenechloride: workers (1940s-80s)	Methylenechloride, 471 mg	724	(918)
0.5	**Wine, 28.0 g**	**Ethyl alcohol, 3.36 ml**	91 I0	(—)
0.4	Conventional home air (14 hours/day)	Formaldehyde, 598 mg	2.19	(43.9)
0. 1	**Coffee, 13.3 g**	**Caffeic acid, 23.9 mg**	297	(4900)
0.04	**Lettuce, 14.9 g**	**Caffeic acid, 7.90 mg**	297	(4900)
0.03	**Safrole in spices**	**Safrole, 1.2 mg**	(441)	51.3
0.03	**Orange juice, 138 g**	**d-Limonene, 4.28 mg**	204	(—)
0.03	**Pepper, black, 446 mg**	**d-Limonene, 3.S7 mg**	204	(—)
0.02	**Mushroom (*Agaricus bisporus* 2.55 g)**	**Mixture of hydrazines, etc. (whole mushroom)**	—	20,300
0.02	**Apple, 32.0 g**	**Caffeic acid, 3.40 mg**	297	(4900)
0.02	**Coffee, 13.3 g**	**Catechol, 1.33 mg**	118	(244)
0.02	**Coffee, 13.3 g**	**Furfural, 2.09 mg**	(683)	197
0.009	BHA: daily US avg (1975)	BHA, 4.6 mg	745	(5530)
0.008	**Beer (before 1979), 257 g**	**Dimethylnitrosamine, 726 ng**	0.124	(0.189)
0.008	**Aflatoxin: daily US avg (1984-89**	**Aflatoxin, 18 ng**	0.0032	(+)
0.007	**Cinnamon, 21.9 mg**	**Coumarin, 65.0 mg**	13.9	(103)
0.006	**Coffee, 13.3 g**	**Hydroquinone, 333 mg**	82.8	(225)
0.005	Saccharin: daily US avg (1977)	Saccharin, 7 mg	2140	(—)
0.005	**Carrot, 12.1 g**	**Aniline, 624 mg**	194C	(—)
0.004	**Potato, 54.9 g**	**Caffeic acid, 867 mg**	297	(4900)
0.004	**Celery, 7.95 g**	**Caffeic acid, 858 mg**	297	(4900)
0.004	**White bread, 67.6 g**	**Furfural, 500 mg**	(683)	197
0.003	**Nutmeg, 27.4 mg**	**d-Limonene, 466 mg**	204	(—)
0.003	Conventional home air (14 hour/day)	Benzene, 155 mg	(169)	77.5
0.002	**Carrot, 12.1 g**	**Caffeic acid, 374 mg**	297	(4900)
0.002	Ethylene thiourea: daily US avg (1990)	Ethylene thiourea, 9.51 mg	7.9	(23.5)
0.002	[DDT: daily US avg (before 1972 ban)]	[DDT, 13.8 mg]	(84.7)	12.3
0.001	**Plum, 2.00 g**	**Caffeic acid, 276 mg**	297	(4900)
0.001	BHA: daily US avg (1987)	BHA, 700 mg	745	(5530)
0.001	**Pear, 3.29 g**	**Caffeic acid, 240 mg**	297	(4900)
0.00I	[UDMH: daily US avg (1988)]	[UDMH, 2.82 mg (from Alar)]	(—)	3.96
0.0009	**Brown mustard, 68.4 mg**	**Allyl isothiocyanate, 62.9 mg**	96	(—)
0.0008	[DDE: daily US avg (before 1972 ban)]	[DDE, 6.91 mg]	(—)	12.5

Table 2.5. (con't) Ranking possible carcinogenic hazards from average US exposures

0.0007	TCDD: daily US avg (1994)	TCDD, 12.0 pg	0.0000235	(0.000156)
0.0007	Bacon, 11.5 g	Diethylnitrosamine, 11.5 ng	0.0237	(+)
0.0006	**Mushroom (*Agaricus bisporus* 2.55 g)**	**Glutamyl-p-hydrazino-benzoate, 107 mg**	.	277
0.0005	**Jasmine tea, 2.19 g**	**Benzyl acetate, 504 mg**	(—)	1440
0.0004	**Bacon, 11.5 g**	**N-Nitrosopyrrolidine, 196 ng**	(0.799)	0.679
0.0004	**Bacon, 11.5 g**	**Dimethylnitrosamine, 34.5 ng**	0.124	(0.189)
0.0004	[EDB: Daily US avg (before 1984 ban)]	[EDB, 420 ng]	1.52	(7.45)
0.0004	Tap water, 1 litre (1987-92)	Bromodichloromethane, 13 mg	(72.5)	47.7
0.0003	**Mango, 1.22 g**	**d-Limonene, 48.8 mg**	204	(—)
0.0003	**Beer, 257 g**	**Furfural, 39.9 mg**	(683)	197
0.0003	Tap water, 1 litre (1987-92)	Chloroform, 17 mg	(262)	90.3
0.0003	Carbaryl: daily US avg (1990)	Carbaryl, 2.6 mg	14.1	(—)
0.0002	**Celery, 7.95 g**	**8-Methoxypsoralen, 4.86 mg**	32.4	(—)
0.0002	Toxaphene: daily US avg (1990)	Toxaphene, 595 ng	(—)	5.57
0.00009	**Mushroom (*Agaricus bisporus*, 2.55 g)**	**p-Hydrazinobenzoate, 28 mg**	.	45
0.00008	PCBs: daily US avg (1984-86)	PCBs, 98 ng	1.74	(9.58)
0.00008	DDE/DDT: daily US avg (1990)	DDE, 659 ng	(—)	12.5
0.00007	**Parsnip, 54.0 mg**	**8-Methoxypsoralen, 1.57 mg**	32.4	(—)
0.00007	**Toast, 67.6 g**	**Urethane, 811 ng**	(41.3)	16.9
0.00006	**Hamburger, pan fried, 85 g**	**PhIP, 176 ng**	4.29c	(28.6c)
0.00005	**Estragole in spices**	**Estragole, 1.99 mg**	.	51.8
0.00005	**Parsley, fresh, 324 mg**	**8-Methoxypsoralen, 1.17 mg**	32.4	(—)
0.00003	**Hamburger, pan fried, 85 g**	**MeIQx, 38.1 ng**	1.99	(24.3)
0.00002	Dicofol: daily US avg (1990)	Dicofol, 544 ng	(—)	32.9
0.00001	**Cocoa, 3.34 g**	**a-Methylbenzyl alcohol, 4.3 mg**	458	(—)
0.00001	**Beer, 257 g**	**Urethane, 115 ng**	(41.3),	16.9
0.000005	**Hamburger, pan fried, 85 g**	**IQ, 6.38 ng**	1.89c	(19.6)
0.000001	Lindane: daily US avg (1990)	Lindane, 32 ng	(—)	30.7
0.0000004	PCNB: daily US avg (1990)	PCNB (Quintozene), 19.2 ng	(—)	71.1
0.0000001	Chlorobenzilate: daily US avg (1989)	Chlorobenzilate, 6.4 ng	(—)	93.9
<0.00000001	Chlorothalonil: daily US avg (1990)	Chlorothalonil, <6.4 ng	828d	(—)
0.000000008	Folpet: daily US avg (1990)	Folpet, 12.8 ng		2280d
0.000000006	Captan: daily US avg (1990)	Captan, 11.5 ng	269d	(2730d)

a = no data in CPDB; (—) = negative in cancer test; (+) = positive cancer test(s) not suitable for calculating a TD_{50}.

b = This is not an average, but a reasonably large sample (1027 workers).

c TD_{50} harmonic mean was estimated for the base chemical from the hydrochloride salt.

d Additional data from EPA that is not in the CPDB were used to calculate these TD_{50} harmonic means.

[Chemicals that occur naturally in foods are in bold.] Daily human exposure: Reasonable daily intakes are used to facilitate comparisons. The calculations assume a daily dose for a lifetime. Possible hazard: The human dose of rodent carcinogen is divided by 70 kg to give a mg/kg/day of human exposure, and this dose is given as the percentage of the TD_{50} in the rodent (mg/kg/day) to calculate the Human Exposure/Rodent Potency index (HERP), i.e. 100% means that the human exposure in mg/kg/day is equal to the dose estimated to give 50% of the rodents tumours. TD_{50} values used in the HERP calculation are averages calculated by taking the harmonic mean of the TD_{50}s of the positive tests in that species from the Carcinogenic Potency Database. Average TD_{50} values have been calculated separately for rats and mice, and the more potent value is used for calculating possible hazard.

Source. Gold et al.

specific for each chemical, makes good evolutionary sense. The reason that predators of plants evolved general defences is presumably to be prepared to counter a diverse and ever-changing array of plant toxins in an evolving world; if a herbivore had defences against only a set of specific toxins, it would be at a great disadvantage in obtaining new food when favoured foods became scarce or evolved new toxins.

(b) Various natural toxins, which have been present throughout vertebrate evolutionary history, nevertheless cause cancer in vertebrates (Ames et al. 1990b; Gold et al. 1997). Mould toxins, such as aflatoxin, have been shown to cause cancer in rodents and other species including humans (Table 2.4). Many of the common elements are carcinogenic to humans at high doses (e.g. salts of cadmium, beryllium, nickel, chromium, and arsenic) despite their presence throughout evolution. Furthermore, epidemiological studies from various parts of the world show that certain natural chemicals in food may be carcinogenic risks to humans; for example, the chewing of betel nuts with tobacco has been correlated with oral cancer world-wide.

(c) Humans have not had time to evolve a 'toxic harmony' with all of their dietary plants. The human diet has changed dramatically in the last few thousand years. Indeed, very few of the plants that humans eat today (e.g. coffee, cocoa, tea, potatoes, tomatoes, corn, avocados, mangoes, olives, and kiwi fruit), would have been present in a hunter-gatherer's diet. Natural selection works far too slowly for humans to have evolved specific resistance to the food toxins in these newly introduced plants.

(d) DDT is often viewed as the quintessentially dangerous synthetic pesticide because it concentrates in the tissues and persists for years, being slowly released into the bloodstream. DDT, the first synthetic pesticide, eradicated malaria from many parts of the world, including the US. It was effective against many vectors of disease such as mosquitoes, tsetse flies, lice, ticks, and fleas. DDT was also lethal to many crop pests, and significantly increased the supply and lowered the cost of food, making fresh nutritious foods more accessible to poor people. It was also remarkably non-toxic to humans. A 1970 National Academy of Sciences report concluded: 'In little more than two decades DDT has prevented 500 million deaths due to malaria, that would otherwise have been inevitable (NAS 1970).' There is no convincing epidemiological evidence, nor is there much toxicological plausibility, that the levels normally found in the environment are likely to be a significant contributor to cancer. DDT was unusual with respect to bioconcentration, and because of its chlorine substituents it takes longer to degrade in nature than most chemicals; however, these are properties of relatively few synthetic chemicals. In addition, many thousands of chlorinated chemicals are produced in nature and natural pesticides also can bioconcentrate if they are fat soluble. Potatoes, for example, naturally contain the fat soluble neurotoxins solanine and chaconine, which can be detected in the bloodstream of all potato eaters. High levels of these potato neurotoxins have been shown to cause birth defects in rodents (Ames et al. 1990b).

(e) Since no plot of land is immune to attack by insects, plants need chemical defences - either natural or synthetic - in order to survive pest attack. Thus, there is a trade-off between naturally occurring pesticides and synthetic pesticides. One consequence of disproportionate concern about synthetic pesticide residues is that some plant breeders develop plants to be more insect-resistant by making them higher in natural toxins. A recent case illustrates the potential hazards of this approach to pest control: when a major grower introduced a new variety of highly insect-resistant celery people who handled the celery developed rashes when they were subsequently exposed to sunlight. Some detective work found that the pest-resistant celery contained 6,200 parts per billion (ppb) of carcinogenic (and mutagenic) psoralens instead of the 800 ppb present in common celery (Ames et al. 1990b).

Misconception No. 8: pesticides and other synthetic chemicals are disrupting our hormones

Synthetic hormone mimics are likely to be the next big environmental issue, with accompanying large expenditures. Hormonal factors are important in cancer (Misconception No. 2). A recent book (Colburn et al. 1996), holds that traces of synthetic chemicals, such as pesticides with weak hormonal activity, may contribute to cancer and reduce sperm counts. The book ignores the fact that our normal diet contains natural chemicals that have oestrogenic activity millions of times higher than that due to the traces of synthetic oestrogenic chemicals (Safe 1994, 1995) and that lifestyle factors can markedly change the levels of endogenous hormones (Misconception No. 2). The low levels of exposure to residues of industrial chemicals in humans are toxicologically implausible as a significant cause of cancer or of reproductive abnormalities, especially when compared with the natural background (Safe 1994, 1995; Reinli and Block 1996). In addition, it has not been satisfactorily shown that sperm counts really are declining (Kolata 1996), and, even if they were, there are many more likely causes, such as smoking and diet (Misconception No. 2).

Misconception No. 9: regulation of low hypothetical risks advances public health

There is no risk-free world, and resources are limited; therefore, society must set priorities based on which risks are most important in order to save the most lives. The EPA reports that its regulations cost $140 billion per year. It has been argued that overall these regulations harm public health (Viscusi 1992), because 'wealthier is not only healthier but highly risk reducing'. One estimate indicates 'that for every 1 per cent increase in income, mortality is reduced by 0.05 per cent' (Shanahan and Thierer 1996). In addition, the median toxin control programme costs 58 times more per life-year saved than the median injury prevention programme and 146 times more than the median medical programme (Tengs et al. 1995). It has been estimated that the US could prevent 60 000 deaths a year by redirecting resources to more cost-effective programmes (Tengs and Graham 1996). The discrepancy is likely to be greater because cancer risk estimates used for toxin control programmes are worst-case, hypothetical estimates, and the true risks at low dose are often likely to be zero (Gold et al. 1992; Gold et al. 1997; Graham and Wiener 1995) (Misconception No. 5).

Regulatory efforts to reduce low-level human exposures to synthetic chemicals are expensive because they aim to eliminate minuscule concentrations that now can be measured with improved techniques. These efforts are distractions from the major task of improving public health through increasing knowledge, public understanding of how lifestyle influences health, and effectiveness in incentives and spending to maximize health. Basic biomedical research is the basis for improved public health and longevity, yet its cost is less than 10 per cent the cost to society of EPA regulations.

Of course, rules on air and water pollution are necessary (e.g. it was a public health advance to phase lead out of gasoline) and, clearly, cancer prevention is not the only reason for regulations. But worst-case scenarios, with their associated large costs to the economy, are not in the interest of public health and can be counterproductive.

References

Ames, B. N., Magaw, R., and Gold, L. S. (1987). Ranking possible carcinogenic hazards. *Science* **236**, 271-280.

Ames, B. N., and Gold, L. S. (1990). Chemical carcinogenesis: Too many rodent carcinogens. *Proc. Natl. Acad. Sci. USA* **87**, *7772-7776.*

Ames, B. N., Profet, M., and Gold, L. S. (1990a). Dietary pesticides (99. 99 per cent all natural). *Proc. Natl. Acad. Sci. USA* **87**, 7777-7781.

Ames, B. N., Profet, M., and Gold, L. S. (1990b). Nature's chemicals and synthetic chemicals: Comparative toxicology. *Proc. Natl. Acad. Sci. USA* **87**, 7782-7786.

Ames, B. N., Shigenaga, M. K., and Gold, L. S. (1993a). DNA lesions, inducible DNA repair, and cell division: Three key factors in mutagenesis and carcinogenesis. *Environ. Health Perspect.* **101**(Suppl 5), 35-44.

Ames, B. N., Shigenaga, M. K., and Hagen, T. M. (1993b). Oxidants, antioxidants, and the degenerative diseases of ageing. *Proc. Natl. Acad. Sci. USA* **90**, 7915-7922.

Ames, B. N., Motchnik, P. A., Fraga, C. G. et al. (1994). Antioxidant prevention of birth defects and cancer. In *Male-Mediated Developmental Toxicity*. (D. R. Mattison, and A.

Olshan, eds). New York: Plenum Publishing Corporation, pp. 243-259.

Ames, B. N., Gold, L. S., and Willett, W. C. (1995). The causes and prevention of cancer. *Proc. Natl. Acad. Sci. USA* **92**, 5258-5265.

Ames, B. N., Gold, L. S., and Shigenaga, M. K. (1996). Cancer prevention, rodent high dose cancer tests and risk assessment. *Risk Anal.* **16**, 613-617.

Bailey, L. B., Wagner, P. A., Christakis, G. J. et al. (1979). Folacin and iron status and haematological findings in predominately black elderly persons from urban low-income households. *Am. J. Clin. Nutr.* **32**, 2346-2353.

Bailey, L. B., Wagner, P. A., Christakis, G. J. et al. (1982). Folacin and iron status and haematological findings in black and Spanish-American adolescents from urban low-income households. *Am. J. Clin. Nutr.* **35**, 1023-1032.

Block, G. (1992). The data support a role for antioxidants in reducing cancer risk. *Nutr. Reviews* **50**, 207-213.

Block, G., Patterson, B., and Subar, A. (1992). Fruit, vegetables and cancer prevention: A review of the epidemiologic evidence. *Nutr. and Canc.* **18**, 1-29.

Blount, B. C., Mack, M. M., Wehr, C. et al. (1997). Folate deficiency causes uracil misincorporation into human DNA and chromosome breakage: Implications for cancer and neuronal damage. *Proc. Natl. Acad. Sci. USA* **94**, in press.

Butterworth, B., Conolly, R., and Morgan, K. (1995). A strategy for establishing mode of action of chemical carcinogens as a guide for approaches to risk assessment. *Canc. Lett.* **93**, 129-146.

Christen, S., Hagen, T. M., Shigenaga, M. K. et al. (1997). Chronic infection and inflammation lead to mutation and cancer. In *Microbes and Malignancy: Infection as a Cause of Cancer*. (J. Parsonnet, S. Horning, eds) Oxford: Oxford University Press, in press.

Colburn, T., Dumanoski, D., and Myers, J. P. (1996). *Our Stolen Future: Are we Threatening our Fertility, Intelligence, and Survival?: A Scientific Detective Story*. New York: Dutton.

Devesa, S. S., Blot, W. J., Stone, B. J. et al. (1995). Recent cancer trends in the United States. *J. Natl. Canc. Inst.* **87**, 175-182.

Doll, R., and Peto, R. (1981). The causes of cancer. Quantitative estimates of avoidable risks of cancer in the United States today. *J. Natl. Canc. Inst.* **66**, 1191-1308.

Feigelson, H. S., and Henderson, B. E. (1996). Oestrogens and breast cancer. *Carcinogenesis* **17**, 2279-2284.

Fraga, C. G., Motchnik, P. A., Shigenaga, M. K. et al. (1991). Ascorbic acid protects against endogenous oxidative damage in human sperm. *Proc. Natl. Acad. Sci. USA* **88**, 11003-11006.

Fraga, C. G., Motchnik, P. A., Wyrobek, A. J. et al. (1996). Smoking and low antioxidant levels increase oxidative damage to sperm DNA. *Mutat. Res.* **351**, 199-203.

Gaylor, D. W., and Gold, L. S. (1995). Quick estimate of the regulatory virtually safe dose based on the maximum tolerated dose for rodent bioassays. *Regul. Toxicol. Pharmacol.* **22**, 57-63.

Gold, L. S., Slone, T. H., Stern, B. R. et al. (1992). Rodent carcinogens, Setting priorities. *Science* **258**, 261-265.

Gold, L. S., Slone, T. H., Stern, B. R. et al. (1993). Possible carcinogenic hazards from natural and synthetic chemicals: Setting priorities. In Cothern, C. R. (ed.), *Comparative Environmental Risk Assessment*. Boca Raton, FL: Lewis Publishers, pp. 209-235.

Gold, L. S., Garfinkel, G. B., and Slone, T. H. (1994a). Setting priorities among possible carcinogenic hazards in the workplace. In *Chemical Risk Assessment and Occupational Health, Current Applications, Limitations, and Future Prospects*. (C.

Smith, D. C. Christiani, K. T. Kelsey, eds). Westport, CT: Greenwood Publishing Group, pp. 91-103.

Gold, L. S., Slone, T. H., Manley, N. B., and Ames, B. N. (1994b). Heterocyclic amines formed by cooking food: Comparison of bioassay results with other chemicals in the Carcinogenic Potency Database. *Canc. Lett.* **83**, 21-29.

Gold, L. S., Slone, T. H., and Ames, B. N. (1997a). Prioritization of possible carcinogenic hazards in food. In Tennant, D. (ed.), *Food Chemical Risk Analysis*. London: Chapman and Hall, in press.

Gold, L. S., Slone, T. H., and Ames, B. N. (1997b). Overview and update analyses of the carcinogenic potency database. In *Handbook of Carcinogenic Potency and Genotoxicity Databases.* (L. S. Gold, and E. Zeiger, eds) Boca Raton, FL: CRC Press, 661-685.

Gold, L. S., Slone, T. H., Manley, N. B. et al. (1997c). Carcinogenic potency database. In *Handbook of Carcinogenic Potency and Genotoxicity Databases* (L. S. Gold, and E. Zeiger, eds) Boca Raton, FL: CRC Press, pp. 1-605.

Gough, M. (1990). How much cancer can EPA regulate anyway? *Risk Anal.* **10**, 1-6.

Graham, J., and Wiener, J., eds (1995). *Risk versus Risk: Trade-offs in Protecting Health and the Environment.* Cambridge, Massachusetts: Harvard University Press.

Hahn, R. W., ed. (1996). *Risks, Costs, and Lives Saved: Getting Better Results from Regulation.* New York: Oxford University Press.

Henderson, B. E., Ross, R. K., and Pike, M. C. (1991). Toward the primary prevention of cancer. *Science* **254**, 1131-1138.

Herbert, V. (1996). Vitamin B-12. In *Present Knowledge in Nutrition* (Ziegler, E. E., and Filer, L. J., eds). Washington, DC: ILSI Press, pp. 191-205.

Hill, M. J., Giacosa, A., and Caygill, C. P. J. (1994). *Epidemiology of Diet and Cancer.* Chichester: Ellis Horwood.

Hunter, D. J., and Willett, W. C. (1993). Diet, body size, and breast cancer. *Epidemiol. Rev.* **15**, 110-132.

Innes, J. R. M., Ulland, B. M., Valerio, M. G. et al. (1969). Bioassay of pesticides and industrial chemicals for tumourigenicity in mice: A preliminary note. *J. Natl. Canc. Inst.* **42**, 1101-1114.

Jacobson, E. L. (1993). Niacin deficiency and cancer in women. *J. Am. Coll. Nutr.* **12**, 412-6.

Ji, B. -T., Shu, X. -O., Linet, M. S. et al. (1997). Paternal cigarette smoking and the risk of childhood cancer among offspring of non-smoking mothers. *J. of Natl. Canc. Inst.* **89**, 238-244.

Kolata, G. (1996). Measuring men up, sperm by sperm. *New York Times*, May 4 E4(N), E4(L), (col. 1).

Kosary, C. L., Ries, L. A. G., Miller, B. A. et al. (eds) (1995). *SEER Cancer Statistics Review*, 1973-1992. National Cancer Institute, Bethesda, MD.

National Academy of Sciences (US). Committee on Research in the Life Sciences. (1970). *The Life Sciences: Recent Progress and Application to Human Affairs, the World of Biological Research, Requirement for the Future.* National Academy of Sciences, Washington.

National Research Council (1994). *Science and Judgement in Risk Assessment.* Committee on Risk Assessment of Hazardous Air Pollutants, Washington, DC.

National Research Council (1996). *Carcinogens and Anti-carcinogens in the Human Diet: A Comparison of Naturally Occurring and Synthetic Substances.* Washington, DC: National Academy Press.

National Cancer Institute Graphic (A). (1996). Why eat five? *J. Natl. Canc. Inst.* **88**, 1314.

Patterson, B. H., Block, G., Rosenberger, W. F. et al. (1990). Fruit and vegetables in the

American diet: Data from the NHANES II survey. *Am. J. Public Health* **80**, 1443-1449.

Peto, R., Lopez, A. D., Boreham, J. et al. (1994). *Mortality from Smoking in Developed Countries 1950-2000*. Oxford: Oxford University Press.

Reinli, K., and Block, G. (1996). Phyto-oestrogen content of foods - A compendium of literature values. *Nutr. Canc.* **26**, 1996.

Safe, S. H. (1994). Dietary and environmental oestrogens and anti-oestrogens and their possible role in human disease. *Environ. Sci. Pollution Res.* **1**, 29-33.

Safe, S. H. (1995). Environmental and dietary oestrogens and human health: Is there a problem? *Env. Health Persp.* **103**, 346-351.

Senti, F. R., and Pilch, S. M. (1985). Analysis of folate data from the second National Health and Nutrition Examination Survey (NHANES II). *J. Nutr.* **115**, 1398-402.

Shacter, E., Beecham, E. J., Covey, J. M. et al. (1988). Activated neutrophils induce prolonged DNA damage in neighbouring cells [published erratum appears in Carcinogenesis 1989 Mar 10(3),628]. *Carcinogenesis* **9**, 2297-2304.

Shanahan, J. D., and Thierer, A. D. (1996). How to talk about risk: How well-intentioned regulations can kill: TP13, Washington, DC, Heritage Foundation.

Steinmetz, K. A., and Potter, J. D. (1991). Vegetables, fruit, and cancer. I. Epidemiology. *Cancer Causes Control* **2**, 325-357.

Steinmetz, K. A., and Potter, J. D. (1996). Vegetables, fruit, and cancer prevention: A review. *J. Am. Diet Assoc.* **96**, 1027-1039.

Subar, A. F., Block, G., and James, L. D. (1989). Folate intake and food sources in the US population. *Am. J. Clin. Nutr.* **50**, 508-16.

Tengs, T. O., Adams, M. E., Pliskin, J. S. et al. (1995). Five-hundred life-saving interventions and their cost-effectiveness. *Risk Anal.* **15**, 369-389.

Tengs, T. O., and Graham, J. D. (1996). The opportunity costs of haphazard social investments in life-saving. In: *Risks, Costs, and Lives Saved: Getting Better Results from Regulation* (R. Hahn, ed.). New York: Oxford University Press, pp. 165-173.

Viscusi, W. K. (1992). *Fatal Trade-offs*. Oxford: Oxford University Press.

Wallock, L., Woodall, A., Jacob, R., and Ames, B. (1997). Nutritional status and positive relation of plasma folate to fertility indices in non-smoking men. *FASEB J. (Abstract)* in press.

Wickramasinghe, S. N., and Fida, S. (1994). Bone marrow cells from vitamin B12- and folate-deficient patients misincorporate uracil into DNA. *Blood* **83**, 1656-61.

Yamashina, K., Miller, B. E., and Heppner, G. H. (1986). Macrophage-mediated induction of drug-resistant variants in a mouse mammary tumour cell line. *Cancer Res.* **46**, 2396-2401.

Zhang, J. Z., Henning, S. M., and Swendseid, M. E. (1993). Poly(ADP-ribose) polymerase activity and DNA strand breaks are affected in tissues of niacin-deficient rats. *J. Nutr.* **123**, 1349-55.

3 Are dietary nitrates a threat to human health?

Jean-Louis L'hirondel

Summary

Two major charges were levelled at nitrates some thirty years ago: infant methaemoglobinaemia or 'blue-baby syndrome', and a greater risk of cancer in adults. These were either presumptions or hypotheses.

The many scientific studies carried out over the last few decades allow us to conclude that neither of those grievances were founded. Dietary nitrates pose no threat to human health.

Nitrates present in blood plasma have two sources; one source of nitrates is exogenous - from food; 80 per cent of these alimentary nitrates come from vegetables and 10 to 15 per cent from drinking water; there is also an endogenous source, providing a similar quantity, of cell origin, involving the amino acid, L-arginine and nitrogen monoxide (NO). In addition to passive urinary excretion, there are two active secretions of nitrate (NO_3) from plasma: colonic secretion and salivary secretion. Salivary secretion reintroduces NO_3 ions a second time in the mouth.

The directives issued in 1962 by the UN's WHO and Food and Agricultural Organisation (FAO), and in 1980 by the EEC are now redundant. The directive on drinking water is in addition very costly. They need repealing; eventually, this will become inevitable.

Introduction

From the twelfth to the nineteenth century, nitrates (NO_3) were used as medicines, sometimes in very large doses, for a wide, and sometimes surprising, range of symptoms.

At the beginning of the twentieth century, the development of aspirin followed by the introduction of corticoids meant the end of the therapeutic use of nitrates. By the 1950s, the growing incidence of infant methaemoglobinaemia in some rural areas of the United States, together with the discovery of the carcinogenicity of many nitrosamines in animals, contributed to bringing dietary nitrates under suspicion and to casting doubt on their innoccuousness.

In 1962 these doubts led the Committee of Experts on Food Additives of the WHO and the FAO to set an acceptable daily intake (ADI) level for man at 3.65 mg/kg of NO_3, and in 1980 the European Community issued a directive setting a limit for NO_3 in drinking water of 50 mg/l, above which water is no longer deemed fit for human consumption.

Some thirty-seven years after the WHO and the FAO's decision and eighteen years after the European Community's directive, the scientific perspective has changed totally. Many studies and experiments have taken place; any suspicions which were legitimate a few decades ago can no longer be justified.

The metabolism of nitrates

Figure 3.1 gives an overview of the metabolism of nitrates. Nitrates are always present in our bloodstream, at levels normally ranging between 1 and 3 mg/l before meals. In normal conditions two sources of nitrates coexist - the exogenous source from food and water, and the endogenous source from cell activity; they each provide 70 to 75 mg/day.

Of the food source, 80 per cent of nitrates ingested by man normally come from vegetables, and 10 to 15 per cent from drinking water. On swallowing, these nitrates pass down into the stomach as NO_3, before being quickly and virtually totally absorbed in the upper section of the small intestine. Less than 2 per cent of all nitrates ingested therefore reach the large intestine (Bartholomew 1984).

The endogenous source of nitrates has been known for some thirteen years, since the research carried out by Stuehr and Marletta in 1985. The metabolic process of the amino acid L-arginine releases a nitrogen atom at cell level which forms a molecule of nitrogen monoxide, NO. Outside the cell, the NO molecules combine with oxygen to form various molecules including nitrates, nitrites and nitrosamines. Many physiological activities, like running and cycling, and pathological conditions, like infections, lead to cell stimulation and thereby contribute to increasing this endogenous synthesis of nitrates.

Thereafter, the fate of plasmatic nitrates, from two different sources - exogenous and endogenous - is rather unusual:

- A small proportion of plasmatic nitrates, approximately 10 per cent of the quantity of nitrates ingested, is eliminated through sweat and tears.
- The largest amount of nitrates leaving the plasma is passively excreted in urine, this excretion being based only on NO_3 plasmatic concentrations.

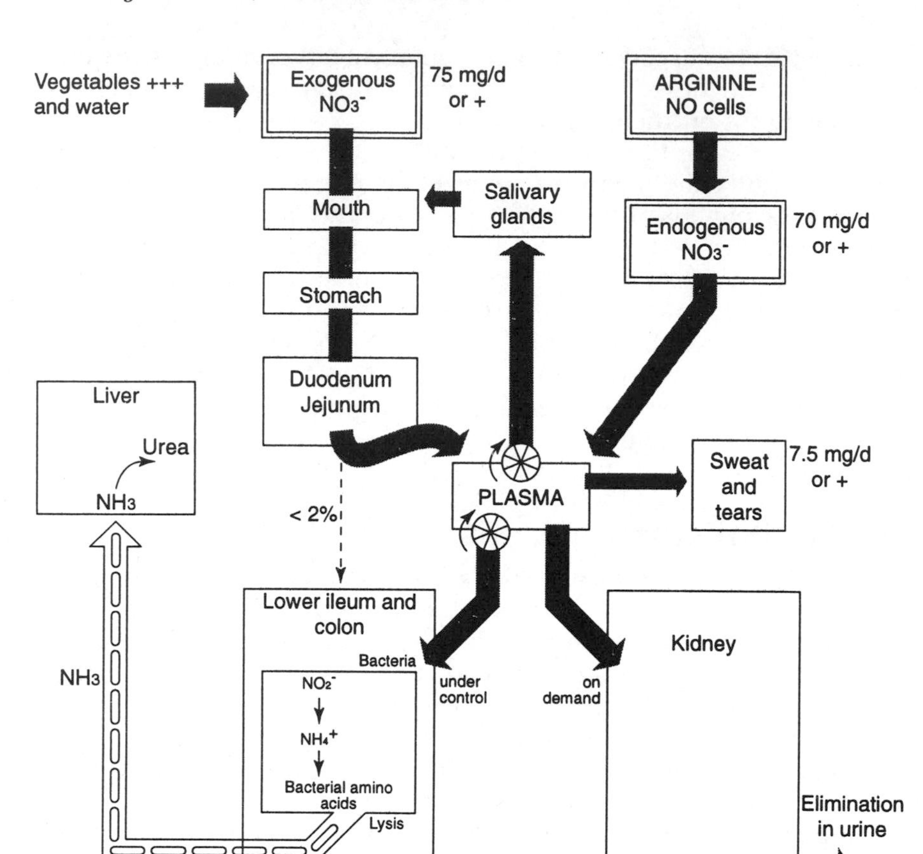

Figure 3.1 The metabolism of nitrates in man

- Two active phenomena also occur, which have a significant impact from a physiological viewpoint: the colonic and salivary secretions of nitrates.
- One of the functions of the columnar cells of the colonic epithelium is to draw NO_3 ions from the plasmatic sector towards the colon's light through an active capture phenomenon. The purpose of this colonic secretion of NO_3 ions is most probably to ensure the nutrition of the colonic bacterial flora.
- The cells of the salivary acinus also actively draw NO_3 ions from the plasmatic sector and release them into the saliva, its secretion product. At times therefore, levels of salivary nitrates are between 6 and 30 times higher than for plasmatic nitrates.

Thereafter, these salivary nitrates, which remain for a while in the mouth, come under the influence of bacterial enzymes produced from a relatively abundant physiological bacterial flora.

Some of these salivary nitrates (NO_3) thereby turn into salivary nitrites (NO_2).

As Figure 3.2 shows, NO_3 ions therefore pass through the mouth twice, the first time as dietary nitrates, the second as salivary nitrates. Only the latter process induces the formation of a certain amount of salivary nitrites, which reach the stomach when the saliva is swallowed.

The role of this salivary secretion of nitrates merits clarification. It is quite possibly a preliminary stage in the digestion of proteins, as salivary nitrates have the ability to make food proteins more sensitive to the subsequent action of proteolytic enzymes (pepsine and trypsine). Benjamin also showed in 1994 that in an acid medium, ingested salivary nitrates, which have therefore reached the stomach, release nitrogen monoxide and thereby destroy organisms such as Candida albicans and Escherichia coli, thereby promoting host defence against ingested pathogens.

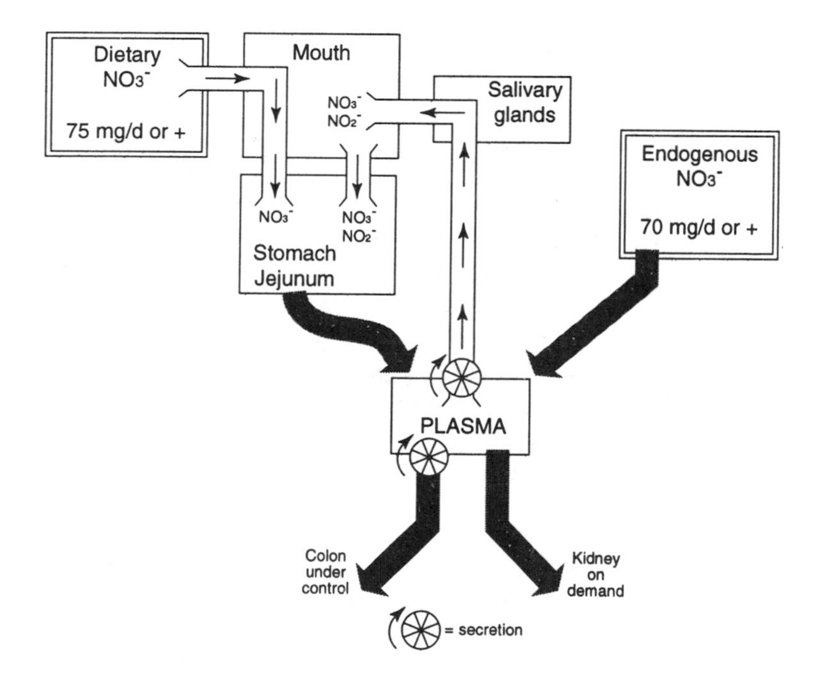

Figure 3.2 The duality of nitrates in the mouth. Dietry nitrates and nitrates secreted by the salivary glands are two distinct entities. Only NO_3^- ions secreted by the salivary glands, as NO_2^- precursors, can induce nitrosamines. By contrast, NO_3^- ions from food reach the stomach intact, intact, i.e. without turning into NO_2^{3-} and therefore without any risk of turning into nitrosamines

Grievances against nitrates and their refutation

In the 1950s and 1960s, alimentary nitrates aroused disquiet on two counts. They were thought to be responsible for methaemoglobinaemia in infants, and people wondered whether they might not also induce the onset of some cancers, in particular stomach cancer. These two presumptions gave rise to two major grievances. The many studies conducted in the last thirty years now allow us to state that both are unfounded.

Dietary methaemoglobinaemia in infants - Blue Baby syndrome

Methaemoglobin is an oxidised derivative of haemoglobin, which loses its ability to carry oxygen molecules. In the physiological state, 1 to 2 per cent of haemoglobin in the red cells are in the form of methaemoglobin. Clinical disorders, in this case cyanosis, appear if methaemoglobinaemia levels exceed the 10 to 20 per cent threshold. Any increase reaching 70 to 80 per cent can be fatal.

The transformation of haemoglobin into methaemoglobin in red cells is due to the action of nitrites rather than nitrates, which are very powerful oxidants. Babies under the age of six months are most at risk as they have not yet fully developed a protective enzyme system (reductase methaemoglobin or NADH-cytochrome b5 reductase). Beyond the age of six months, the risk of pathological methaemoglobinaemia no longer exists.

Prior to 1984, it was generally thought that nitrates in feeding bottles turned into nitrites in the baby's colon, after contact of these nitrates with the colon's large bacterial flora, as only bacterial enzymes are capable of reducing NO_3 nitrates into NO_2 nitrites. However, since Bartholomew's studies in 1984 showed that 98 per cent of alimentary nitrates are absorbed in the upper section of the small intestine, this explanation is no longer valid.

Some scientists then wondered whether the nitrates-nitrites transformation could not take place in the baby's stomach, as a result of a colonisation of the stomach by micro-organisms of enteric origin, under a hypochlorhydric effect of the gastric juices. However, studies on this subject are not conclusive, as Walker showed in 1990; the secretion of gastric acid in infants is actually sufficient to prevent any significant bacterial colonisation.

In reality, as some authors (Knotek 1964, Simon 1966, Dupeyron 1970, J. L'hirondel 1971) had indicated twenty-five to thirty years ago, the nitrates-nitrites transformation, responsible for infant methaemoglobinaemia, occurs in the feeding bottles, if basic hygiene rules are not observed when the bottles are being prepared, thereby causing microbe pullulation.

A number of clinical observations have confirmed this; these include the sudden and unexpected nature of cyanosis, when large amounts of nitrated foods (carrot soup, spinach) have been ingested during the days or weeks preceding the incident without giving rise to the slightest clinical anomaly, the rapid onset of cyanosis 15 to 20 minutes after feeding, and the lack of correlation between the level of methae-moglobinaemia and the amount of nitrates ingested with food.

To limit nitrate levels at 50 mg/l of NO_3 in drinking water does not constitute an appropriate preventive response with regard to infant methaemoglobinaemia.

Nitrates will continue to be a potential element of infant food via the intake of water used to prepare formula milk, and vegetables. The only effective preventive solution consists in acting on the bacterial element: all risks of bacteria pullulation must be avoided in feeding bottles, whether they contain nitrates or not, by following a few basic hygiene rules when the bottles are being prepared. In the case of carrot soup, it should be boiled for a few minutes and fed to the baby shortly thereafter, and in no circumstances should the soup be left to stand at ambient temperature for more than six to eight hours.

What causes methaemoglobinaemia in infants therefore is not alimentary nitrates, but rather the nitrites formed in the feeding bottles after the reduction of nitrates into nitrites as a result of an unfortunate microbe pullulation in the bottles. It is the latter phenomenon which has to be prevented at all cost. Infant methae-moglobinaemia caused by food has been virtually eradicated in developed countries, where people are familiar with basic hygiene rules for preparing bottles of formula milk.

Cancer

As shown above, the salivary glands draw nitrate ions from the plasmatic sector; salivary nitrates partly turn into nitrites in the mouth, and, on swallowing, these salivary nitrites reach the stomach.

Salivary nitrites then react with various amines in the stomach to form nitrosamines. Ninety per cent of nitrosamines tested in experiments are known to be carcinogens in animals. It was deduced from this that nitrates have a potential carcinogenic power, and this presumption has now been hanging over them for almost forty years.

On analysis, however, this suspicion proves unfounded:

1. As Figures 3.1 and 3.2 show, nitrites in the stomach do not come directly from alimentary nitrates; they come from plasmatic nitrates on which the salivary glands have had a very specific action.
2. The amount of nitrosamines thereby formed in the stomach through the metabolism of nitrates is very tiny.

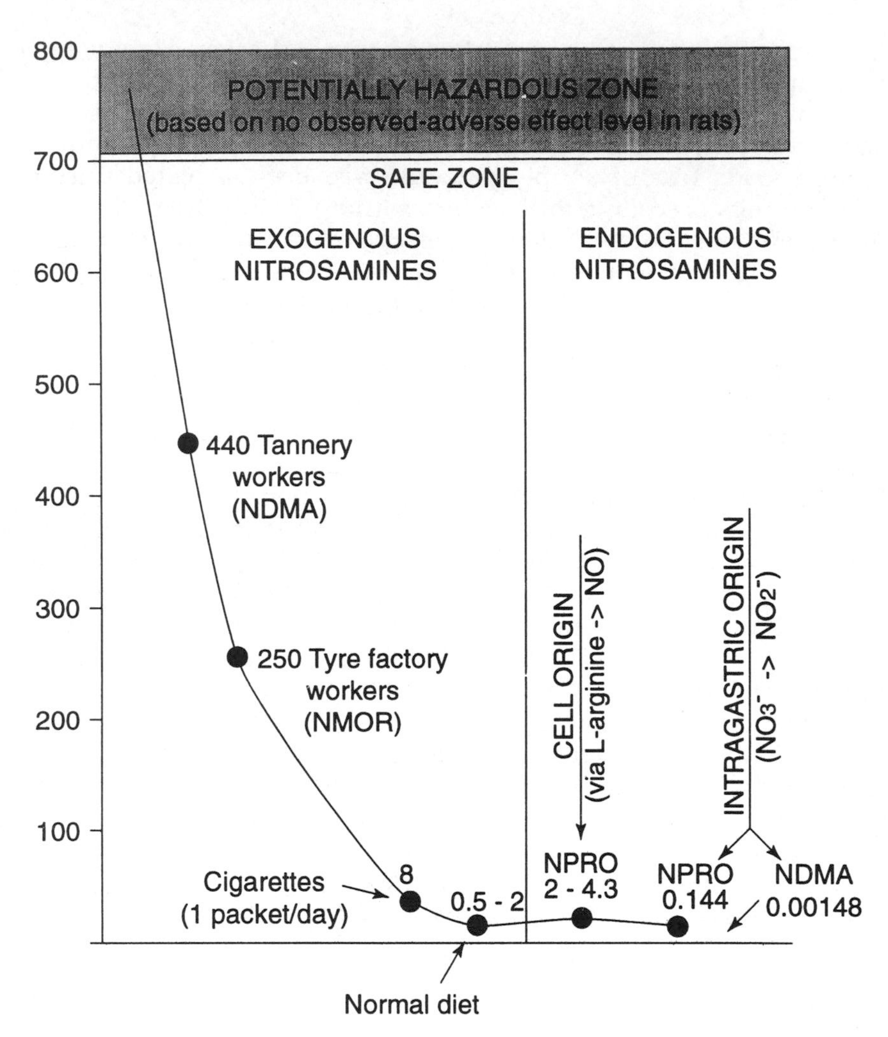

Figure 3.3 Comparison of endogenous syntheses and exogenous intakes of nitrosamines. NDMA: nitrosodimethylamine, NMOR: nitrosomorpholine, NPRO: nitrosoproline

Figure 3.3 compares levels between nitrosamines produced from endogenous synthesis, i.e. via the body's cells through the metabolic process involving L-arginine, those brought in by direct exogenous intake via food, those brought in exogenously by extra-dietary means (tobacco, tyre factories, tanneries), and the no observed-adverse effect level, in animals.

Many foods (beer and seasoned cooked meat in particular) contain nitrosamines. Levels of intake by direct dietary means are several tens or hundreds of times higher than for nitrosamines formed in the stomach through the metabolism of nitrates. If the precautionary principle were to be used, it would have to be applied in priority to nitrosamines of direct dietary origin, which would involve introducing restrictive measures for a number of foods. In fact, such restrictive measures are not necessary as these endogenous and exogenous nitrosamines remain confined to very tiny amounts compared to the theoretical toxicity threshold.

The level of direct dietary nitrosamines intake is several hundreds of times lower than the potential toxic level, and the amount of nitrosamines formed in the stomach during the metabolism of nitrates is several tens of thousands of times smaller than the potential toxic level (J. L'hirondel and J. L. L'hirondel 1996).

3. All nitrates-cancer experimental studies have proved negative in animals. Not one study conducted on rats or mice has succeeded in showing that even a considerable and prolonged intake of nitrates results in an increase in the incidence of cancers.
4. Since 1945, some twenty epidemiological studies have attempted to clarify the possible correlation in man between nitrate intake and the incidence of stomach cancers. Only two out of twenty show a positive correlation. Seven out of twenty even point to a statistically significant negative correlation. Such a negative correlation should not surprise us at all; the favourable impact of vegetables on the incidence of cancer pathology in general is universally acknowledged (WHO, 1990), and, as we know, 80 per cent of ingested nitrates come from vegetables.

   Therefore, as stated by the European Commission's Scientific Committee for Food in its 'Opinion on Nitrate and Nitrite' (22nd September 1995): 'Epidemiological studies thus far have failed to provide evidence of a causal association between nitrate exposure and human cancer risk'.

In conclusion therefore, the amounts of nitrosamines formed in the stomach during the metabolism of nitrates are actually very tiny; in no way are they capable of increasing the incidence of cancer pathology in man.

Other grievances

Other, less serious charges have been levelled at dietary nitrates: an increase in the risk of foetal death, an increase in the risk of congenital malformation, a tendency towards enlargement of the thyroid gland, and an early onset of arterial hypertension.

There have been few studies on these issues, and some of them contain a number of methodological flaws. As a result, these ancillary grievances cannot be sustained legitimately as they lack a sound, documented scientific basis.

Conclusions

Whether they are considered major or secondary therefore, no grievances against dietary nitrates in food can stand up to analysis. Scientific knowledge leads to the following conclusion: in the short, medium and long term, nitrates from food and from drinking water have no negative impact on human health.

Consequently, the directive drawn up in 1962 by the Committee of Experts on Food Additives of the WHO and the FAO on an acceptable daily intake (ADI) level for man is now redundant; so is the directive from the European Community dated 15th July 1980 on the quality of water intended for human consumption (80/778/EEC), which set a permitted limit of 50 mg/l for NO_3, above which water is no longer deemed fit for human consumption.

The implementation of this latter directive on drinking water is particularly costly for the citizens of the European Community. Its repeal is necessary and inevitable.

References

Bartholomew, B., Hill, M. J. (1984). The pharmacology of dietary nitrate and the origin of urinary nitrate, *Fd. Chem. Toxic.*, 22, 789-85.

Benjamin, N., O'Driscoll, F., Dougall, H. et al. (1994). Stomach NO synthesis, *Nature*, 368, 502.

Dupeyron, J. P., Monier, J. P., Fabiamni, P. (1970) Nitrites alirnentaires et methemoglobinemie du nourrisson, *Ann. Biol. Clin.*, 28, 331-6.

European Commission (1980). Directive on the quality of drinking water for human consumption, Council Directive 80/778/EEC OJNI L229, 30-8-1980, 11-26.

European Commission (1995). Scientific Committee for Food Annex 4 to document III/15611/95. Opinion on nitrate and nitrite (expressed on 22 September 1995), 20.

Knotek, Z., Schmidt, P. (1964). Pathogenesis, incidence and possibilities of preventing alimentary nitrate methemoglobinemia in infants, *Pediatrics*, 34, 78.

L'hirondel, J., Guihard J., Morel, C. et al. (1971). Une cause nouvelle de méthémoglobinémie du nourrisson: la soupe des carrottes, *Ann. Pediat.*, 18, 625-32.

L'hirondel, J., L'hirondel J. L. (1996). *Les nitrates et l'homme, le mythe de leur toxicité.* Editions de l'Institut le l'Environnement, 142 p.

Simon, C. (1966). L'intoxication par les nitrites après ingestion d'épinards (une forme de méthémoglobinémie) *Arch. Fr. Ped.*, 23, 231-8.

Stuehr, D. J., Marletta, M. A. (1985). Mammalian nitrate biosynthesis: Mouse macrophages produce nitrite and nitrate in response to *Escherichia coli* lipolysaccharide, *Proc. Nati. Acad. Sci.*, 82, 7738-42

Walker, R. (1990). Nitrates, nitrites and N-nitrosocompounds: a review of the occurrence in food and diet and the toxicological implications, *Food Add.* Contam. 7, 717-68.

WHO (1962). Evaluation of the toxicity of a number of antimicrobials and antioxidants. Sixth report of the joint FAO/WHO Expert Committee on Food Additives. World Health Organization, Technical Report Series 228, 76-78, Geneva.

WHO (1990). Régime alimentaire, nutrition et prévention des maladies chroniques. Rapport d'un groupe d'étude de l'OMS. Série de rapports techniques 797.

4 Farmyard follies: the end of antibiotics on the farm?

Roger Bate

Summary

New strains of bacteria, such as E coli and tuberculosis have politicans to analyse the problem of bacterial resistance to antibiotic drugs. Most health experts agree that the overwhelming reason for this antibiotic resistance is misprescription of drugs and dirty hospitals. Nevertheless the political urge to be seen to act, and the massive cost of targeting the main problem, has led to calls for a ban on the use of antibiotics in routine use on farm animals. Although this makes food more expensive, leads to more pollution and may even harm human health, the EU has enacted a ban from 1st July on four 'growth promoters'. Powerful Scandinavian interests, who have already banned the use of such promoters, used alarmist tactics based on the BSE scare to alarm the public. They claim, with no scientific proof, that antibiotic resistance may jump the species barrier from chickens to humans.

It is likely that as EU farmers become uncompetitive internationally following the ban, there will be calls to ban imports from producers who still use these growth promoters. Since there are no good grounds for such a ban a trade war (probably with the US) could be the result.

Introduction

At the time of writing, EU farmers are about to be prohibited from using four antibiotics that have been commonly used as digestive enhancers in animals destined for the table. This move was prompted by a fear that such regular use might lead to antibiotic resistance in animals, which might then be passed on to humans. The EU looks set to go ahead with the ban, due to come into force on July 1st 1999, despite the fact that official scientific investigations failed to find any evidence to warrant this fear. The cost of such a ban to European producers and consumers has been estimated at $2.5 billion per year.

Officials and farmers in Scandinavian countries and especially Sweden, where routine use of certain antibiotics has been banned for some years, campaigned vigorously for other countries to join them in their abnegation. They had support from many groups, including various government authorities, consumer groups and doctors, who all relied upon the precautionary principle – which rests upon the assumption that if uncertainty lies ahead, no step should be taken towards it – as justification for a ban.

Bacterial resistance

Fears that human antibiotic resistance may be increasing are well founded as tough new strains of salmonella, E.coli, gonorrhoea and tuberculosis have recently appeared. However, many of these have been bred by misuse of human antibiotics and no doubt several have evolved naturally (Bates 1997). The treatment for tuberculosis, for example, lasts for over a year and is extremely onerous, and completion of the drug regime after the symptoms have abated is probably considered by many to be a nuisance. Furthermore, many people with tuberculosis are homeless and so do not have a regular pattern in which to fit their drug regimen. After all, how many of us stop taking the pills for our chest infection once the coughing has stopped? Every time we fail to complete a course of antibiotics, we run the risk that those bacteria not killed off will have a greater chance of developing resistance.

Bacterial resistance to antibiotics is as old as antibiotics themselves. Some bacteria are naturally resistant to certain antibiotics, which is why both physicians and veterinarians culture an infection to better determine which antibiotic to use. Bacteria also reproduce rapidly and random changes in their genetic structure can pass on drug resistance to future generations (Levy 1998). In order to counter this, drug companies constantly develop new antibiotics so that they remain one step ahead of the bacteria.

One theory of acquired human resistance is that some antibiotic-resistant bacteria in animals might be transferred to people through improperly cooked food. The food-borne bacteria could then make a person sick, manifesting itself in a mild stomach-ache or diarrhoea. Some people with compromised immune systems could become so sick that they require antibiotics to treat the food-borne infection. The fear is that subsequent treatment could be unsuccessful because the bacteria had already grown resistant to the drug due to exposure on the farm (AHI 1998).

However, most experts seem to be of the opinion that human use of antibiotics has caused far more drug resistance than animal use (Levy 1998). According to a 1997 editorial in the *Washington Post*, more than 50 million antibiotic prescriptions made in the USA each year are not necessary (cited in AHI, 1998).

Nevertheless, political and media attention has focused on animal use of antibiotics. In 1998, the House of Lords Select Committee on Science and Technology recommended that animal growth promoting antibiotics should be phased out, rather than suggesting that human use of antibiotics should be restricted. Its Chairman, Lord Soulsby, argued that a precautionary approach of eventually banning animal antibiotics was essential. He was quoted in *New Scientist* on April 25th 1998 as saying, 'Humans are the most important consideration'.

New Scientist magazine further explained in a lengthy article that drug resistance primarily came from over-prescription by doctors and dirty hospitals. Yet it called on the UK Government to follow the Danish and Swedish lead and ban animal growth promoters immediately.

But how frequently are resistant microbes actually transferred to people from animals? It is obvious that bacteria can be transferred from food to people, (see the Chapter by Craven & Johnson in this volume) but there are no conclusive data showing that animal-to-human transfer of organisms resistant to antibiotic growth promoters causes resistance in people. Matt Ridley, in his book *Disease*, explains that the idea is still theory. After 35 years' research 'the evidence linking such growth promoters to antibiotic resistance does not so far exist'. For example, there is no documented human case of an antibiotic proving ineffectual because of resistance developed from use of the antibiotic in animals. A letter in the *New England Journal of Medicine* (Van den Bogaard 1998) suggests transfer resistance from turkeys to humans, but this has not been substantiated. So, even if transfer of resistance from animals to humans does occur, it is at most a tiny fraction of total resistance to antibiotics in humans. Addressing human antibiotic usage is therefore far more important than limiting any drug use in animals.

Antibiotics on the farm

Since the 1950s, farmers have been using antibiotics to treat and control numerous animal diseases such as pneumonia, salmonellosis and dozens of other infections. Without antibiotics, there would be an increased chance that these diseases could pass to people through food and through contacts with animals. Animals also grow faster when they are free from infection and disease (Nefato 1997).

That infection management increases productivity and drives down costs is obvious, but the welfare benefit to the animals is less apparent. Dairy cows commonly suffer from udder infections and reproductive problems. On an ordinary farm, this means that every year about 30 per cent of the dairy herd are culled, but on organic farms where antibiotics are not used, the cull rises to between 50 and 60 per cent (AHI, 1998b).

One effect of animals suffering from mild infection is that they eat more per pound of weight gained, as diseased intestines are usually less efficient at absorbing nutrients. They also produce more waste. One estimate from the UK National Office of Animal Health put the figure at 6 per cent more manure from animals not on a drug regime, or 7 million cubic metres for the EU as a whole. That is enough to submerge Birmingham a metre deep. This manure would include 78,000 tonnes of nitrogen and over 15,000 tonnes of phosphorous being released into the environment each year. These by-products of less healthy animals probably contribute to pollution problems (NOAH, 1998a).

The EU's own respected Scientific Committee for Animal Nutrition (SCAN), which reported in July 1998, largely condemned the Danish ban of virginiamycin. It called the data presented by the Danish government in favour of a ban, 'misleading' and 'speculative'. It concluded that the use of the virginiamycin growth promoter did 'not constitute an immediate risk to public health'. Professor Michael Pugh, a member of SCAN called the proposed EU-wide ban 'contrary to scientific evidence' and said 'there is not yet any evidence which proves that these substances have caused adverse effects to humans and therefore the proposal for an immediate ban is disproportionate' (NOAH 1998c).

Danish political pressure for a ban continued even after the SCAN report. The Danish Agriculture Minister, Dam Kristensen, cryptically said that the debate: 'is not only about science...this looks like the avoparcin case, where we also lost the first round'. Avoparcin is another antibiotic that a previous SCAN report had also said had posed no health threat. Yet under political pressure it was eventually banned across the EU from January 1997. Kristensen's remarks imply that a precautionary ban of other antibiotics may occur even though it is not backed by the science, since a precedent was set with avoparcin.

As to what considerations other than science Kristensen thought should play a part in the debate is not clear. Perhaps he was concerned to placate the various stakeholders that had demanded action. Perhaps he wished to be seen to be taking some kind of action in response to these stakeholder demands, in order to avoid damage to his reputation amongst the wider voting public. Perhaps he was implying that economic factors were important – although for the EU as a whole the economic implications of a ban are all bad, as Danish and Swedish farmers have found to their cost.

But why have there been such insistent demands to ban animal antibiotics when the real risk is human-induced drug resistance and there is no evidence of resistance spreading from animals? There seem to be at least three possible (and not mutually exclusive) explanations.

One explanation is political blame avoidance (see Craven and Johnson). Large veterinary drug companies and intensive farmers make easy scapegoats and soft targets, especially for governments that need to be seen to act for the public health.

A second explanation is self-interest on the part of certain government agencies. One curious feature of this scare is that although transfer of antibiotic resistance has been acknowledged as a potential problem for over a decade, only now, when ironically there is more evidence indicating that the problem is not likely to be significant, is a fuss being made. A reasonable inference from this is that either politicians have only just woken up to the issue, or someone's funding has been threatened. One can only speculate, but it is difficult to believe that it is mere coincidence that the UK Public Health Laboratory Service (PHLS), which identifies the resistant bugs, has recently had its funding cut. Diana Walford, Director of the PHLS, interviewed alongside Lord Soulsby in the *New Scientist* said, 'We're going to have to have additional funds if we're going to [combat antibiotic resistance] properly' (*New Scientist*, 25th April 1998). Hence calls for a Swedish/Danish style ban continued to make media headlines.

A third possible explanation is that the Danes and Swedes took a brave but emotional decision that placed them at an enormous economic disadvantage, and that they are now looking for a way out while saving face.

Swedish failure

For two reasons the Swedes were less enthusiastic supporters of the benefits of an EU-wide ban than were the Danes. First, the ban only affects four antibiotics, allowing four to remain on the market, leaving Sweden with its complete ban at a continued disadvantage. Sweden has made an official request to the EU to ban the remaining four products. Second, the Swedish ban had been in place a decade longer than the two-year-old Danish ban, and Swedish officials must be acutely aware of the costs. While the Swedish Government rhetoric has been that their ban has been a success, many Swedish farmers disagree with them. The strategy appears to have been that the Swedes want a level playing field for all EU farmers, and hence want an EU-wide ban, but dare not shout too loudly about their own ban in case it is shown up for the costly mistake it has become. An influential insider attacked the 12-year-old Swedish ban. Dr Berndt Thafevelin, retired head of the Swedish Animal Health Service, claims the ban backfired. Thafevelin said that zinc oxide, used as a replacement for growth promoters, was less effective in suppressing disease, and more antibiotics were subsequently needed to deal with outbreaks of infection (see Figure 4.1). Since it also led to environmentally damaging zinc contamination of manure, that too is being restricted, leaving pork producers struggling for existence.

The costs had been significant as well. The Swedish ban came into force before EU membership, and their pork market was completely closed by high import tariffs. Now, Sweden is flooded with cheaper imports from the rest of Europe, and the Swedish policies have pushed up farmers' costs making their products uncompetitive, which could explain why they and the Danish pig farmers have been squealing for a level playing field (Viane 1997). The effect of increasing costs to these levels across Europe could be to close the export market. According to the European Federation of Animal Feed Additive Manufacturers (FEFANA) the costs of an EU-wide ban would be significant, reaching as much as $2.5 billion per year for consumers and farmers (FEFANA 1997). Furthermore it takes more than a decade and millions of pounds to complete the necessary testing and approval processes to gain permission to market a new veterinary or human antibiotic. Banning the four most widely available animal antibiotics would reduce the incentives for pharmaceutical companies to invest in discovering new drugs. The EU would probably respond by protecting its own market. Experience suggests that following a ban it would not be long before an EU-wide ban was proposed on meat imports from countries that still used growth promoters.

The Swedish experiment should have been a warning to those nations contemplating a ban on antibiotic animal growth promoters (see Figure 4.1). Sweden has less than 2 per cent of the European meat production market and it has struggled with higher production costs than other EU farmers – many of whom are uncompetitive internationally. Sweden also has to import meat to satisfy domestic demand. In a competitive international market, EU farmers cannot afford to adopt the Swedish meat model.

Copenhagen coup

As part of its campaign to persuade the rest of Europe to restrict antibiotics, Denmark's Health, Food and Agriculture Ministries hosted a conference in Copenhagen in September 1998. The stated aim of the conference was to discuss the latest research on antibiotic resistance in man and the link, if any, between animal antibiotic resistance and human antibiotic resistance. However, the conference was called 'The Microbial Threat' and political wrangling quickly overshadowed scientific debate. The Danish Government was accused of planting alarmist stories in newspapers, while the large Scandinavian contingent lobbied delegates to support a ban. The newspaper article in the daily Danish newspaper *Berlingske Tidende* on the first day of the conference (8.9.98) claimed that a Danish woman who died from salmonella (strain DT 104) had been infected by Danish pork. DT 104 is resistant to many antibiotics, and it was implied that animal use of antibiotics as growth promoters had encouraged the resistance. According to Vibeke Rosdahl, of the Danish Serum Institute, the salmonella was resistant to fluorquinolone but sensitive to other human antibiotics. Quinolones are not used for growth promotion, indeed, they are not related to any compounds that are used as growth promoters, but the story was out, the damage done.

Those delegates who were hoping for a proper scientific discussion of the dangers to humans of animal antibiotics were dismayed at the overtly political nature of the meeting. Tony Mudd of FEFANA (representing Europe's feed additive industry) was particularly worried that Europe was moving towards a mentality of banning without proper risk assessment (Yeo 1998, 405).

His fear was confirmed by Richard Smith (editor of the *British Medical Journal*) who called for a vote in a plenary session dominated by human health experts and Scandinavians, on whether there should be an immediate ban of growth promoters or a risk assessment first. That the immediate ban without evaluation won the vote, even that the question should be framed in that way, showed that science took a back seat at the conference. According to one attendee I spoke to, there was widespread consternation about the use of voting at a supposedly scientific conference. Although it was supposed to be merely a symbolic vote, it was widely reported in international newspapers, including the *Financial Times*, and strengthened the political will to ban.

The more prescient commentators at the Conference considered that a ban was likely not just because of the Scandinavian pressure but because of the economic realities in farming in the EU. In 1998 there was vast over-supply in the pig market, and anything that resulted in a reduction in production was welcome for farm ministers.

Ban details

Following the Copenhagen meeting, which became known in business circles as the 'Scandinavian Stitch-up', EU Agricultural Commissioner Franz Fischler called for a suspension of authorisation of four growth promoters (including the widely used virginiamycin and tylosin) under Directive 70/524. Although there was no evidence of transmission of antibiotic resistance from animal to humans, Fischler considered a suspension was appropriate as a precautionary measure. Fischler's support for a ban was more remarkable because on 17th November 1998 he said that decisions would be based on the EU scientific studies that were underway. A week later, however, he performed an abrupt U-turn and started proceedings for the ban. However, Fischler did not obtain a majority in favour of his proposal in the Standing Committee, composed of representatives of Member States. In accordance with EU procedure, the decision was passed to the EU Council of Ministers, which met on 14th December 1998. Scandinavian lobbying continued unabated and the result was predictable. Three countries abstained, the remaining 12 voted in favour and the suspension of authorisation for the growth promoters was passed. Because the approved antibiotic list (in an annex to Directive 70/524) is EU-wide, no new national legislation was required; hence the suspension could have been instantaneous. However, accepting that alternative arrangements would have to be found, the EU Council of Ministers decided that the suspension should not come into effect until 1st July 1999.

Reaction to the ban

Opposition to the ban came from many sources, including pharmaceutical firms, large pig farmers outside Denmark, several farming groups, and veterinarians. Even the British organic farming lobby organisation, the Soil Association, now headed by the Prince of Wales, recognised the value of antibiotics for treating sick animals and said that the evidence regarding the impact of animal antibiotics on humans was 'scientifically inconclusive' and the ban 'essentially a political gesture'. The independent Heidelberg Appeal Netherlands Foundation conducted a comprehensive study of the published scientific literature on the subject and concluded that the ban was 'unsupportable' (Bezoen et al. 1998).

One of the more unexpected negative reactions to the proposed ban came from within the European Commission. This was because the ban pre-empted two key scientific studies that were in progress. The first was a multidisciplinary scientific committee established by EU Directorate General XXIV to investigate the relationship between resistance and the use of antimicrobials in human and veterinary medicine, animal husbandry, and plant protection due to report in March 1999. The other project was a joint EU Commission-Member States/industry surveillance programme to monitor resistant bacteria to feed additive antimicrobials, which was originally planned to last until December 1999. Understandably, the scientists involved were annoyed at the EU ban because it curtailed their surveillance programme, probably making the data collected inadequate for assessing any resistance. As one scientist put it, 'political expediency has once again over-ridden science'.

It is possible that the ban will be delayed in view of the research being conducted by the Commission. The French Government is calling for a minimum of two years for the phase-out of antibiotic growth promoters, which is probably the only hope for a temporary reprieve from the ban.

Discussion

EU health officials need to be seen to act for the public health, but the situation with regard to antibiotics is really not so bad. Resistance to antibiotics is not undermining treatment of major diseases; new antibiotics are being developed, but in any case will be displaced over the next few decades. The golden age of antibiotics, begun when Alexander

Fleming literally opened the window for penicillin to fly in, may soon be over, but monoclonal antibodies, haemopoietins and maganins are antibacterial chemicals we have barely begun to develop (Ridley, 1997). DNA vaccines are on the horizon, and may work on viruses as well as bacteria. They have the advantage that they should be cheap to produce, store and transport, unlike antibiotics. But even though antibiotics will probably be superseded, we should continue to use them, in animals and humans, until they become obsolete.

Bans on animal antibiotic growth promoters appear to be quite unwarranted on the evidence so far available. Although a potential adverse effect has been described, its size has not been measured; indeed, it may never be accurately ascertainable. In particular, the flow of genes and bacteria between humans and other animals is not known. The fact that vegetarians carry more resistant enterobacteria than meat-eaters vividly illustrates the complexity of the situation (Bories 1998).

The ban may even backfire by encouraging a black market in antibiotics. As one farmer put it to me before the Copenhagen conference, farm incomes are at an all-time low at the moment. A further dive in profits caused by a drug ban may lead some desperate farmers to buy growth promoters from whatever source they can, undermining the proposed legislation.

Another effect is that alternative, unlicensed materials will be promoted to farmers. Not only do many of these have their own antimicrobial properties – and hence resistance implications – but they will not have been through any licensing process. The result will be to replace proven, licensed products with unknown and unquantified risks to human and animal health.

Perhaps the most perverse effect of the ban comes from the fact that imports from outside the EU will not have to comply with it. A spokesman for EU Directorate General XXIV said there was 'no case for banning imports on safety grounds'. How long that position will last when British, German, French and Dutch pig and poultry farmers start losing huge numbers of animals to bacterial infections is uncertain. But the pressure to ban imports may eventually be overwhelming, and another idiotic piece of legislation in one country will lead to more EU-wide protectionism.

Figure 4.1 Swedish antimicrobial usage since 1986

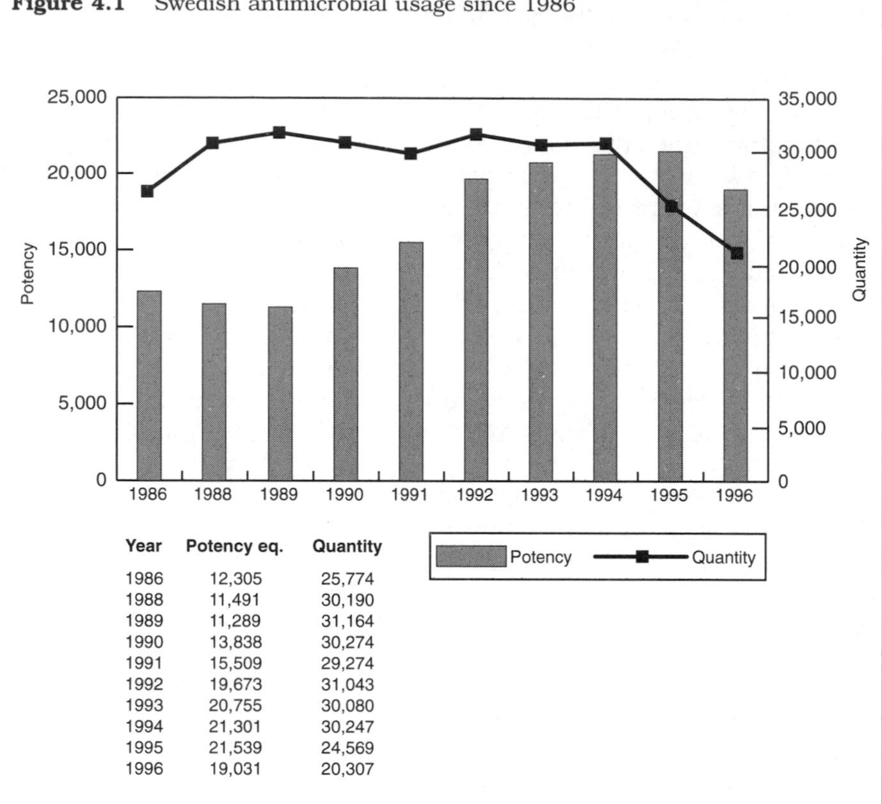

Year	Potency eq.	Quantity
1986	12,305	25,774
1988	11,491	30,190
1989	11,289	31,164
1990	13,838	30,274
1991	15,509	29,274
1992	19,673	31,043
1993	20,755	30,080
1994	21,301	30,247
1995	21,539	24,569
1996	19,031	20,307

In 1986 Sweden banned the use of antibiotic growth promoters and Swedish authorities have claimed that as a result the total quantity of antimicrobials has fallen in Sweden. This is strictly true, but as antibiotics were still used to treat infection in Sweden, and as more animals became sick as a result of the ban on preventative use, there was an increase in therapeutic use. The reason that the total quantity of antibiotic fell overall was merely technical – the drugs had become stronger.

According to a study reported in the journal *Feedstuffs* in October 1998, if the use of antibiotics is translated into potency equivalents (the power of the antibiotic multiplied by the amount used) then total use over the period has increased. As the graph above shows, over the ten-year period for which data were available, potency of use (column) increased by over 50%, whilst the quantity of antibiotic (line) fell by 20%.

Swedish farmers compensated for not using growth promoters by using more potent therapeutic antibiotics on sick animals. The result was an increase in the potency of antibiotic use. Furthermore growth promoters were replaced by large-scale use of zinc oxide at levels far in excess of those permitted elsewhere in the EU.

References

AHI (1998): *Antibiotics Briefing File*, Alexandria, VA: Animal Health Institute, April 22.

Bates, J. (1998) Epidemiology of vancomycin-resistant enterococci in the community and the relevance of farm animals to human infection. *J Hosp Infection*, 37, 89-101.

Bezoen, A., van Haren, W. Hanekamp, J. C. (1998) *Emergence of a Debate: AGPs and Public Health. Human Health and Antibiotic Growth Promoters (AGPs): Reassessing the Risk*. Heidelberg Appeal Netherlands Foundation, Amsterdam.

Bories, G. and Louisot, P. (1998): *Report on the use of antibiotics as growth promoters in animal nutrition*, Paris: Inter-ministerial and Inter-professional Commission on Animal Nutrition (CIIAA), High Advisory Committee on Public Hygiene.

FEFANA (1997): Fact Sheets on *Antimicrobials as Feed Additives*, Brussels: European Federation of Animal Feed Additive Manufacturers.

Health & Hygiene Focus on Europe (1998): *Pig Farming*, July.

Levy, S. B. (1998) The challenge of antibiotic resistance. *Scientific American*, March 46-53.

Ministry of Health, Ministry of Food, Agriculture and Fisheries Denmark (1998): *The Microbial Threat. Health of the population: Strategies to prevent and control the emergence and spread of antimicrobial-resistant micro-organisms*, Copenhagen, September.

Mudd, A.J., Lawrence, K. and Walton, J. (1998): 'Study of Sweden's model on antimicrobial use shows usage has increased since 1986 ban,' *Feedstuffs* October 26.

Nefato journaal (1998) The Netherlands Issue (5).

NOAH (1998a): *Environmental Benefits of Dietary Enhancing Feed Additives*, Briefing Document No. 7, Middlesex: National Office of Animal Health.

NOAH (1998b): *Precautionary ban on antibiotic growth promoters not backed by scientific data*, Press Briefing, Middlesex: National Office of Animal Health, November 30.

NOAH (1998c): *EU vote puts UK meat producers at risk*, Press Briefing, Middlesex: National Office of Animal Health, December 3.

Ridley, M. (1997): *Disease*, London: Viking.

Van den Bogaard, A. E. J. M., Jensen, L. B., Stobberingh, E. E. (1997) Vancomycin-resistant enterococci in turkeys and farmers. *New England J Medicine* 337 (1): 1558-1559.

Viane, J. (1997) The Swedish animal production system. Could it be applied across the EU? University of Ghent.

Yeo, A. (1998): 'The Microbial Threat Conference: Little fish off the hook – but for how long?' *Animal Pharm.* 405, September 25, Richmond, Surrey.

5 Genetic modification in context and perspective

Michael A. Wilson, John R. Hillman and David J. Robinson

Summary

Biotechnology has been shown to be safe and reliable to the satisfaction of numerous scientific inquiries. Research and development is extremely well monitored and regulated. It holds the promise of increased productivity at reduced costs, of hardy and robust crops which can stand extreme conditions and climates, among many other beneficial features. Many environmental problems associated with agriculture can be reduced by using GMOs, which need lower chemical and energy inputs, as well as less land, than traditional strains. While Britain and the EU are expressing fear and confusion, Australia, Canada, China and Japan have joined America in the research and commercial growth of GMOs. Arguments against GMOs offer little scientific evidence, relying on shock and alarm to carry their case.

Introduction

Rational, unambiguous, clear and credible presentations of the costs, benefits and risks are essential to inform consumers and restore public confidence in crop biotechnology in general, and in genetic modification (GM) in particular. Political and economic judgements and regulatory decisions need to be based on sound science rather than propaganda and prejudice. Scientific progress will create opportunities to protect the environment and remedy past environmental abuses; protect public and private investments and future national economic potential; improve agronomic procedures, and enhance global food security. Public concerns and confusion must be answered with scientific facts, which is a tall order but one that becomes increasingly urgent as technical advances and economic advantages accrue to competitor nations. As so often in the past, UK and European consumers and taxpayers face the prospect of having to import, at premium prices, the improved products of new technologies which we ourselves first discovered and developed. Moreover, we risk exporting our leading young scientists to those parts of the world that respect and appreciate their innovations and discoveries.

> There is a serious problem in striking a socially responsible balance between the influence of the articulate and the evidence of the informed.
>
> Professor John Marsh, Centre for Agricultural Strategy, University of Reading, in *The Agronomist* 1997.

> Biotechnology offers new breeding strategies for plants and animals to improve products and give better resistance to pests and environmental stresses.
>
> *DTI Biotechnology Foresight*, December 1996.

> It is not the strongest of the species that survive, nor the most intelligent, but the one most responsive to change.
>
> Charles Darwin

> Just as the food requirements of today's population of nearly 6 billion people could not have been met by the techniques of the 1940s, we cannot assume that current practices will feed the population of 8 billion expected by 2020. New approaches are needed in addition to the continued improvement of existing methods of crop and animal husbandry and food processing.
>
> Statement on 'Genetically Modified Plants for Food Use' by The Royal Society of London, September 1998.

Genetic modification in context and perspective

Archaeological data show that humans ceased being nomadic hunter-gatherers and settled on fertile riverside and coastal land to domesticate crop plants and animals about 10,000 years ago. Today, 95 per cent of our rapidly expanding global population relies for its daily food on only 15 plant and 7 animal species. We also rely on plants for most of our fibre, fuel and shelter materials. 'Natural' selective breeding of crops with desirable traits progressed gradually over the first 9,900 years, by a process that is best described as 'choose two of the best-performing parents, cross them, and hope for the best'. Until comparatively recently, most human beings were self-sufficient (subsistence) farmers who engaged in non-intensive agriculture, and were continually exposed to food insecurity through climatic and environmental fluctuations (abiotic stresses), or crop depredations by pests and diseases. Local cropping practices and crop varieties changed imperceptibly and, for better or worse, were a tradition transferred between successive generations. Such forms of agriculture, with their grinding poverty, high labour inputs, low yields, poor wealth creation and dependence on western aid still pervade the world's economy. Nonetheless, there are ideologists who, from the comfort of a western lifestyle, glorify subsistence, 'traditional value' agriculture and seek to constrain universal human improvement. Even in Europe, 200 years ago, primitive medicines, scarce food of low nutritional quality, and the rigours of survival meant that average human life expectancy was less than half that of today. The quality of life and food security that many humans enjoy today is directly attributable to scientific and technological advances over the past 100-200 years. Not until studies on the inheritance of traits in peas by Gregor Mendel (1866), and the rediscovery of his conclusions about 100 years ago, was any systematic genetic approach to crop improvement feasible.

Traditional breeding of crops and animals is risky, slow and never predictable. This is because tens of thousands of genes, representing 50 per cent of each parent's DNA, each with unknown effects, especially from the so-called 'unselected' parent, are mixed randomly into the new hybrid variety. The unwanted DNA then has to be bred out over 10-15 years of backcrossing with the original selected parent, and cycles of trial and error to create a new crop variety with one or a few desirable new traits and without any overtly deleterious characteristics. Nevertheless, many genes from the unselected parent will remain, and it is these which can confer unpredicted characteristics on the newly created cultivar under certain environmental conditions, or when the crop is exposed to a new pest or pathogen. Given our present-day global dependence for food on so few crop species, of such limited genetic diversity, created over the past century by only a handful of commercial breeders, our faith in and reliance on such haphazard natural biological processes is unnerving. Our rapidly accumulating genetic knowledge and modern biotechnological methods, including but not restricted to GM techniques, provide a far more robust, rapid, targeted, predictable and quantifiable route to new, locally adapted crop varieties. Mankind has to face up to the industrialisation of agriculture, horticulture and food production. Urban and suburban humanity will not revert to a subsistence-level existence, devoid of ambition.

Over the last few millennia as world population expanded, urbanisation encroached on the most fertile agricultural land. To feed the urban population, proportionately fewer farmers produced more food with successively more intensive production methods. New varieties of plants became higher yielding and more amenable to mechanisation, but were also more susceptible to pest and disease epidemics. Traditional breeding methods of introducing naturally resistant genes, even where such genes existed, were not enough to guarantee crop protection and food security. The Green Revolution of the 1950s was sparked by the parallel development of higher yielding semi-dwarf cereal varieties alongside synthetic agrochemicals such as fungicides, insecticides, nematicides, herbicides, growth promoters or retardants and ripening promoters or retardants. Each year, global agriculture spends more than $34 billion on fungicides and uses large amounts of insecticides and soil fumigants such as 70,000 tonnes of methyl bromide, to protect existing non-GM crop cultivars from pests and pathogens.

Over the coming 50 years a second green revolution will be required to feed the projected increase of 90-100 million every year, equating to 200 babies every minute. The new revolution must find a more environmentally sustainable approach than simply increasing 'spraying and praying'. A further handicap is that the world's most fertile and accessible land, amounting to 14 per cent or about 8 million square miles, is already under the plough. It is often impractical to recruit new land for agriculture because most is too dry, wet, salty, or prone to erosion. Meanwhile, high quality agricultural land continues to be lost to urban expansion. Future food production targets, even at current levels of agrochemical inputs, will rely on less productive marginal land or wilderness areas. It has been estimated that an additional 15 million square miles or 25 per cent of the total land area of Earth, equivalent to 4-times the area of the USA or Brazil, must be converted to intensive crop production to feed a burgeoning population with more westernised, carnivorous dietary demands. Irrespective of the consequent loss of global biodiversity, such a target is simply unachievable. Therefore, we must speed up the development of new crops and cultivars for higher yields with lower inputs using land already under intensive farming to ameliorate the pressure to slash, burn and till poorer wilderness areas (Avery, 1998). Biotechnology in its many forms – enhanced crop breeding, GM technology, smart agrochemicals, intelligent machines, IT, and mathematical modelling will all contribute to solving this major challenge confronting our species. These views have been validated by international scientific and technology foresight programmes.

So far, a variety of more-or-less invasive laboratory techniques, including tissue culture methods, have been used to attain new plant characteristics. 'Traditional' breeders frequently use chemical mutagens or irradiation to increase genetic variation. Golden Promise, the most successful malting barley in the 1970s-80s was a mutant produced by gamma-irradiation of seed. Contrary to sentimental notions, crop production practices over the last 100 years have been neither natural, rapid, nor predictable. Nor are they risk-free.

Darwinian evolution predicts a perpetual genetic 'cat-and-mouse game' in which pests and disease agents (pathogens) can, and do, mutate to overcome natural resistance genes. Moreover, there are many pests and pathogens for which no natural resistance genes exist, so farmers must fall back on the zealous and costly application of sprays, fumigants and other toxic agrochemicals to guarantee a viable crop yield.

Since 1983, it has become possible to introduce, quickly and precisely, only those genes for certain characteristics. This means that existing, desirable plant characteristics are unaffected and no unwanted DNA is brought in as part of the package. Moreover, GM crops express their new traits, such as effective resistance to existing and emerging pests and pathogens, by creating biodegradable natural molecules in the form of RNA, or proteins. These are present at extremely low levels within the susceptible cells and tissues of the crop itself, where they will be most effective. In contrast, only 5 per cent of a crop spray may actually reach its intended target, while timing the application so that the pest or pathogen is accessible and/or susceptible is critical and very dependent on the weather.

What is genetic modification?

Since the early 1970s, scientists have known the techniques for cutting, transferring and pasting fragments of the genetic material, DNA, from any source, whether living or dead, plant, viruses, animal or microbial cells, or even entirely synthetic DNA made in the laboratory, into the heritable chromosomal DNA of another living cell or organism, whether plant, animal, bacterium or virus, no matter how unrelated they may be. This is possible because the chemical structure of DNA, first deduced in 1953, is common across all life forms on Earth no doubt as a consequence of incremental divergent evolution. Moreover, much of the precise genetic information and language held in the common genes – their subtle regulatory mechanisms and the molecular biological basis of cellular metabolism, are shared by what appear to be very distantly related organisms. Thus it is possible to transfer and express genes in a subtle, compatible and predictable way across all nature.

Ever since its first discovery, this recombinant DNA technology has undergone intensive self-monitoring, stringent regulation and risk assessment, even for contained laboratory applications. With the accumulation of extensive knowledge and experience, hazards that were originally perceived as demanding particular levels of physical or biological containment emerged as manageable risks, and the regulations were amended accordingly. This iterative process continues to inform the evolution of reasonable, effective and safe international GM containment and release regulations. In the UK alone there are now seven statutory or advisory Government Committees concerned solely with biotechnological issues, and a further nine Committees with an indirectly biotechnological remit. We are not short of regulations.

GM crop technology is a relatively young science. In 1983, the requisite skills and methods in molecular biology – direct DNA transfer systems, plant tissue culture, transformant selection and verification and whole plant regeneration, merged, since when the science has quickly developed. The first examples of transformed plants with novel and useful agronomic traits, such as resistance to tobacco mosaic virus, phytophagous insects or the herbicide glyphosate, appeared around 1986. For technical reasons, each new characteristic was conferred by a single dominant coding sequence, or gene. More recently, multiple genes can either be added directly or combined, simply by crossing GM plants with single traits and selecting suitable progeny, as in traditional breeding.

The foreign DNA is almost always introduced into the heritable chromosomal DNA contained in the nucleus of the cell, rather than into the smaller amounts of DNA of ancient bacterial origin found in chloroplasts and mitochondria in the plant cell cytoplasm. By 1983, biologists had tamed a soil bacterium, *Agrobacterium tumefaciens*, to act as a vector to carry foreign genes. *Agrobacterium* species naturally transfer a piece of their own DNA into the chromosomal DNA of a plant cell. Therefore, by simply co-cultivating plant cells or cut leaf, stem or root pieces in a special culture medium with *Agrobacterium* cells containing the foreign DNA, a few plant cells received the new gene and became genetically transformed. In order to select only the transformed cells from the vast majority of unaltered plant cells, a second gene was included which made the rare transformed cells resistant to the toxic effects of an

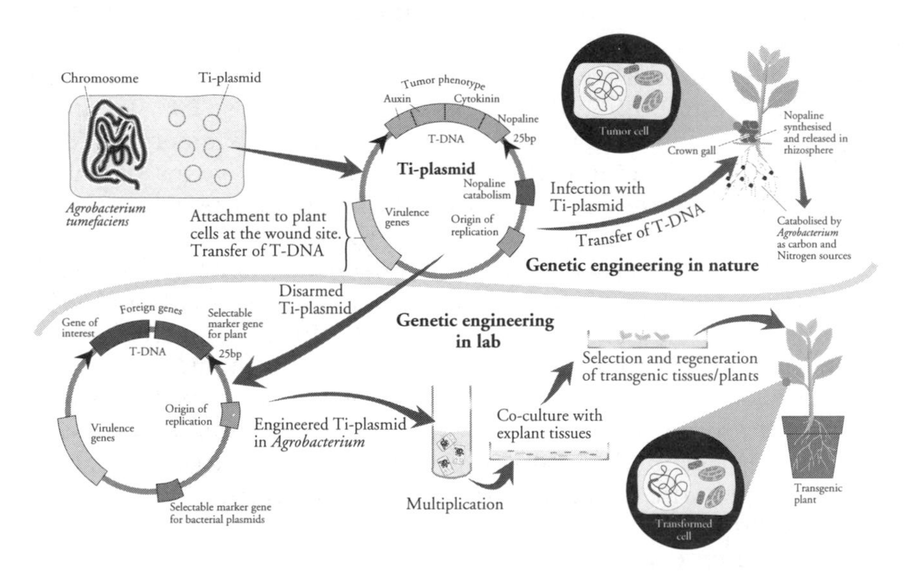

Figure 5.1 General features of the natural Ti-plasmid and the disarmed Ti-plasmid vector used in the process of DNA transfer into plant cells by the soil bacterium, *Agrobacterium tumefaciens*

antibiotic or a herbicide added to the culture medium. By vegetatively propagating shoots and roots from the surviving cells, small transformed, or transgenic, plantlets could then be grown to maturity. Resistance to the toxic antibiotic or herbicide often remained, however. While antibiotic resistance had no economic or biological advantage, herbicide tolerance was frequently a useful additional agronomic trait in the resulting GM crop plants. The presence and expression of the newly added gene(s) are then verified respectively by DNA fingerprinting or marker techniques and by analytical methods which detect the specific new RNA or protein molecules. Finally, the GM plant is tested for functional expression of the new trait in contained glasshouse or laboratory trials, then in open field trials under different conditions and at increasing scales. Since the mid-1980s, there have been more technical advances than can be described in detail, but the processes summarised above have culminated in the licensing and commercial-scale cultivation of more than 50 different GM crop-gene combinations, all outside the EU, and the introduction of GM crop products in many parts of the world. For example, according to the ISAAA, 69.5 million acres of transgenic crops were cultivated commercially (excluding China) in 1998 (James, 1998).

The intense public and media debate, scientific scrutiny, regulatory controls and precautionary risk assessments which have accompanied the development of GM crops since 1983 are without precedent. Yet plant biotechnology has an unblemished safety record. There have been over 25,000 closely monitored experimental field trials at all scales, involving billions of individual GM plants; more than three years' commercial scale sowings covering 4.3 million acres in 1996, 27.5 million acres in 1997 and 69.5 million acres in 1998 (James, 1998). During all this time no predicted or unpredicted hazard has emerged. The UK Health and Safety Executive concluded after 25 years of extensive scrutiny, 'GMO technology has proven to be one of the safest yet developed'.

Since 1994, the UK has held two Consensus Conferences (on Plant Biotechnology, and on Food) and New Zealand also held one on Plant Biotechnology. Each had a lay panel of 14 to 16 volunteers from all socio-economic groups and ages. When they interrogated experts from all sides of the debate and were given accurate information on the processes, risks, and alternatives available, each panel recognised and accepted the reasonable need and promise of agricultural biotechnology and GM crops, and rejected the technophobic propaganda. Sadly, the recommendations and conclusions of these conferences did not command high media attention nor much political recognition. It would seem that sensational misinformation, corporate or Government conspiracy theories, fear and speculation sell more newspapers and raise radio and TV ratings than does calm and considered analysis.

GM technology is the only method currently available that can create genetic resistance to many devastating pests and agents of crop disease. Paradoxically, it also allows us to harness the wealth of natural genetic biodiversity in wild species which could never be crossed with our domesticated crops. Without such biotechnology, plant breeders are restricted to the same gene pools they have been recording and reassorting for over a century. Biotechnology also provides the precise methods to quantify, catalogue and deliver information on species biodiversity.

What is feasible now, and in the future?

Most commercial GM crops carry dominant, single gene traits, such as resistance to a particular insect, virus, fungus or bacterium, tolerance towards a specific herbicide for better weed control, altered fruit ripening properties or a new flower colour. Some simple biochemical pathways have also been altered, such as one that produces biodegradable plastic from oilseed rape, or one that changes the spectrum of plant oils to substitute for products otherwise obtainable only from fossil fuels.

Future generations of GM crops will carry multigenic traits, such as tolerance to drought, salt or high UV-B irradiation, or which accomplish complex alterations in physiology or metabolism to create renewable raw materials for industrial processes. They will also be designed to produce foods for better human or animal nutrition, such as high starch potatoes, which absorb less fat when fried, and better tasting

Figure 5.2 Control (right) and transgenic (left) strawberry plants grown in soil contaminated with root-damaging vine weevils and not treated with conventional pesticides. The transgenic strawberry plant contains a gene from cowpea plants which expresses an inhibitor of a digestive enzyme (trypsin) in the gut of the weevils.

fruit achieved by slower ripening.

Until recently, most foreign DNA sequences introduced into GM crops were expressed continuously, or constitutively, and/or in every part of the plant. In recent years, and increasingly in future, the trigger sequences which switch on the foreign gene will work only in the appropriate part of the plant, not necessarily the edible parts, or to be turned on only in response to a stress signal from a pest attack, a cold snap or high UV-B. For example, strawberries containing a gene for a protein which deters vine weevils from devouring the roots of plants need only express the protective protein in their roots and not in the fruit.

Many new disease-resistance transgenes operate without expressing any protein at all, obviating issues such as potential allergenicity of a foreign protein in a food crop. Moreover, selectable marker genes are replacing the antibiotic resistance genes formerly used in GM plant production, a method which has caused some popular consternation. Likewise, there are now molecular systems which silence or remove marker genes once they have done their job, and before the crop is harvested.

There are several ways that biotechnology has increased farm productivity, making it possible for the land area under cultivation to be reduced in those regions outside Europe where new methods have been tried.

It is far more effective to treat weeds when they are in leaf than before a crop is planted, therefore crops which tolerate a weedkiller while they are growing need less treatment. This uses less herbicide, which is cheaper and less environmentally intrusive.

The benefits of many current GM crops are remote from the consumer which causes problems in public perception and acceptance. Over the last three years, field data from commercial farmers of tens of millions of acres of GM crops in the USA, Canada, Asia, South America and elsewhere, have confirmed industry predictions of significant reductions in pesticide and herbicide use, increased yields from existing arable land areas and better environmental protection. This relieves pressure on marginal land and wilderness zones, which may be pressed into agriculture.

Glyphosate, a weedkiller familiar to gardeners and farmers alike, is a broad spectrum herbicide as it kills all plants, yet it has a very favourable environmental profile, being broken down rapidly by soil bacteria. The manufacturer Monsanto, which produces glyphosate under the brand name Roundup™, has developed crops known as Roundup Ready™ to be resistant to its effect. The first crops to be grown commercially were soybeans in 1996. In 1997, 75 US seed companies distributed all available Roundup Ready seed to farmers, 90 per cent of whom said it exceeded expectations and 88 per cent said they would buy it again. By 1998, 36.25 million acres of herbicide-tolerant soybean were grown worldwide. In 1997, farmers reported lower

Roundup use on the GM soybean variety than on adjacent non-GM fields; reductions ranged from 39 per cent in the Southeastern USA to 9 per cent in the central USA, and over 75 per cent of farmers said they applied the herbicide only once to control weeds over the whole growing season. In the USA, GM soybean yields per acre were 5 per cent higher than non-GM soybean on average. Further, indirect savings on labour and energy inputs, and reduced soil erosion were achieved by removing the need to till for weed control between sowings.

The reduced need for ploughing will help conserve soil microfauna and microflora and reduce soil erosion and fuel consumption. For example, herbicides are usually cultivated into the soil before conventional crops are sown, but GM herbicide-tolerant crops can be direct-drilled into relatively undisturbed soil. Herbicide is then applied much later, post-emergence, when crop and weeds have sprouted, a far more effectual method as it is absorbed through the leaves directly into the roots.

Small-scale UK field trials in 1998 on herbicide-tolerant sugarbeet also showed that herbicide applications were reduced from the customary 5-7 sprays with cocktails of up to 9 selective herbicides, to only 2 applications of Roundup per season. With no need for pre-emergence spraying, and by allowing weeds to grow for longer than usual, until they began to compete with the crop, the GM sugarbeet plots preferentially attracted and supported more aphids and predatory ladybirds than the adjacent weed-free non-GM crop sustained by excessive spraying. Eventually, the mulch of dead and dying weeds in the GM crop provided a better environment for insects than the normal bare earth, while conserving water and reducing soil erosion. These findings run counter to the claims of environmental groups that GM crops will encourage the use of more herbicides, leave fields barren of wildlife and promote the growth of 'superweeds' immune to herbicide. The benefits for the farmer were even more clear-cut. The costs of insecticide and herbicide spraying were reduced to £24 an acre, compared with up to £140 an acre in fields growing ordinary sugarbeet. If weeds were left completely untreated in a GM, or a non-GM crop, sugarbeet yields would be reduced by at least 80 per cent. Moreover, herbicides can be applied to tolerant crops over a wider time period so that there may be more flexibility in weed control and a more targeted approach to weed management practices.

A further advantage of GM herbicide-tolerant crops is that broad spectrum herbicides, such as glyphosate, remain active for relatively short periods of time and may replace persistent soil-acting herbicides such as atrazine. Atrazine is routinely used in maize crops, where it is applied to the soil before any weeds are observed. The agricultural use of atrazine and other triazine herbicides is banned in the UK because of contamination of aquifers.

Insect-tolerant plants can dramatically reduce the use of conventional, wide-acting insecticides, as well as overcome the dependence of pest control on the vagaries of the weather and traditional application methods.

A development likely to be important to developing countries is a crop with built-in resistance to specific pests, as the need for expensive pesticide is drastically reduced. It has been estimated that insect damage results in crop losses of 13 per cent worldwide (BBSRC, 1998). For example, cotton growers in the USA use 20-25 per cent of annual pesticide applications, and despite spending $50-100 per acre on insecticides, they still suffer losses. In the USA in 1996, 5600 growers planted BOLLGUARD™ cotton which resists pest attack by expressing a gene from *Bacillus thuringiensis* (Bt) for a toxin which selectively kills bollworm larvae. Most of the crop (70 per cent) required no insecticide sprays, and the remainder was sprayed only once instead of 4-6 times. The economic benefit from using the GM cotton was estimated at $86 million (James, 1998). Between 1996 and 1998, total insecticide applications to GM cotton were reduced by over 3 million litres compared to non-GM cotton. In 1998 alone, overall insecticide use on cotton throughout Alabama was reduced by 80 per cent.

In 1997, 75,000 acres of a related GM crop, INGUARD™ insect resistant cotton, were grown in Australia. The Cotton Research and Development Corporation reported, midway through the season, that the GM crop required 68 per cent less insecticide spray than non-GM cotton.

Colorado beetle-resistant GM varieties of potato which express a different Bt toxin are also being grown widely and similar statistics on reduced pesticide use can be quoted (James, 1997, 1998).

Provided insect-resistant GM crops have minimal effects on non-target insects, the consequent reduced use of insecticides should encourage the establishment of beneficial insects in the crop and in the field margins, which in turn may favour other forms of wildlife. The specificity and internal location of a Bt toxin in a GM crop should provide more efficient and effective pest control than widespread, periodic spraying of pesticides or even of Bt itself. Moreover, Bt GM crops offer new opportunities for integrated pest management, because it is feasible to contemplate additional biological pest control strategies using beneficial insect pest predators and/or parasitoids in a crop which requires minimal pesticide application, something that could not be done in a non-GM crop needing sprays.

Anti-viral GM crops

In 1986, transgenic tobacco and potato plants that expressed a plant viral coat protein gene, which conferred laboratory and field resistance against the cognate virus, were among the first GM crop species reported with enhanced and potentially valuable new traits. Since then, the efficacy and commercial utility of plant virus-derived sequences as stable, heritable, single dominant resistance transgenes has been confirmed in many hundreds of cases. Examples include members of most genera of plant viruses and all types of gene sequence, expressing either functional or defective proteins, or no protein at all. In 1994, the first virus-resistant transgenic plant (ZW20 squash) expressing the coat proteins of both courgette yellow mosaic potyvirus and watermelon mosaic virus 2, was approved by the USDA for widespread, unmonitored commercial release and sale to consumers. The levels of viral coat proteins expressed from the transgenes were about 1000-times less than the amount of viral coat protein ordinarily consumed in non-GM, and hence commonly virus-infected, squash.

The Hawaiian (Oahu/Puna) papaya industry was destroyed 30 years ago by a virtually uncontrollable virus (papaya ring spot virus, PRSV), but has now been rescued by taking a commercial variety ('Sunset') and introducing the coat protein gene of a mild strain of PRSV. The new, patented PRSV-resistant GM variety, called 'Rainbow', was deregulated by the USDA Animal & Plant Health Inspection Service in November 1996. 'Rainbow' was approved by the US Environmental Protection Agency and the Food & Drug Administration in September 1997, licensed to the Papaya Administrative Committee, and, after lengthy and complex legal negotiations, finally distributed free of charge between May-September 1998, to commercial growers on approximately 1000 acres, to restore their businesses.

In Mexico in 1998 it was confirmed that GM technology preferentially benefited, in both relative and absolute terms, small-scale potato farmers who tend to use farm-saved seed (tubers). The latter carry latent virus infections between successive generations of stock. Using proprietary technology and training donated by Monsanto, and with funding from the Rockefeller Foundation, local scientists transformed three Mexican varieties of potato (Alpha, Rosita and Nortena) for resistance to potato viruses X and Y, and to potato leafroll virus (PLRV). While these GM varieties decreased unit production costs on large-scale farms by up to 13 per cent, costs to small-scale farmers were cut by up to 32 per cent (Quaim, 1998).

Anti-nematode GM crops

Nematodes cause crop losses of around $60 billion worldwide, including some £30 million in the UK potato crop (House of Lords, 1999). Over 80 per cent of the damage is due to female cyst and root-knot nematodes. GM crops that express a gene which inhibits sexual maturation of the worms could provide control and be an effective and more sustainable alternative to chemical control agents and soil fumigants such as methyl bromide, 70,000 tonnes of which is currently used worldwide. Although shorter lived in the atmosphere than CFCs, methyl bromide is more damaging to ozone while it is present. Alternatively, in work at the

John Innes Centre, a rice gene which codes for an enzyme that inhibits the digestive enzymes of nematode pests has been modified to increase its biological activity, then re-introduced into rice plants.

There are countless other present-day targets for GM crop and food research in public and private laboratories around the world. Although several hundred illustrative case studies are possible, a few examples of GM projects being developed for the future to improve agricultural practices, environmental sustainability, crop yield and quality, and food safety may be helpful.

- Cornell University researchers recently claimed that we are wasting most of our plant genetic resources by cross-breeding existing plants. We should now be mapping and isolating the best gene(s) for each trait from all the genes in all species. Then they hope to construct the ideal crop plant 'from the (genetic) ground up'. To illustrate this, they have added genes from two wild rice relatives to the best Chinese rice hybrids and are getting 20-40 per cent higher yield. In tomato, they are getting 48 per cent more yield and 22 per cent higher solids by introducing, through GM, some genes from wild relatives which cannot interbreed with cultivated tomatoes.
- The International Rice Research Institute (IRRI) in the Philippines is re-designing the rice plant to achieve 30 per cent higher yield by redirecting 10 per cent of the plant's metabolic energy into the grain while simultaneously reducing the number but increasing the strength of the stalks. Since 1993, IRRI has also used GM technology to introduce new resistance genes against the important pests and diseases endemic to rice in Asia; for example, rice tungro virus, which causes annual yield losses estimated at 7 million tonnes. Thus, GM crops have the proven ability to increase yield and provide food security in developing countries.
- Aluminium is the most abundant metal on Earth, and is highly toxic to plant roots, especially under acidic soil conditions where the metal becomes soluble. Over one-third of the world's arable land suffers from acidity and hence from aluminium toxicity, especially in the tropics. Production losses of up to 80 per cent occur in corn, soybean, cotton and field bean. Farmers add lime to neutralise the acid soil, but this adds a further economic burden on poorer farmers and also causes pollution of run-off water. In Mexico recently, researchers introduced a gene for an enzyme, citrate synthase, from a bacterium (*Pseudomonas aeruginosa*) into tobacco and papaya. The gene caused the roots to make and secrete 4-times normal levels of citric acid into the soil. The aluminium citrate formed could not then enter the roots, so making the plants tolerant of the metal and producing higher yields. Work is underway on rice and corn to allow them also to be grown on otherwise toxic acidic soils.

- Although seedless grapes have been bred by traditional methods, other fruits are less amenable. Japanese and Australian scientists are researching the possibility of developing pip-free (seedless) fruits by introducing a gene that switches off seed development.
- Cassava is an important source of dietary energy for 400 million people, mostly in the developing world. However, cassava and sorghum are also among the few food crops in nature whose roots and leaves contain highly toxic cyanogenic glycosides, such as linamarin and lotaustralin. These are associated with diseases known as goitre and konzo (paralysis of the lower limbs), which occur when the roots of bitter and highly cyanogenic cassava varieties are insufficiently processed before consumption. In 1999, a Danish research team aims to create an acyanogenic variety of cassava by switching-off endogenous expression of two closely-related, probably duplicated, native genes in the allopolyploid genome of cassava. To do this they will introduce anti-sense versions of two cassava genes, which they have already isolated, that are essential for the production of cyanogenic glycosides.
- Genetic engineering can increase the content of vitamin E, a powerful anti-oxidant, in plants nine-fold (Shantini and DellaPenna, 1998). Vitamin E detoxifies free radicals which have been linked to many ailments, as well as to ageing. Food contains four types of vitamin E, the most active but least abundant being alpha-tocopherol, a fat-soluble vitamin which reduces the destruction of low-density lipoprotein (LDL) cholesterol, and is essential for many important body functions such as forming red blood cells, muscles, lung and nerve tissue. It is also claimed that vitamin E bolsters immunity, lowers the risk of cardiovascular disease and prostate cancer, and slows the progress of arthritis and Alzheimer's disease. Alpha-tocopherol (7 per cent of the total vitamin E in soybean oil) is made by an enzyme (gamma-tocopherol methyl transferase, gamma-TMT) found in oilseeds which adds a methyl (CH_3) group to gamma-tocopherol (a more abundant, but less active form of vitamin E). In the developed world, our diet provides the minimum recommended daily allowance (RDA) of vitamin E (12-15 IU) from greens, vegetable oil and nuts. But a balanced diet cannot supply therapeutic levels of vitamin E (400 IU). To achieve this one would have to eat 930 almonds (an extra 6600 calories) or 1.5 kg of spinach. Moreover, vitamin E is degraded when foods are stored for long periods or cooked at high temperatures. Using a seed-specific promoter to over-express the gene for gamma-TMT, Shintani and DellaPenna (1998) altered the ratio of alpha- to gamma-tocopherol such that there was 80-times more of the more active form of vitamin E present. Thus, weight-watchers could restrict their intake of oils and nuts, but still receive their RDA of vitamin E. Likewise, therapeutic levels of the vitamin could be ingested (or extracted) from less plant material.

- GM cotton varieties are being developed that are coloured, softer, wrinkle-resistant and shrinkage resistant. Coloured cotton does not need to be dyed, which reduces industrial water usage and highly toxic chemical waste. Additionally, fabric made from such cotton is less susceptible to fading (Dutton, 1999).

The list of examples of applications of GM technology to improve crops and to produce functional foods (nutraceuticals), or therapeutic or prophylactic medicines is already vast and growing rapidly. For example, successful clinical trials are underway using antibodies produced in GM plants against dental caries or genital herpes (Travis, 1998). As with all other so-called 'designer genes' for endogenous pest or pathogen control or for resistance to abiotic stresses, the repertoire of candidate sequences and functional proteins is limited only by imagination, ingenuity, a knowledge of natural biological control systems and funding.

However, not all the stories have happy endings. After the civil war in the newly independent state of Georgia, the national economy and infrastructure were in ruins. Their main agricultural export business had been providing early season, high-health seed potatoes for Russia. It required urgent assistance to restore some wealth-creating capacity to the economy. Ill-able to afford the chemical sprays needed to protect their seed potato stocks from insects carrying devastating viruses such as PLRV, a programme was devised through US Agency for International Development (USAID) for a multinational company to provide, free of charge, GM potatoes resistant to PLRV. Sadly, a European branch of an international environmental activist organisation sent members to recruit a Georgian chapter of eco-warriors, who then destroyed the GM potato stock material. Thus thwarted, Georgia must now import agrochemicals and/or virus-free seed potatoes from the same country that so helpfully provided the eco-warriors.

Financial and economic aspects

New niche markets can be generated by biotechnology and GM crops, with high added value, such as the production of veterinary or human therapeutic or prophylactic medicines, even in crops such as tobacco. Opportunities like these, in addition to allowing reduced chemical inputs into high-volume food and commodity crops, are vital to sustain a competitive agricultural sector, particularly in Europe when faced with the end of the Common Agricultural Policy and no more subsidies. Agricultural biotechnology now provides intellectual property protection on both processes and varieties. After more than a decade of exhaustive risk analysis and media/consumer/political obstacles, many global players now expect to see improved returns on their investments. The technology is certainly giving a competitive advantage (lower inputs, and higher, better quality outputs) to those farmers and nations that use it.

As with any crop, economic returns to the grower will vary year-by-year, depending upon local weather conditions and the incidence of diseases, weeds or pest infestations. Nevertheless, because of data from the huge areas of GM crops grown and closely scrutinised in the USA and Canada in 1996, 1997 and 1998, certain general conclusions can be drawn concerning benefits to growers. Clearly, any reduced use of agrochemicals, no matter how smart and biodegradable, must be advantageous for the environment and non-target wildlife. In 1996 in the USA, the economic benefits to growers alone from GM crops were conservatively estimated at: $61 million for Bt cotton; $19 million for Bt corn; and $12 million for herbicide-tolerant soybean. In 1997, the values were: $81 million for Bt cotton; $119 million for Bt corn; $109 million for herbicide-tolerant soybean; $5 million for herbicide-tolerant cotton; and more than $1 million for Bt potato. In Canada, herbicide-tolerant canola (oilseed rape) saved $5 million in 1996, and $48 million in 1997, plus $5 million for Bt corn. Hence, over just two years in the USA and Canada alone, growers of GM varieties saved $465 million.

Global sales of GM crop products have grown rapidly and by 20-fold (James, 1998) from 1995 ($75 million), through 1996 ($235 million) and 1997 ($670 million), to 1998 (estimated at $1.2-1.5 billion). The market is projected to increase to $3 billion in 2000, $6 billion in 2005 and $20 billion in 2010. The numbers of countries growing GM crops commercially has increased from one in 1992, to six in 1996, to nine in 1998 and is expected to reach 20-25 by the millenium. The speed of uptake of this technology is a clear indication of its potential economic and agronomic benefits.

Even by 1997, twenty-seven GM crop varieties, nutritionally, functionally and environmentally equivalent to the respective non-GM crop, had permission to be grown commercially and enter the food chain in the USA, compared to 23 in Canada, 15 in Japan, but only 4 in the EU. US Food and Drug Administration (FDA) policy states that a GM product or ingredient label is needed on a novel food only if there is a safety issue or a nutritional or compositional difference, and not for 'right-to-know' information. Three-quarters of the US public agree with this.

Thus we must recognise and accept that while UK and EU environmental pressure groups continue to call for moratoria on growing GM crops, and a ban on GM soya and maize imports from the USA, these and many analogous products have already passed through stringent US and international safety regulations and food chains from 'plough to platter' without any health incidents, social revolution, disaster or consumer rejection. In the USA, when FDA, EPA and USDA Committees say that extensive field trials, nutritional, compositional and safety tests show these products to be safe for human or animal consumption, they are believed and trusted, which then allows technology and industry to make progress.

In Europe, on the other hand, despite regulations and procedures based on equally sound scientific principles, the public have lost (or been encouraged to lose) trust in Government agencies, private- and public-sector scientists, regulators, supermarkets and farmers, as well as the food processing, agrochemical and seed industries. In fact, the anti-technology propaganda campaign in Europe has been waged by classical methods that encourage public fear, distrust and confusion through shocking but erroneous imagery. *Non sequiturs* and misinformation are used to undermine detailed scientific knowledge, precise compositional data and the extensive, already prolonged risk assessments undertaken on GM crops and foods. The result has stigmatised GM research and GMOs, and has marginalised Europe in the eyes of much of the rest of the biotechnological world.

Much EU regulation aimed at 'harmonisation' is based on out-moded science and reflects unjustified concerns. The complexity and confusion in EU actions, edicts and regulations on GM processes and products elicit incredulity, amusement and even some anxiety among foreign observers, depending on their point of view. Many are seen to be taken in response to public pressure groups rather than being based on sound scientific evidence and facts. All EU and UK legislation should be regularly reviewed and redrafted, if necessary, in the light of the most recent scientific data. Intrinsic ethical, moral and religious issues may be valid, and merit debate in their own right, but they have no place in public health and safety regulations.

It is startling to contrast the negative perceptions of GM crops and food with the positive media hype, individual hope and public approval generated for human gene therapy. Yet, not a single beneficial positive result has emerged from more than 200 clinical trials of gene therapy, involving thousands of patients, conducted since 1990.

Since 1982, over 50 medicines derived from GMOs have been marketed worldwide. GMOs are now used to produce over 25 per cent of the top 20 drugs, for example insulin, growth hormone, several hepatitis B vaccines, and monoclonal antibodies to treat cancer. Clearly, public attitudes to GM-derived oral and injectable medical products are more pragmatic, utilitarian and less amenable to technophobic propaganda than is the case with GM crops and foods.

Although scientists are often rightly accused of failing to communicate clearly about the early development phase of GM crops, the problem is not entirely of their own making. The slow, incremental and provisional nature of scientific knowledge, its methods, theories and counter-theories during the research and development phase are not newsworthy and do not lend themselves easily to informed public debate, or even to short-term attention by the media. Sadly, only when the final products or 'breakthroughs' such as Dolly or Roundup

Ready™ soybeans emerge, are they seized upon by reporters, with a customary 'spin', usually involving conspiracy theories, boffin imagery, and meaningless, science fictional tags such as 'X-files', 'Doomsday', Frankenstein Food or Gene Food. By then, of course, any public debate is prejudiced and the original rationale for the work is overlooked. Innovative scientists are deeply discomforted by having their motives or quality of work questioned by unscientific sensation-seeking reporters. Scientists conducting important experiments with GM crops, particularly food crops, should recognise the potential sensitivity of their data, especially very preliminary findings, far from market. There is a desperate need for carefully validated, peer-reviewed data with proper, statistically robust experimental designs and analyses. Scientists in this field must not operate primarily through the tabloid media as it is difficult to undo the damage wreaked on years of study. Researchers should always remember that for the media, 'good news is not news'.

It is vital, for public acceptance, that GM crops offer clear, demonstrable and significant benefits which substantially outweigh any perceived risk to consumers and/or the environment. The onus is on the biotechnology industry to develop products with such benefits. The primary benefit of most GM crops commercialised to date has been indirect or remote from consumers. A benefit to a US farmer, seed or agrochemical company is not going to win acceptance of a GM crop product by a European consumer unless there is also a real benefit to his/her health, wealth or to the environment. Facts must then be provided showing better environmental protection, improved yields from existing arable land without further intensive use of agrochemicals, or loss of marginal land, wilderness and consequent biodiversity.

If process labelling of GM ingredients is demanded, then, in all fairness, relevant equivalent information on current non-GM food production methods, composition, storage and safety data should also be provided, to allow the public to make properly balanced judgements and informed choices. Labels should not be designed to stigmatise a safe and approved product, in order to selectively undermine, for example, the GM process. After all, many organically-approved, old-fashioned natural control agents are far more environmentally toxic and nutritionally undesirable than most modern agrochemicals, for example Bordeaux mix which contains high levels of copper, Bt toxin and even arsenic are approved for organic use, but these facts escape the public right-to-know. In short, consumers and opinion formers need to conduct more sensitive, thorough and rational risk-cost-benefit analyses.

The key paradoxes

Those who raise speculative risks, promote public fear and media misinformation, and demand moratoria to insist on largely unnecessary experiments to prove that their own imagined, improbable hazards could never happen (it being scientifically impossible to prove a negative), simultaneously encourage, sponsor and approve the actions of self-appointed eco-warrior groups who terrorise farmers, scientists and regulators alike, while destroying the very same experiments designed to address the issues posed by their own unscientific propaganda.

Those who claim that the (now proven) benefits of GM crops such as up to 80 per cent reductions in pesticide and herbicide use, less ploughing and soil erosion, increased yields and greater field biodiversity, accrue only to US farms and farmers while UK/EU consumers derive no benefit, but are forced to eat GM ingredients because of WTO rules and a major multinational corporate conspiracy, are the self-same activist groups who oppose commercial farm-scale releases of any GM crop in the UK/EU, thereby denying our farmers and rural environment any of the above.

A final word

Genetic modification for crop enhancement is a vital new technology whose use must not be rejected, as happened with food irradiation, through ignorance and irrational fears fuelled by anti-technology lobby groups. GM crops hold the most important key to solving future problems in feeding an extra 5 billion mouths over the next 50 years, and to using less land without causing agrochemical overkill and freshwater pollution. The indiscriminate and inefficient use of agrochemical sprays cannot continue at present levels, on even greater acreages, if the environment and biodiversity are to be protected, while simultaneously

Table 5.1. Some real or perceived benefits of GM crops

- Lower chemical and energy inputs
- Sustained or improved production from existing areas of arable land (thus protecting wilderness and biodiversity)
- Encourage use of biodegradable, smart agrochemicals
- Faster process, more predictable, with resistance targeted to emerging pests and pathogens
- Unlimited opportunity to alter designer resistance genes in successive crops to reduce selection pressure on pathogen to evolve counter-resistance
- Provide resistance genes against disease agents for which conventional resistance is not available
- Better quality food, improved shelf-life, fewer pesticide residues, lower price, better texture, taste etc
- Reduce need for ploughing, thus helping to conserve soil microfauna and reduce soil erosion

catering for a 100 per cent (or greater) rise in food demand by a global population which, hopefully, will peak at only 10.8 billion. Large-scale organic farming is not an efficient, economic or phytosanitary option. Thus blanket ideological objections to GM crops and other modern improved cropping practices effectively condemn us to continued dependence on agrochemicals alone; an ironic finale for so-called 'environmental protectionist' groups.

References

Avery, D.T. (1998) 'Saving the Planet with Pesticides, Biotechnology and European Farm Reform' 24th. Bawden Lecture, 1997.

James, C. (1998) Global Review of Commercialised Transgenic Crops: 1998, Ithaca, New York: ISAAA Briefs No. 8 (www.isaaa.cornell.edu).

Quaim, M. (1998) Transgenic Virus Resistant Potatoes in Mexico: Potential Socioeconomic Implications of North-South Biotechnology Transfer, Ithaca, New York: ISAAA Briefs No. 7 (www.isaaa.cornell.edu).

Shintani, D. & DellaPenna, D. (1998) Elevating the vitamin E content of plants through metabolic engineering. *Science* **282,** 2098-2100.

Travis, J. (1998) Scientists harvest antibodies from plants. *Science News* **154,** 359.

James, C. (1997) Global Status of Transgenic Crops in 1997, Ithaca, New York: ISAAA Briefs No. 5. (www.isaaa.cornell.edu)

Dutton, G. (1999) Transgenic cotton boosts agbio. *Genetic Engineering News*, 15 January 1999.

House of Lords (1999) Select Committee Report.

BBSRC (1998) In-gene-ious exhibition.

Science Museum and BBSRC (1994) Final Report: UK National Consensus Conference on Plant Biotechnology.

Straughan, R. and Reiss, M. J.(1996) *Ethics, Morality and Crop Biotechnology*, Swindon: BBSRC.

Ringe, C. (ed.) (1998) *Old Crops in New Bottles?* Warwickshire: Royal Agricultural Society of England.

Neal, M. and. Davies C (1998) *The Corporation Under Siege: Exposing the Devices Used by Activists and Regulators in the Non-Risk Society*, London: Social Affairs Unit

Bruce, D., Bruce, A, A., Appleby, M. et al. (1998) Engineering Genesis – The Ethics of Genetic Engineering in Non-Human Species' Church of Scotland Society, Religion & Technology Project. Earthscan Publications. IS BILLION 1 85383 570 6

Robinson, D. J., Davies, H.V., Birch, A.N.E. et al. (1998) Development, release and regulation of GM crops, *1997-98 SCRI Annual Report*, pp. 44-53.

Hillman, J. R. and Wilson, T. M. A.(1995) Delivering thepPromise – plant biotechnology, *Science in Parliament* **52,** 7-12.

6 Packaging and food: interconnections and surprises

Lynn Scarlett

Summary

Packaging of foods, whether for preservation or distribution, has been intrinsic to man's development from archaeological time to the space age – and beyond. Today, manufacturers, shippers, packagers and fillers, wholesalers, retailers and consumers demand many functions from packaging under economical and ecological constraints. The disadvantage of waste resulting from discarded packaging is balanced against the advantages of preventing food waste and reducing use of energy and resources. The disadvantages are typically overvalued and the advantages undervalued.

Introduction – packaging through the ages

Dates, plums, and olives await consumers in a London store. The presence of these packaged foods would be unremarkable but for one detail – they are products left behind in a store that served London under Roman rule over 1,100 years ago (Wood 1987).

Food packaging is not new to human settlements. The oldest known clay containers appeared in Japan 14,000 years ago; the oldest known baskets came 1,000 years later (Diamond 1997). Archaeological evidence from as far back as the Neolithic Age (8000 BC-4000 BC) shows the presence of earthenware in many human settlements (Lox 1992). Sumerians, in 3000 BC, used glass containers, and Egyptian pyramids of 2700 BC entombed pottery, papyrus leaves, and wooden containers apparently filled with goods. Historian Herodotus (480 BC-430 BC) described the amphorae – two-handled clay containers – of Persians;

others described their use by Phoenicians and Greeks to transport foods (Alexander 1993).

Earlier primitive settlements left behind evidence of skins, wood, clay, dried peels, and other vessels used as food containers. Stone Age tribes that survive into the 20th century subsist without metal-making technology or wheels, but they have food containers and 'packaging.' Anthropologist Napoleon Chagnon describes isolated, indigenous South American Yanomamo tribes that use leaves as 'packages' in which to cook food. Leaves also serve the Yanomamo for transport packaging as they squeeze 'honey-soaked leaves onto a pile of broad leaves and wrap it up to take it home' (Chagnon 1992).

Nor is the challenge of managing packaging waste an exclusively modern trial. Floors in Bronze Age Troy were littered with debris, including the remains of food containers (Blegen 1958). But another legacy of past eras is food waste. American archaeologist William Rathje recounts an incident in which Native Americans some 6,500 years ago apparently 'stampeded a herd of *Bison occidentalis* to their death into an arroyo 140 miles south-east of what is now Denver' (Rathje 1992). Without means of preserving this feast for many months, the Native American hunters took away meat that could supply the needs of 150 people for several weeks. Of the 200 bison killed, the remains of some 50 were left to rot.

Even into 20th century, food waste presented major sanitation challenges. Unused food was often discarded in streets and alleys as 'slop' for pigs; the practice persisted in towns and cities into the early 20th century (Melosi 1981).

As human societies evolved in form, so too did packaging. Nomadic tribes, reliant solely upon human locomotion, could not transport large amounts of goods; their use of packaging was modest and largely restricted to containers for cooking or short-term storage. But the advent of animal (and later mechanical) locomotion made long-distance travel and trade possible. With trade came the need for packaging and preservation techniques to maintain food quality during lengthy journeys and under varying shipping conditions. Salting, smoking, and drying extended the 'shelf-life' of foods; amphorae, glass jars, leather pouches, baskets, and wooden crates offered means to transport large quantities of food.

But through most of the 18th century, preservation of many foods in edible form remained a challenge – so much so that Napoleon offered a prize to anyone who could discover a way to preserve wholesome fruits and vegetables for the duration of long military marches and campaigns. With this incentive, Frenchman Nicholas Appert invented the first 'canned' goods by cooking under pressure foods packed in glass jars, introducing the first modern revolution in packaging (Lox 1992).

People waited at least another century before other types of food could be preserved – or even made possible at all – through the introduction of new packaging technologies. Modern critics of packaged goods sometimes yearn for the era of the 'cracker' barrel described as common features of 18th century dry goods stores. Customers dipped into these barrels to extract biscuits – no fuss, no packaging

This idealised picture of the old cracker barrel leaves out critical details. These crackers were not the crisp delicate biscuits modern consumers would recognise; they were, instead, 'a type of ship's biscuit so teeth-cracking hard they earned the name 'crackers.' They were dispensed to patrons by the handful.... The hard surface of the crackers offered one practical advantage; it made it easier for consumers to brush off rodent droppings before serving the biscuits to their families' (Alexander 1993).

It was also not until the 20th century that packaging and the techniques to mass-produce packages essentially democratised food consumption by making available a wide variety of food products at very low cost. Nineteenth-century canning systems could produce just 200 containers of canned food during one 10-hour workday; a single production line by the 1990s could churn out 36,000 containers each day, with dramatic implications for costs to consumers. In today's currency, the price for a can of corn made using old techniques might reach $10.00; instead, a can of corn costs well under $1.00.

This abbreviated retrospective on packaging offers a reminder that packaging in itself is not waste. It is an integral part of preserving, storing, and transporting food, and preparing it for final consumption. The history of packaging is one in which people have refined and improved the ways in which these packaging functions are accomplished. Those improvements have not only made possible advances in human diets; they have also dramatically reduced food waste and the energy required to produce and deliver a given 'unit' of food to consumers. Over time, packaging innovations have also generally performed given food-delivery functions with fewer materials for each food product delivered to the consumer. Finally, and perhaps most notably, packaging innovations have given consumers a wider array of options on how they might use their time and meet their food-consumption needs. Greater safety, healthier diets, more convenience, less food waste, and better food quality all have depended on the evolution of packaging.

Complexity of packaging choices

American archaeologist William Rathje examines modern landfills rather than ancient middens to better understand human cultures. After decades of poking through trash, Rathje concludes that the modern waste stream is a 'vast, interconnected, impossibly complex system' that results from millions of choices made by interconnected lives (Rathje 1992).

What is true of trash in general is equally true of packaging choices.

These choices are neither simple nor independent of the broader social and economic context in which they are made. These choices are also intimately linked to the many functional purposes that packages serve. Needs and resource constraints both shape packaging choices. The needs themselves vary among manufacturers, shippers, packers and fillers, wholesalers, retailers, and consumers but fall into three main categories: protection, marketing requirements, and control or tracking requirements (see Table 6.1).

Even among these categories, complexities abound. Packages must protect products, especially food products, against a wide spectrum of hazards – and do so while minimising costs and weight as well as resource and labour requirements. Temperature, light, moisture, bacteria and other animal pests, atmospheric pressure, gases, mechanical shocks, and vibrations during handling and transport all can undermine product quality or even destroy a product before it reaches the intended end user (see Table 6.2).

Packaging manufacturers and fillers must, then, design packages that take into account permeability, shock absorption, light transmission, thermal conductivity – the list goes on and on. Sometimes a package may need to prevent 'the transmission of moisture vapour into dry milk or corn chips' (Alexander 1993). At other times, a package may be required to prevent transmission of moisture vapour out of the package (as, for example, with some baked goods or frozen foods). Former American packaging company executive Judd Alexander (1992) points out that:

> The amount of protection needed by various products differs. Generally, the more barrier, the higher the cost, so packagers choose the best-value material to serve specific needs. Some packages need to hold vacuums (coffee, canned foods, poultry) or artificial atmospheres (refrigerated pasta), or vent carbon dioxide released by products (ground coffee, cheese, some produce), and bar oxygen from entrance to prevent mould or rancidity. Carbonated beverages and aerosol products can build up pressures of a hundred pounds per square inch in transit or handling, so containers must protect against rupture and possible injury.

Yet these packaging qualities are often invisible to consumers, who, though they want packaging that ensures delivery of a good-quality product, have an additional set of criteria that they demand from packaged products (see Table 6.3).

These criteria include attention to labels, appearance, safety, weight, product and packaging convenience, and price, for example. Even odour, tactile attributes, and opacity (or transparency) can matter to consumers.

Meeting all these requirements is neither simple nor merely a matter of satisfying modern predilections for convenience. Belgian packaging

Table 6.1. Packaging requirements needed by various interested parties from production through Distribution to Consumption

	Manufacturer	*Packer*	*Transport Sector*	*Warehousing (stacking)*	*Distribution*	*Consumer*
Packaging cost	X	X			X	X
Nature of packaging material						
a. relative to physical, chemical, and mechanical properties	X	X	X	X	X	X
b. relative to preserving the quality of the product		X		X	X	X
Feasibility of automatic packaging		X				
Good printabilitv (labeling)		X				
Information about the product					X	X
Feasibility of stacking (palletization)			X	X	X	
Appearance and emotional elements					X	(X)
Ease of handling			(X)	(X)	(X)	X
Goods coding by means of packaging	(X)	X	X	X	X	
Ecological aspects of packaging	X				X	X

X: major interest (X): minor interest

Source: Frans Lox, *Packaging and Ecology*, Table 1.10, p. 23.

Table 6.2. Packaging protective function against external factors

External factors	Functional elements and required properties for packaging and packaging material
Mechanical shocks, vibrations and compressive loads	Resistance (brittleness) Bending strength Shock absorption Tensile strength (elasticity) Compressive strength (resistance to stacking) Seal resistance
Gases (oxygen, nitrogen, carbon dioxide), water vapor, odors such as aromas and flavors and their mixture	Permeability Porosity and density
Liquids (water, oil, etc.)	Solubility, absorption
Light intensity and radiation energy (ultraviolet light, visible and infrared (IR) light)	Light transmission characteristics Light absorption Reflection (brilliancy)
Thermal content and temperature	Thermal conductivity
Biological factors	Resistance or sensitivity

Source: Frans Lox, (1992) *Packaging and Ecology*, Table 1. 11, p. 27.

expert Frans Lox (1992) points out that 'generally light accelerates oxidation and/or decomposition by atmospheric oxygen.' Through this process, fats may become rancid or sour. Milk can lose substantial vitamin content within a few days; some products experience reductions in nutritional value and may even become toxic. Packaging that results in carbon dioxide build up can accelerate rotting (Lox 1992).

No single remedy can address any one of the hazards that confront food producers, shippers, wholesalers, retailers, and consumers. What will work depends on the product contained within the package and the primary function of the package. In addition, packages come in many forms, depending on their purpose, and these forms will constrain the options available to address certain hazards (see Table 6.4).

The trash tango: packaging and waste

Consumers in most industrialised nations expect to have access to a wide variety of foods at modest cost. They also expect food to be undam-

Table 6.3. Consumer attitudes

In a 1992 survey of female shoppers AcuPoll Research asked the following question:

'The packaging of foods and drinks can provide many benefits. How do you rate the importance of each of the benefits listed below?'
(0=Not important; 10=Very important)

• Keeping product in good condition up to time you buy it:	9.2
• Keeping remaining product in good condition after package has been opened:	9.2
• Has nutritional information package:	8.8
• Easy to reclose:	8.5
• Provides 'enough' packaging but not more:	8.4
• Easy to open:	8.1
• Made with recyclable materials:	7.4
• Right size for freezer, cupboard etc.:	7.4
• Made with recycled materials:	7.2

Source: Richard Lawrence, 'Trends in Packaged Goods Innovations,' presented at *Packaging Magazine*'s 1993 Packaging Forecast & Planning Seminar, Sept. 22, 1992.

Table 6.4. Main groups of packaging forms

Boxes and trays (carton, corrugated fibreboard, wood, tinplate, plastic)
Wrappers, foils, and films (paper, metal, plastic)
Bottles and jars (glass, ceramic, plastic)
Ampoules (glass)
Tubs (glass, paper, plastic)
Sleeves and pipes (metals, paperboard, plastic)
Containers (metal plastic)
Tubes (metal, plastic)
Thermoforms (blisterpak plastic)
Bags and sacks (paper, plastic, natural and synthetic yarn
Space filling (paper, wood wool, cotton wool, plastic)
Bowls (Paper, paper board, plastic)

Source: Frans Lox, (1992). *Packaging and Ecology*, Table 1-2, p. 11.

aged at the point of purchase, and they expect packaged foods to have a reasonable shelf life, which can vary from days to months, or sometimes even longer, depending on the particular product. Cost, quality, and shelf life are visible to consumers – though they often fail to understand the role packaging plays in achieving these goals. Even less apparent to consumers is what happens to foods *before* they reach the supermarket, and how packaging innovations affect the movement of food from farm to supermarket.

The US Department of Agriculture has estimated that 50 per cent of food production worldwide fails to reach the final consumer. Rodents, insects, mildew, and other purveyors of rot damage foods – sometimes before they even leave the field. A simple comparison of food losses in India, the former Soviet Union and the United States illustrates the consequences of these losses in individual nations (Alexander 1993). In India and the former Soviet Union, where use of food packaging was limited in the 1980s, food production losses were 70 per cent and 50 per cent respectively, while losses in the United States were 17 per cent. Alexander (1993) tells of a conversation with the Soviet Minister of Agriculture in the 1970s who lamented, 'Our problem is not producing food, it is getting it to market in edible condition'.

The efficient food distribution system in the United States, made possible through a combination of sophisticated packaging and transportation networks, translates into lower cost foods to the consumer. For modern industrialised nations, including Japan, the United States and those in Europe, food costs absorb 12 to 23 per cent of consumer expenditures; for India and the former Soviet Union, by comparison, food costs amount to 56 per cent and 38 per cent respectively of consumption expenditures (Alexander 1993).

Packaging used to collect, ship, store, and sell food makes up about 60 per cent of all packaging discards in the United States and other modern industrialised nations. This packaging is part of a total distribution system that offers five key benefits, including lower food costs, reduced spoilage, greatly increased food variety and improved diets, extended shelf life, and greater convenience. Without this packaging system that makes possible food shipments and trade, most nations would have highly restricted diets, with fruit and vegetable consumption being limited by season and much more limited in variety. Individual nations would also be much more vulnerable to food crises and incidents of hunger resulting from natural disasters and cyclical weather patterns.

Packaging used for food shipments has, in effect, helped to 'democratise' food consumption by widening access to different foods and reducing costs. The disadvantage of the food shipment system in the form of packaging waste is modest, since much of this packaging is reused or recycled. Pallets and crates made of wood or corrugated paper are among the most highly recycled packages in the marketplace. In the

United States, for example, the National Wood Pallet and Container Association reported a recycling and reuse rate for wood pallets of some 80 per cent; the recycling rate for corrugated containers exceeded 60 per cent (Alexander 1993). New transport packaging systems, such as returnable plastic pallets, offer further potential for reducing food spoilage and packaging waste. Damage rates for shipping tomatoes using returnable plastic pallets declined from 15 per cent to 3 per cent; shipping costs dropped 25 per cent; and the pallets can be reused repeatedly (*Cheap Produce News* 1998). Moreover, improved systems for packing and shipping foods have reduced the need for warehousing space from 40 per cent of a self-service supermarket in 1960 to under 10 per cent just a decade later (Lox 1992). Finally, the prospect of food trade, made possible by transport and other packaging systems, reduces pressures within individual nations to farm sensitive lands and less-productive lands.

William Rathje and colleague Michael Reilly looked at the relationship between food packaging and food waste by comparing Mexican households and US households. The latter had far greater reliance on processed and packaged food. Though US households generated ten times as much waste from fresh produce than their Mexican counterparts within similar income groups during the study period, their total food-waste generation was far less than in Mexico because packaged foods, upon which Americans relied, generate the least amount of waste (Rathje 1992). Rathje and Reilly (1985) noted that Mexican households, on average, throw out three times as much food waste – skins, rinds, peels, tops, and other inedible parts discarded in food preparation and portions of edible food discarded. They estimate that the Mexican daily household discards represent over half the amount of food (by weight) necessary to provide an adult with a nutritionally sound diet for one day (though, of course, the mix of discards may not yield a balanced diet).

Another way to look at the role packaging plays in preventing food waste at the household level is to compare waste from fresh foods to waste from processed and packaged foods. Through examining what people actually throw out, Rathje (1992) estimates that:

> The waste figure for the much-maligned prepared-food category (which includes microwave dinners, take-out food and packaged soups and stews) is a relatively modest 4 to 5 per cent. The food category that undoubtedly has far and away the most positive public profile – fresh produce – is also far and away the biggest contributor to food waste: Produce accounts for from 35 to 40 per cent of total edible-food discards by weight. This figure does not include thrown-away portions of produce that aren't really waste – rinds, peels, skins, and so on which constitute a considerable category unto themselves. By weight, the inedible part of an avocado is some 24 per cent of the total; of a banana, 32 per cent; of lemon or grapefruit, 59 per cent.

Former chief executive of Coca-Cola Foods, Harry Teasley, points out that no package is not necessarily the best package, as some waste hawks have suggested. Many US households buy frozen orange juice, delivered to them in a spiral paperboard carton with metal ends, rather than squeezing fresh oranges into juice. Though users of fresh oranges have no household packaging waste, they do generate waste in the form of discarded rinds. Rathje and Reilly noted that the typical Mexican household, which relies more on juice squeezed from fresh oranges, tosses out 10.5 ounces of orange peels each week while the typical American household throws out only 2 ounces of cardboard or aluminium from a frozen concentrate container (Scarlett 1991).

Even that comparison fails to capture the full and complex set of trade-offs between packaging and food waste. Looking more closely at this question, Harry Teasley noted that to yield the same quantity of orange juice, a consumer uses 25 per cent more oranges than does an industrial processor. This means that fresh oranges require about 25 per cent more fertiliser, water, fuel and other resources to produce a given quantity of juice. The industrial processor does not discard the rinds but, rather, uses them in animal-feed preparations, perfumes, and other products. Moreover, fresh oranges are transported on containers requiring nearly nine times more corrugated cardboard waste at the retail level than the 12-ounce frozen concentrate (though 70 per cent of corrugated cardboard is then recovered for recycling). Finally, it takes 6.5 times more truckloads of fresh oranges to produce equal quantities of orange juice – resulting in 6.5 times more energy consumption and 6.5 times greater production of pollution (Scarlett 1991).

The orange juice example is not meant to suggest that consumers abandon purchasing fresh fruits and vegetables. They are nutritious and often tasty; food waste is not the sole criterion that shapes food-consumption choices.

In the aggregate, the low levels of food waste associated with packaged products illustrate the interdependent relationship between packaging choices and total waste generation. This relationship is not static. Packaging innovations during the 20th century have contributed to dramatic additional reductions in food waste over time.

Alexander (1993) recounts the success story of a humble new packaging system for lettuce. In the 1960s, lettuce in the United States generally was shipped from farms to supermarkets in lightweight wooden boxes – so-called lugs. Lettuce losses through spoilage amounted to 20 per cent – waste that wholesalers and retailers had to discard. Packagers then developed a new, coextruded, multi-material plastic film that could be wrapped around the heads of lettuce within seconds of their harvesting. Alexander describes the plastic wrapper as the perfect package for lettuce: it kept moisture in the lettuce, prevented oxygen from reaching the lettuce, and allowed carbon dioxide releases from the lettuce to be released into the atmosphere. Moisture-retention maintained freshness; the oxygen barrier prevented decay from oxygen-loving bacteria; carbon dioxide release slowed rotting processes.

Other unglamorous packaging innovations yielded similar reductions in waste: Reclosable plastic bread bags replaced waxed paper wrappers, reducing both packaging waste and food waste by extending the shelf-life of sliced bread. In fact, William Rathje's (1992) careful investigation of trash show sliced bread to generate almost no food waste.

Meat packaging, too, has been transformed since the 1970s. Through the mid-70s, packing houses sent meat to wholesalers and supermarkets in half-carcass 'sides' covered in a cotton shroud and refrigerated. Shipped this way, the meat lost moisture, resulting in weight losses of as much as two per cent per day. These sides of beef have been replaced by ten- and twenty-pound beef that is bagged in special air-evacuated, sealed plastic film. Much of the fat, gristle, bone, and other inedible parts of the beef are removed during this packaging process by the packer, so that neither supermarkets nor customers need to handle and discard it (Alexander 1993).

The relationship between packaging and waste is complex. Critics of modern, single-serve, pre-packaged foods have viewed these products as inherently wasteful. That conclusion ignores the subtle (and often individual) trade-offs that surround packaging choices. Four trends lie behind the demands for single-serve and ready-to-eat packaged food in industrialised nations: 1) an increasing number of single-person households, 2) an increasing percentage of two-person households, 3) a decline in the number of children per household, and 4) an increasing average age of the population (Franklin Associates 1992). Single-serve, pre-packaged portions allow these smaller households to avoid food waste associated with purchases of food in 'bulk' packaging.

Though single-serve packages often introduce somewhat more packaging waste at the household level, they sometimes generate less waste at the retail level and allow more-efficient transportation:

> For example, a set of 12-ounce beverage containers, which delivers 64 ounces of drink, yields 0.0415 pounds of retail waste and 0.0917 pounds of consumer waste. By contrast, to deliver 64 ounces of drink in 6-ounce containers made of the same materials results in 0.028 pounds of retail waste and 0.12 pounds of consumer waste. In this comparison, one container produces more consumer waste, the other more retail waste. The set of larger containers produces 0.012 pounds less total waste by weight, making it appear to be the

Table 6.5. Composition of Municipal Solid Waste (by weight) for European countries

Country	*Waste Type*	*Year*[a]	*Paper/ board (%)*	*Plastics (%)*	*Glass (%)*	*Metals (%)*	*Food/ garden (%)*	*Textiles (%)*	*Other (%)*	*References* [b]
Austria	MSW	1990	21.9	9.8	7.8	5.2	29.8	2.2	23.3	1.2
Belgium	MSW	1990	30.0	4.0	8.0	4 0	45.0	–	9.0	2
Bulgaria	MSW	1990	8.6	6.9	3.8	4.8	36.7	–	39.2	2
Czech Republic	MSW	1990	9.5	5.9	7.6	6.4	7 2	–	63 4	2
Denmark	MSW	1985	29.0	5.0	4.0	13.0	28.0	–	21.0	1
Finland	Household	1985	51.0	5.0	6.0	2.0	29.0	2.0	5.0	3
France	MSW	1990	31.0	10.0	12.0	6 0	25.0	4.0	12.0	2.4
Germany	MSW	1990	17.9	5.4	9.2	3.2	44.0	–	20.3	2
Greece	MSW	1990	22.0	10.5	3.5	4 2	48 5	–	11 3	2
Hungary	MSW	1990	21.5	6.0	5.5	4.5	–	–	62.5[c]	2
Iceland	MSW	1990	37.0	9.0	5.0	6.0	15.0	–	28.0	2
Ireland	MSW	1992	34.0	15.0	5.0	4.0	24.0	3.0	15.0	5
Italy	MSW	1990	23.0	7.0	6.0	3.0	47.0	–	14.0	6
Luxembourg	MSW	1990	17.0	6.0	5.0	3.0	–	–	67c	2
Netherlands	MSW	1990	24.7	8.1	5.0	3.7	51.9	2.1	4.5	2.7
Norway	MSW	1990	31.0	6.0	5.5	4.5	30.0	–	23.0	2
Poland	MSW	1990	10.0	10.0	12.0	8.0	38.0	–	22.0	2
Portugal	MSW	1990	23.0	4.0	3.0	4.0	60.0	–	6.0	6
Spain	MSW	1992	20.0	7.0	8.0	4.0	49.0	1.6	10.4	8
Sweden	MSW	1990	44.0	7.0	8.0	2.0	30.0	–	9.0	2
Switzedand _	MSW	1990	31.0	15.0	8.0	6.0	30.0	3.1	6.9	2.9
Turkey	MSW	1990	37.0	10.3	9.0	7.0	19.0	–	18.0	2
U.K.	Household	1992	34.8 11.3	9.1	7.3	19.8	2.2	15.5	10	

[a] Where more than one source is used, year gives latest date.
[b] References: 1, Carra and Cossu (1990); 2, OECD (1993); 3 Etalla (1990); 4, Barres, et al. (1990); 5, ERL/UCD (1993); 6, Elsevier (1992); 7, Beker (1990); 8, MOPT (1992); 9, Gandolla (1990); 10, Warren Spring Laboratory (see Table 5.9).
[c] Includes food/garden waste
Source: P.R. White, M. Franke, and P. Hindle, (1995). *Integrated Solid Waste Management: A Lifecycle Inventory*, Table 5.5, p. 72.

preferable choice from a waste standpoint. But the set of larger containers requires more trucks than delivering a similar quantity of beverage in the smaller containers, and this requirement translates into more fuel and greater emissions associated with the larger containers (Scarlett 1994).

Food packaging: the big picture

No category of items in the waste stream has attracted more critical attention in recent years than packaging. Some subcategories, like fast-food packaging, have inspired especially vigorous criticism. The focus on packaging is understandable to some extent. In the United States, packaging composes about a third of the total municipal solid waste stream which includes household, retail, and some institutional waste (Franklin Associates 1997). In Europe, estimates of the contribution of packaging to the waste stream range from 27 per cent (by weight) in the United Kingdom to 35 per cent or more in Germany (White et al. 1995). Inconsistencies among reported figures often are attributable to use of different definitions for municipal or household waste. Identifying how much of this packaging waste is food packaging is also difficult, since waste stream data often report composition by material rather than by product type (see Table 6.5).

However, in the United States, over 50 per cent of packaging is estimated to be food packaging (see Figure 6.1). One estimate puts the food packaging portion of total packaging waste at 59 per cent (see Table 6.6).

In the United Kingdom, Warren Spring Laboratory looked at house-

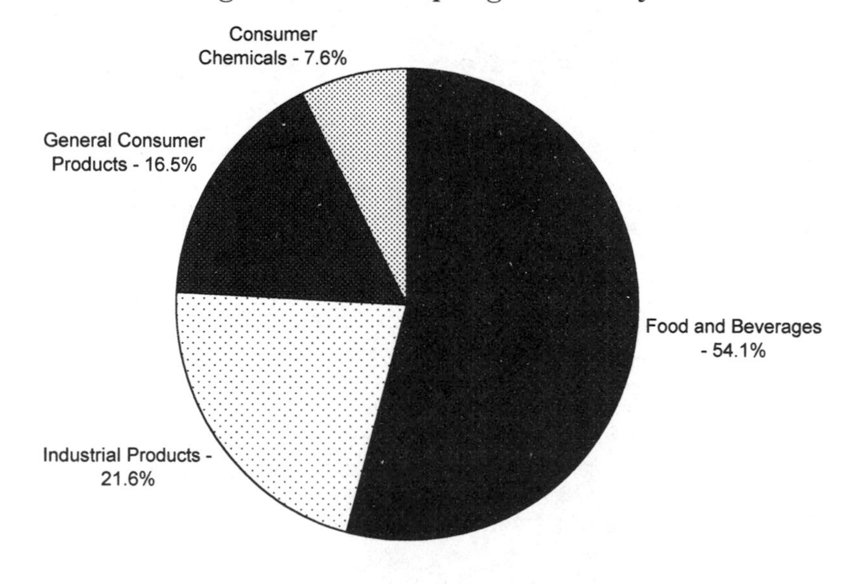

Figure 5.1 Sales by End Uses – Source Rauch Associates Inc. *Source:* Scarlett (1994)

Table 6.6. Food and beverage: share of MSW packaging discards—1998

Package	*Total discards (millions tons)*	*Food and beverage Share*	*Food and beverage (million tons)*
Glass containers	9.9	98%	9.7
Steel	2.4	90%	2.2
Aluminum	1.0	98%	1.0
Shipping Boxes	12.6	35%	4.4
Other Paper Packaging	9.3	46%	4.3
Plastic	5.5	63%	3.5
Wood	2.1	21%	0.4
Miscellaneous	0.2	15%	0.0
Total	43.0	59%	25.5

Source: Judd H. Alexander, (1993). *In Defense of Garbage*. Table 6-1, p. 67.

Table 6.7. Resin properties and uses

	Resin charactenstics	*Uses*
PP	Tough, hard, doesn't scratch, good tensile strength, good resistance to heat and chemicals, high melting point	Extruded into fibers, filaments Auto batteries
HDPE	Translucent, milky white, resists most chemicals, durable, UV light barrier	Blow-moulded non-durables
LDPE	Light, flexible, translucent or opaque, takes high or low gloss, durable	Extruded into packaging film
LLDPE	High tensile strength, puncture resistance, tear qualities, less clear than LDPE	Packaging film
PS	Solid or foam: as solid it is hard, smooth, brittle, scratch-resistant, clear or opaque; as foam it is light, shock absorbent, poor chemical resistance when foamed, insulator	Trays, cups, package stuffing, containers
PET	Clear, lightweight, shatter-resistant, good chemical resistance, effective gas barrier	Bottles
PVC	Tough with smooth finish, scratches easily, flexible, easy to shape into handles, cuts without stretching, strong resistance to reactive materials	Tubing, film, bottles, sheeting

Source: Lynn Scarlett, (1997). *Packaging, Recycling, and Solid Waste*, Table PL-7, p. 122.

hold waste, dividing it into 33 categories. Interpreting these categories indicates that about 20 per cent of household waste is food packaging (White et al. 1995). This assessment appears to include only household waste and not retail or other commercial waste that might be part of a municipal waste stream. Since corrugated packaging makes up a significant portion of commercial waste, and some of this corrugated packaging transports food, the Warner Spring Laboratory figures probably understate the total amount of food packaging waste.

Though precise estimates of the amount of packaging in the waste stream vary – both within and among nations – the overall portion is substantial enough to attract attention. The general public is largely correct in their impression that packaging is abundant in the waste stream. Far less accurate are their estimates of how much each individual packaging type contributes to municipal waste.

In the United States and Europe, fast-food packaging and plastic packaging have been vilified, with the public wildly overestimating their contribution to total waste. For example, the non-profit Audubon Society held a meeting at which participants were asked to estimate the volume of fast-food packaging and expanded polystyrene foam (plastic) waste. Participants estimated that fast-food packaging composed 20 to 30 per cent of landfilled waste (Rathje 1992). The same group guessed that expanded polystyrene foam (used for cups, food clamshells, and so on) took up 25 to 40 per cent of landfill space. Combined, these estimates put these two categories at between 45 and 70 per cent of landfilled waste. Yet, in examining landfill waste, William Rathje found that fast-food packaging represented less than half of one per cent of the waste (by weight) and not more than a third of one per cent of the waste by volume. Polystyrene foam estimates were equally far off, leading Rathje to conclude that: 'plastic is surrounded by a maelstrom of mythology'.

In a 1990 US survey of consumers, respondents believed that plastics made up 60 per cent of the waste stream, although they actually make up 21 per cent by volume and 10 per cent by weight of total municipal waste in the United States (Ottman 1991). Not only do consumers overstate how much plastic is in the waste stream, they also underestimate the role plastic has played in reducing overall packaging weight. A German research organisation estimated what would happen to the municipal waste stream if all plastic packaging was eliminated and replaced by more traditional packaging forms. Under this hypothetical scenario, they estimated that total materials usage by weight would increase fourfold; energy consumption in packaging would almost double; packaging costs would more than double; waste would increase by 256 per cent (GVM 1987).

Plastic packaging, contrary to its poor public reputation, offered food processors and distributors distinct opportunities to improve the quality and shelf life of some foods at very low cost. Lox (1992) notes that plastics are especially suitable for 'fast, automatic pre-packaging', a trend closely tied to the rise of self-service supermarkets. Benefits came to consumers not only in the form of improved shelf life and reduced waste. Pre-packaged foods, initially celebrated as 'maid-service in a package', reduced the need for routine, weekly home baking, canning, and preparation of other items such as baby food, soup, and so on (Alexander 1993). These packaged foods also dramatically reduced consumer costs – in time and/or money – to obtain these items. Many of these pre-packaged foods contained some plastics.

In the 1967 film, *The Graduate*, the young hero Ben, anxious about his future career, is soberly informed: 'Ben, I want to say one word to you, just one word – plastics.' The advice rightly anticipated future trends. Use of plastics climbed at a rate of 10 per cent per year in the United States between 1958 and 1988 (Scarlett et al. 1997). This jump in plastics consumption resulted largely from the many attractive properties the dozens of different plastic resins offer (see Table 6.7).

Plastic technology enabled the development of flexible packaging, which is sometimes composed entirely of plastic and sometimes combines plastic, paper, and aluminium foil in thin, laminated layers. The modern snack chip bag, for example, uses thin laminated layers of nine lightweight materials, each of which serves a distinct function in assuring overall product integrity and consumer utility. The US Office of Technology Assessment (1992), examining this package, concluded that the package is 'much lighter than an equivalent package made of a single, recyclable material and provides longer shelf life, resulting in less food waste'. Overall, though snack food consumption in the United States increased over a fifteen-year period from 1972 to 1987 by 43 per cent, the total weight of packaging associated with delivering these snack foods declined nine per cent, largely due to 'lightweighting' made possible by use of plastic and laminated packages. In fact, according to the consulting group, Franklin Associates (1992), 'the average weight of primary packaging for 100 pounds of product decreased 36 per cent, from 11.3 pounds in 1972 pounds in 1987'.

High-technology laminated packaging also made possible the aseptic package – often known to consumers as the 'juice box,' though it is used for dairy products, soups, tomato sauce, and even water. The package provides a sterile internal environment, hence its designation as the aseptic package. This sterile package enabled the packaging of milk without refrigeration, providing shelf-stable storage.

The package offers environmental advantages, as well, though it is more costly to recycle than other, simpler packaging forms. A lifecycle analysis by the Coca-Cola Company of seven orange juice packaging and delivery systems showed that the aseptic container consumes less energy than the six alternatives, largely because all the other alternatives require refrigeration, which constitutes 45 to 75 per cent of total energy requirements in juice delivery systems. Even at high levels of recycling for glass or plastic containers, energy requirements for these containers still exceed that of the aseptic box because of refrigeration requirements (Scarlett 1994).

Just as plastic packaging has been the subject of a maelstrom of mythology, so too has fast-food packaging. Consumers, as Rathje has pointed out, vastly overstate its presence in the waste stream – overestimating its contribution to landfilled waste by 80-fold or greater. But consumers also misunderstand its benefits as a food-delivery system. Disposable packaging at fast-food restaurants began to replace reusable ware in part due to health considerations. Comparisons of the bacterial content on disposable and reusable ware have shown that 'the probability of microbial contamination was found to be 50% greater with the reusables than with disposable items used in the same establishment' (Felix et al. 1990).' Moreover, the same researchers found that 15 per cent of the reusables had microbial counts above the maximum recommended level of 100 colonies per utensil.

Reflecting on the pressure against McDonald's fast-food packaging in the early 1990s by young children, Rathje (1992) recalls the complex nature of waste. He laments that the protesting student 'is probably unaware ... that he or she throws away three-and-a-half ounces of *edible* food a day; looked at another way, the average elementary-school student every month throws away the equivalent by weight in edible food of 300 Big Mac foam clamshells'.

Understanding these complexities, Rathje concludes that material objects are 'partners in a dynamic relationship'. Consumption choices, including packaging selection, involve a complex and dynamic interaction between the consumer and the object. The availability of new materials and packaging technologies allow new lifestyle choices; and the emergence of changing lifestyles and demographic patterns drive the search for new packaging options to meet ever-wider performance criteria. 'Packaging serves many masters', observes Alexander (1993), among which are the protection of public health and the delivery of high-quality, diverse food to the household.

Trends for the future

Packaging does help reduce food waste – from field to factory to supermarket to household. But it also makes up about one-third of the modern municipal waste stream. However, two points are worth noting. First, overall per capita waste over the past century has been relatively stable, though its composition has changed. Second, trends in packaging waste are toward less, not more, packaging per unit of consumption. In some instances, the trend is even toward less absolute amount of packaging waste.

Studies that measured household waste in the United States in 1959, and again 20 years later, found that discards remained relatively steady at 1.9 pounds per person per day (Rathje 1992). Harvey Alter, former environmental researcher at the US Chamber of Commerce, identified those few cities that actually measured the amount of solid waste disposed of over selected periods of time and found no general increase. Alter (1990) also found that the amount of solid waste generated in the United States is declining in relation to gross national product. That is, despite consumption at the household level of more goods, the amount of trash produced per dollar of GNP is declining, a trend also observed by Franklin Associates (1992).

Even turning the historical page back further, to the early 1900s, waste generation by households reached levels similar to current amounts – but the composition of that waste was decidedly different. One study estimated that 'between 1900 and 1920 each citizen of Manhattan, Brooklyn, and the Bronx annually produced about 160 pounds of garbage, 1,231 pounds of ashes, and 97 pounds of rubbish' (Melosi 1981). Packaging, paper, and other discards found in a modern waste stream were fractions of what they are today; but coal and wood burning for heat produced substantial amounts of ash that required disposal.

The second trend – toward a net reduction in packaging material per unit of production – has contributed to the stabilisation of per capita waste generation within industrialised countries at the same time that incomes have increased and overall consumption has increased. These net reductions occurred through a combination of source reduction – lightweighting, use of bulk packaging, elimination of outer packaging – and market-driven recycling that predate the politically induced recycling of the 1990s. For example, in the United States a typical basket of US grocery items fell from over 2,750 pounds of packaging per gross production unit in 1989 to approximately 2,100 pounds in 1993-94 (Scarlett et al. 1997). These source-reduction trends combined with increases in market-driven recycling to produce absolute declines in grocery packaging as a percentage of total municipal solid waste. In the United States, where data are broken down into different packaging categories, grocery packaging declined between 1980 and 1993 from 15.3 per cent to 12.1 per cent, despite a 14 per cent population increase over the same time. Restated, grocery packaging discards declined 26 per cent between 1980 and 1993 in absolute terms, due to a combination of source reduction and recycling.

A few simple examples demonstrate how these reductions occurred:

- Plastic milk jugs weighed 95 grams in the early 1970s; by 1990 the same jug weighed just 60 grams.
- Plastic grocery bags were 2.3 mils thick in 1976; by 1989 they were 0.7 mils thick.
- New materials, such as flexible packaging, allowed large reductions by weight in packaging for a number of products. For example, the

plastic frozen food bag resulted in an 89 per cent weight reduction and an 83 per cent volume reduction for some frozen foods previously packaged in waxed wrap cartons (Scarlett forthcoming).

For the consumer, these changes are often invisible, as are their overall waste-reduction implications. For example, one juice manufacturer reduced the cubic dimensions of a juice package by 16 per cent and the size of the label by 10.7 per cent. This change saved nearly 20,000 pounds of materials, more than 500 truckloads of outgoing freight, 20,000 shipping pallets, 7,000 pounds of stretch wrap, and 250,000 square feet of chilled warehouse space (Scarlett, 1994). This sort of reduction in packaging waste has been accompanied with improved shelf life, better product quality, and greater convenience to the consumer, a win–win enterprise.

Along with source reduction, market-driven recycling also moderated net packaging discard levels, despite population and consumption increases. For example, between 1972 and 1987, recycling of corrugated cardboard in the United States increased from 21 per cent to 41 per cent. A few years later, that rate had climbed above 50 per cent. On a per capita basis, Franklin Associates (1992) points out that 'generation of corrugated cardboard waste increased 31 per cent, and discards decreased by three per cent'. Market-driven recycling in Europe prior to introduction of packaging ordinances requiring specific levels of recovery was also robust and helped moderate the impact of packaging on the waste stream without imposing costs on either consumers or municipalities. Glass recycling, for example, ranged from 20 to over 50 per cent among different European countries in the late 1980s (White et al. 1995).

These trends toward net reductions in packaging per unit of output will probably continue, though there are two caveats in this general success story.

First, the growing attempt to politicise packaging choices, dominated by an emphasis on recycling, may distort and slow source-reduction trends over time. Many source-reduction achievements result from developing composite and laminated packages. These packages permit the delivery of food using very little overall material, but they pose recycling challenges since they are neither easy to isolate from the waste stream nor are they easy to process into uniform, uncontaminated high-quality materials, attributes that are prerequisites to efficient recycling.

Second, in the press to eliminate packaging waste, increases in food waste may result. 'Packaging,' as anthropologist William Rathje notes, 'must not be looked at in isolation ... it serves larger purposes – purposes that involve such things as efficient resource management, product protection, the prevention of tampering, and maintaining public health (as well as, of course, turning a profit).'

References

Alexander, J. (1993). *In Defense of Garbage* London: Praeger, pp. 64-65.

Alter, H. (1990). The Future of Solid Waste Management in the United States Washington, D.C.: US Chamber of Commerce.

Blegen, C.W. (1958) *Troy* Princeton: Princeton University Press.

Chagnon, N. (1992). *Yanomamo* Fort Worth, TX: Harcourt Brace Jovanovich, p. 62.

Diamond, J. (1997). *Guns, Germs, and Steel: The Fates of Human Societies* New York: W.W. Norton & Company, p. 261.

Felix, C.W., Parrow, C and Parrow, T. (1990). Utensil sanitation: A microbiological study of disposables and reusables, *Journal of Environmental Health*, **2**.

Franklin Associates, (1992). *Analysis of Trends in Municipal Solid Waste Generation: 1972-1987*, report prepared by Procter & Gamble, Browning Ferris Industries, General Mills, and Sears, Roebuck, and Co. Prairie Village, KS.

Franklin Associates, (1997). *Solid Waste Management at a Crossroads Prairie Village*, KS, December.

GVM (Gesellschaft fur Verpackungsmarktforschung) (1987). Packaging without plastic: Summary report prepared for Verban Kunststofferzeugende Industrie, Frankfurt, Germany.

Lox, F. (1992). *Packaging and Ecology* (Surrey, U.K.: Pira International, p. 8.

Melosi, M. (1981). *Garbage in the Cities Chicago*: Dorsey Press, p. 9, 113-14.

Ottman, J. (1991). *Environmental Consumerism*, J. Ottman Consulting.

Rathje, W. (1992) *Rubbish: The Archaeology of Garbage* New York: HarperCollins, p. 39.

Rathje, W. and Reilly, M. (1985). *Household Garbage and the Role of Packaging* Tuscon, AZ: University of Arizona, July.

Scarlett, L. (1991). *Consumer Guide to Environmental Myths and Realities* Dallas, TX: National Center for Policy Analysis.

Scarlett, L. (1994). Packaging, Solid Waste, and Environmental Trade-offs, *Illahee: Journal for the Northwest Environment*, Spring.

Scarlett, L. (forthcoming). Dematerialization: Unsung Environmental Triumph, in Ron Bailey, Earth Report 2000: *The True State of the Planet*, 2nd edition. New York: McGraw-Hill.

Scarlett, L., McCann, R. Anex, R. et al. (1997). *Packaging, Recycling, and Solid Waste Los Angeles*: Reason Public Policy Institute.

US Office of Technology Assessment, (1992). *Green Products by Design* Washington, D.C.: Government Printing Office.

White, P.R., Franke, M. and Hindle, P (1995). *Integrated Solid Waste Management: A Lifecycle Inventory* London: Blackie Academic & Professional.

Wood, M. (1987). *In Search of the Dark Ages* New York: Facts on File Publications, pp. 23, 29.

Section 2:

If it's not part of the solution, it's part of the problem

7 Dietary disarray: guidelines with a pinch of salt*

David M. Warburton, Eve Sweeney and Neil Sherwood

Summary

Dietary guidelines from around the World were examined, using information gathered from Government and non-governmental organisations, books and journals, the Internet and expert sources at Universities and other institutions. The research reveals that countries and organisations differ in which dietary components they feel require comment, and shows up to a ten-fold difference where recommended maximum intakes are specified. In an attempt to explain these discrepancies, the scientific basis upon which such guidelines are based was examined and found wanting. An increasing amount of evidence shows that recommendations may be ill-founded, non-productive and even positively harmful. This suggests that while gross contradictions remain between professional bodies, it is little wonder that the layman is confused and remains sceptical of dietary advice.

> A man seldom thinks more earnestly of anything than he does of his dinner.
>
> Dr Samuel Johnson

Introduction

Governmental agencies are trying to encourage us all to consume less of this and more of that on a population-wide basis. We should eat more fruit and vegetables and restrict our fat consumption. In other words, we are being advised to be more like Jack Sprat, who ate no fat. While his wife, who could eat no lean, is being told to change her eating habits completely. This population approach ignores a fundamental fact – we

*Research for this paper was funded by ARISE (see www.arise.org)

are all different. But, not only is Mrs Sprat being advised to eat more like Mr Sprat but this advice extends to the little Sprats. 'The Population Approach', of the National Cholesterol Education Programme, aims to prevent Coronary Heart Disease (CHD) by lowering levels of cholesterol in children and adolescents through changes to their diet.

In any event, are today's guidelines clearer? Based on scientific research? And is the consumer given all the facts to make educated decisions for themselves regarding their lifestyle?

From many reports, CHD is presented as the biggest cause of concern worldwide. In 1992, the UK government published a White Paper 'The Health of the Nation' and 'Our Healthier Nation' in 1998. In the latter Report, it is stated that in the 1990s around 90,000 people die each year before they reach their 65th birthday and of these people more than 25,000 die of heart disease, stroke and related illnesses (Secretary of State 1998). A briefing pack from this White Paper (Fact Sheet – Coronary Heart Disease/Stroke 1997) states that 88 per cent of men and 90 per cent of women have at least one of the four major risk factors for CHD – high blood pressure, high cholesterol, lack of exercise and smoking. Being overweight is said to increase the risks from cholesterol and blood pressure.

The Surgeon General's Report 'Nutrition and Health' in 1988, states that the causes of CHD are a combination of multiple environmental, behavioural, social, and genetic factors. In 1981, there was already a list of 246 (Hopkins and Williams 1981) associated 'risk factors' for CHD and James McCormick of Trinity College, Dublin has found over 300 (McCormick 1996). However health campaigns have been focused almost exclusively on lifestyle. 'Drink less alcohol'; 'eat less saturated fats', 'especially cholesterol'; 'avoid salt'; 'don't smoke' and 'take much more exercise'.

One would think that the consumer is being presented with the information which is required for making balanced and informed choices about their lifestyle. From the certainty of the dietary advice, it would be reasonable for us to believe that there was a strong scientific basis for the recommendations and that there would be a worldwide consensus on the recommended levels of consumption. But we would be wrong.

The dietary guidelines

Alcohol

A good example of the variability in recommended levels of consumption can be seen for beverage alcohol in Table 7.1, which has been expanded from a table, which was published in *Alcohol in Moderation* (1996, 5 (2): 11). First, it can be seen from Table 7.1 that countries cannot even

Table 7.1. Recommended maximum alcohol consumption

Country	Authority	Sub-Group	Men		Women		Unit (g)
			Max g/day	Max g/wk	Max g/day	Max g/wk	
France	Academy of Medicine	wine	60	420	36	252	12
Austria	Bundesministerimn für Gesundheit und Konsumenschutz	beer & wine	50	249.9	33.3	166.6	11.9
Italy	Ministero della Sanita	non elderly	40	280	30	210	10
Portugal	Conselho Nacional De Alimentacao E Nutricao (CNAN)[5]	wine	37.3	261.1	18.6	130.6	14
Australia	National Health and Medical Research Council[1]		40	200	20	100	10
Japan	Ministry of Health and Welfare[1]	men only	39.5	197.5			19.75
Denmark	Sundhedsstyrlsen[1]		36	252	24	168	12
Romania	Health Ministry	beer	32.5	227.5	17.3	121.1	
UK	Department of Health[1]		32	224	24	168	8
Italy	Ministero della Sanita	elderly	30	210	25	175	10
Spain	Ministry of Health and Consumption		30[4]	210	20[4]	140	10
New Zealand	Alcohol Advisory Council		30	210	20	140	10
USA	Departments of Agriculture/Health and Human Services		28	196	14	98	14
Canada	Addiction Research Foundation[1,2]		27.2	190.4	27.2	190.4	13.6
Austria	Bundesministerium für Gesundheit und Konsumenschutz	spirits	26.5	132. 3	17.6	88.2	6.3
Ireland	Department of Health		24	168	16	112	8
Czech Republic	National Institute of Public Health		24	168	16	112	
Finland	Oy ALKO AB		23	161	16	112	11
Romania	Health Ministry	wine	20.7	144.9	10.8	75.6	
Sweden	Systembolaget (Low Risk)	3	18	126	13.6	95	
Sweden	Systembolaget (Not dangerous)	3	6.7	47	5.4	38	
Germany	DGE	6					

Table reproduced from AIM Digest 1997 with data for Portugal and Germany added

Note The data presented above are estimated and listed in descending order, according to the daily recommendations for men. Where daily advice alone is given it has been extrapolated to weekly advice (and vice versa), for the purposes of comparison. In some cases different advice is given to sub-populations (such as pregnant women)

1 Drink free days are clearly advised and may be reflected in the estimates

2 Endorsed by the Royal College of Physicians and Surgeons of Canada, the Canadian Medical Association, the College of Family Physicians of Canada and the Canadian Medical Society on Alcohol and Other Drugs. No distinction by sex; 'adult man of average build' is used as a reference.

3 Systembolaget publishes its advice in terms of 40% spirits, and notes approximately equivalent values for beer and wine. The chart lists figures for 'not dangerous' and 'low risk' catergories

4 Suggests a maximum of double this value 'on any one occasion'

5 Calculated on average glass of wine being 150ml

6 No recommendation given

Table 7.2. Recommended maximum fat intake

Country	*Authority*	*Total Fat (% of total calories)*	*Saturated Fat*	*Polyunsaturated Fat*	*Monounsaturated Fat*	*Cholesterol (mg)*
UK	COMA	35	10%	10%	–	245
Netherlands	Netherlands Nutrition Council	30-35	10%	–	–	–
USA	American Dietetic Association*	30	–	–	–	–
	American Academy of Pediatrics*	30	10%	–	–	300
	American Heart Association*	30	8-10%	10%	15%	300
	National Cholesterol Education Program*	30	10%	10%	10-15%	300
	US Dept Agriculture & University of Nebraska*	30	10%	–	–	–
Italy	Istituto Nazionale della Nutrizione[1]	15-30	7-10%	7-10%	10-15%	–
France	Apports Nutritionnels Conseillés pour la population Française	30	10%	7%[2]		
Canada	Scientific Review Committee	30	10%			
Czech Reoublic	REGA	30	10%	–	–	300
Portugal	Conselho Nacional De Alimentacao E Nutricao (CNAN)	30	10%	–	–	300
Belgium	Conseil National de la Nutrition	30	10%	3-7%	–	300
Slovak Republic	Research Institute for Nutrition	27.5	–	–	–	–
Germany	DGE	25-30[3]	–	–	–	–
Singapore	Food & Nutrition Dept. Ministry of Health	20-30	1/3rd	1/3rd	1/3rd	300
Poland	National Food & Nutrition Institute	30				
Sweden	'Food Circle'	30-40g[4]				

* – recommended for children over the age of 2

1 Italy specifically recommend for small children 40% total fat

2 recommends be balanced between Linoleic and Linolenic acids

3 recommended for over 25's with a suggested maximum of 40

4 30g per day for women and 40g per day for men – edible fat

agree on what constitutes a standard unit; it is 19.75g of alcohol in Japan and 6.3g in Austria. Second, while there is a consensus that 'moderate' consumption is permissible, there are wide variations for the different countries.

For example, the recommended limits for alcohol consumption range from 60g a day of absolute alcohol for men in France to 6.7g a day for men in Sweden. It is obvious that high levels of consumption are not good for health, but no-one knows what constitutes 'safe' levels. It should also be remembered that the values are derived from the association between illness and self-reported consumption to a doctor, a figure which is typically an underestimate.

Fat

The recommended levels of fat intake show that the majority of countries copy the USA (Table 7.2) and limit total fat intake to 30 per cent of total calories with cholesterol intake no more than 300mg.

Yet, as Professor Walter C Willett (1995) of The Harvard University School of Public Health states, the dietary recommendations regarding fat consumption are not based on facts. No research has ever proved an ideal amount of fat to aim for.

In addition, the abundance of information on fat is confusing to the average person. The American Heart Association (AHA 1997) states that saturated fats (those found in animal products) are the main culprits for raising blood cholesterol so people should not eat butter and substitute saturated fats with monounsaturates and polyunsaturates. However, in June of the same year, the AHA revised this view by stating that some margarines raise serum cholesterol as much as butter; those which contain trans-fatty acids, that is, those that are hydrogenated (Lichtenstein 1997). In order to help the consumer through the fat quagmire, the Food and Drug Administration (FDA) created their own morass in the form of a 5-page pamphlet 'A Consumer's Guide to Fats' (FDA 1994). The majority of the guide refers to cholesterol, with no clear distinction between dietary cholesterol and dietary saturated fats. However, it did try to clarify everything by pointing out that, in addition to saturated and monosaturated and polyunsaturated fats, there are triglycerides, trans-fatty acids, omega-3 and omega-6 fatty acids to be monitored and fretted about.

Eggs

If one read the recommendations on egg consumption, one would believe that an egg is shell-to-shell cholesterol, although there does not seem to be agreement on the amount of cholesterol in one egg; 213mg (AHA 1997) to 447mg (Readers Digest 1999). Therefore, one would think it would be easy to give a simple recommendation for egg consumption.

Not so.

Table 7.3 shows that the recommendations for consumption of eggs vary 10-fold worldwide. In the United Kingdom, the Committee on Medical Aspects on Food Policy, with the unfortunate acronym COMA, recommended only one egg per week in their sample diet, (Department of Health 1994) whereas the British Heart Foundation, talking about the same British population, recommended four (Readers Digest 1999). Recommendations from the American Heart Association were only three eggs in 1995 but now four for 1997. The World Health Organisation (WHO) has given the highest recommendation of no more than 10 per week (Readers Digest 1999).

Salt

Even if the authorities cannot get it right for eggs one might assume that there would be some similarities for the recommendations for something as simple as salt. However, recommended salt/sodium intake (Table 7.4) also varies worldwide with Germany recommending the highest at 10g per day whilst the British Heart Foundation recommend 1.6 to 6g per day. Some countries, such as France and Poland, do not specify a maximum figure, and overall the recommendation is 'just reduce salt intake', as if everyone was overconsuming and at risk.

The scientific evidence

Clearly, the available guidelines look as if they are numbers plucked out of the air. Thus, it is instructive to look at the 'scientific evidence' which allegedly backs them up. One might expect that, although the guidelines were not exactly the same, the data on which they were based were sound. One would anticipate that there would be a body of evidence which supported the need to limit consumption, although some agencies might be more conservative than others. However, an examination of the studies do not give any reassurance.

Alcohol

The issue of the health effects of alcohol consumption continues to provoke controversy. Some groups, like the WHO (Anderson 1996; WHO 1993), refuse to accept the J-shaped curve showing that moderate alcohol consumption can have a beneficial effect on diseases such as CHD (Peele and Grant 1999). The American Cancer Association considers that the benefits to CHD of moderate alcohol consumption may be outweighed by a higher risk of cancer, especially in men over 50 and women over 60 (Mezzetti et al. 1998). They claim that even a few drinks a week may increase the risk of breast cancer, even though a 85,709 person study of nurses in 1995 found that the risk of breast cancer

Table 7.3. Recommended maximum egg consumption

Country	*Authority*	*Per week*
	WHO[1]	10
USA	American Egg Board	7-14
USA	Dietary Guidelines for America (4th Ed)	7
USA	American Heart Association (1997)	4
UK	British Heart Foundation[1]	4
Germany	DGE	3[1]
Czech Republic	REGA	3[2]
Slovak Republic	Research Institute for Nutrition	3
Italy	Istituto Nazionale della Nutrizione	2-3
UK	COMA	1[3]

[1] From – [http://foods/readersdigest.co.uk/extracts/pages/eggs/html]

[2] allowance 13 eggs per month

[3] suggested as part of a hypothetical ideal diet

decreased with light to moderate drinking (Fuchs et al. 1995). The British Medical Association (BMA) in 1995 stated that the guidelines for the UK of up to 14 units for women and 21 units for men per week should remain unchanged as they constitute a 'low risk' range of drinking to which individuals should aim to adhere. The risk is the development of 'addiction'.

Salt

The issue of salt consumption is similar to that of fat and cholesterol. Typical of the controversy surrounding intake of salt and the link with hypertension, Dr Graham MacGregor, of St George's Hospital Medical School, London, recently published a study in *The Lancet* (Cappuccio et al. 1997) in which he stated that even a slight reduction in salt intake lowers blood pressure in older people and thus significantly reduces the risk of a heart attack. This change, he calculated, would 'save' 34,000 lives a year.

In contrast, a meta-analysis of 56 studies concluded that reduction of salt intake was unnecessary unless the person was over 45 years old AND suffering from high blood pressure (Midgley et al. 1996). They stated that it was unwise to blindly assume that a low-salt diet was harmless for the population in general, because a study of hypertensive men showed evidence of an increased risk of heart attack with a low-salt diet (Alderman et al. 1992). Another problem associated with salt reduction is observed in people working or exercising in hot climates, that of salt depletion. The symptoms include, fatigue, headache, weakness, nausea, and occasionally vomiting (Weatherall, Ledingham and Warrell 1989).

Table 7.4. Recommended maximum daily intake for sodium/salt

Country	Authority	Sodium per day*	Equivalent amt salt*
Germany	DGE	4,000mg	10g
USA	The American Dietetic Association	2,000–4,000mg	5–10g
Netherlands	Netherlands Bureau for Food & Nutrition Education	3,600mg	9g
Belgium	Conseil National de la Nutrition	3,500mg	8.75g
Italy	Istituto Nazionale della Nutrizione	600–3,500mg	1.5–8.75g
USA	Dietary Guidelines for America (4th ed, 1995)	2,400mg	6g
USA	American Heart Association (1997)	2,400mg	6g
Portugal	Conselho Nacional De Alimentacao E Nutricao (CNAN)	2,400mg	6g
Czech Republic	REGA	2,400mg	6g
UK	COMA	2,400mg	6g
UK	British Heart Foundation	640–2,400mg	1.6–6g
USA	American Academy of Family Physicians	2,300mg or less	5.75g
Singapore	Ministry of Health	2,000mg	5g
Sweden	Swedish Nutrition Recommendations	800mg	2g
France	Apports Nutritionnels Conseillés pour la population Française	[1]	
Poland	National Food & Nutrition Institute	575mg	1.44g

NB – where recommendations are given for sodium or salt alone, the appropriate figure for the other was calculated at 400mg of sodium = 1g salt.

* Not all agencies give salt and sodium figures – many countries recommend 'reduce salt intake' without specific guidelines.

[1] No recommendation given.

[2] This is minimum intake – no maximum recommendation given.

Dietary fat

The simple dietary advice is drastically curtail fat consumption and eliminate heart disease. The serious consequences of the propaganda, Le Fanu (1986) thinks, is that the public is misinformed about the complexity of diseases. For example, McMichael (1979) does not believe atheroma is caused by diet – but that it is a complex process of heredity, infection and other unknown factors, but most of all believes it is a 'wear and tear disease'. Social class variations for CHD mortality have been reported by COMA (Department of Health 1994), with lower than expected mortality in the professional occupations and higher than expected in the unskilled group. A recent study by George Davey-Smith and his colleagues reported an association with higher socioeconomic status and better health, in other words, wealth equals health (Davey-Smith et al. 1996). Thus, why should we expect that changing a single factor, like diet, would be beneficial to everyone in the treatment of a multifactorial disease (McNamara 1986), especially when some would say that for 'multifactorial' we should really read 'unknown' (McCormick and Skrabanek 1988)?

Indeed, the justification of dietary recommendations relies on selective, partial, obsolete information or is even the antithesis of the conclusions reached in studies (Le Fanu 1986, Reiser 1984). Committees looking at the studies have failed to point out that the level of heart disease does not invariably follow the pattern of food consumption (Le Fanu 1986). The most quoted paper is that of Keys (1970) which examined a seven-country study and reported a relationship between dietary fat and heart disease. But this study is not representative of the international comparisons. For example, Le Fanu (1986) found a random relationship between death from heart disease and dietary fat intake in 15 countries, which argues strongly against even a small causal relationship between dietary fat and heart disease. Certainly, in the US, Canada and Australia, fat consumption has steadily increased since the 1970's, whilst ischaemic heart disease deaths have been falling sharply. Japan shows the greatest and fastest increase in fat consumption anywhere in the world but death from heart disease has been falling steadily since 1950 (Atrens 1994).

Part of the explanation for the changes in CHD incidence is that it is a disease of old age. Many previous causes of death, such as infectious disease, are now controlled which means there has been a tripling of the proportion of elderly in the population. Now, 82 per cent of deaths from 'heart attack' occur in persons aged 65 years or more (Chew 1995). Death rates increase rapidly with increasing age for most diseases of adults and Harper (1983) points out that the countries with the highest rates of heart disease are also those with the longest life expectancy. Of the 22 countries where life expectancy for men is 72 years or longer, seven consume low-fat diets and 15 consume high-fat diets. In addition, the average life expectancy is 73 for men and 6 years more for women in both groups (Robb-Smith 1967).

McCormick and Skrabanek (1988) think that 'to base population strategies on unproven hypotheses seems unreasonable'. Results of studies will be disappointing to those that believe a healthier diet – not eating things they enjoy such as cream, chocolate, cakes and chips- will protect them from an early death (Browner et al. 1991). Indeed, the fact that many dietary recommendations are not founded on scientific evidence may result in the public being ignorant of the merits and possible faults of any dietary changes being able to prevent CHD (Ahrens 1979) and the time may not be ripe for national dietary guidelines. The only scientifically defensible position is that the public should be kept fully informed of the questions remaining unanswered, as well as any progress that is being made. They should also understand that science rarely proceeds by breakthroughs, but by the slow accumulation of evidence.

Cholesterol reduction

Over $1,000,000,000 has been spent in primary prevention trials for heart disease based on reducing cholesterol whilst the increasing amount of evidence shows that this may in fact be ill-founded, non-productive and even harmful (Atrens 1994, Ahrens 1985, Ramsey et al. 1991, Reid and Mulcahey 1987).

Evidence that the premise may be ill-founded has come from a number of epidemiological studies. First, the British Regional Heart Study found no association between total cholesterol level and CHD mortality (Shaper et al. 1981). Second, COMA (Department of Health 1994) also reported that, although the level of CHD varied between different regions of the UK, cholesterol levels showed little regional variation, although they still concluded that the evidence clearly indicates a reduction in cardiovascular disease as well as a net benefit to health! Third, there is a lack of any association between cholesterol level and CHD in women (Dunnigan 1993). Women have lower cardiovascular mortality and yet show higher levels of cholesterol compared to men, particularly after the menopause. Yet total CHD mortality in the UK for men is 30 per cent whilst only 23 per cent for women. Fourth, amongst the Masai in Africa death from cardiovascular disease is almost unknown and yet they have a very high intake of fat and substantial atherosclerosis (Atrens 1994). Five, while there has been a great reduction in ischaemic heart disease mortality since the 1950's, cholesterol levels have remained constant or even increased (Bonneux and Barendregt 1994). Some positive clinical trials have used patients with familial hypercholesterolaemia (FH). As Mann (1994) points out, to extrapolate findings from these unusual people (1 in 500 white Americans) to the general population, is absurd. FH individuals are the 'least appropriate' patient if we are trying to treat an acquired disease by diet not an inherited disease. Certainly, they have confused and exaggerated the relationship between serum cholesterol and atherogenesis. Six, a Finnish study (Jousilahti et al. 1996) showed cholesterol levels increase with age even when saturated fat intake remains constant. The authors then state that a reduction in saturated fat intake would prevent the rise of serum cholesterol levels with age! Further they suggest that the effect of dietary changes cannot be proven to be linked to change in cholesterol level by epidemiological analysis or cross-sectional studies – this can only be done by clinical studies.

Evidence that the strategy will be non-productive or have trivial benefits has come from estimates of the mortality savings. These estimates have not been encouraging. A model which was developed by Taylor et al. (1987) predicted one to two months increased life expectancy beginning with cholesterol lowering at age 20. Browner et al. (1991), used a different model which was based on the current guidelines of a restriction of dietary fat to 30 per cent total energy. It employed a 'best case' analysis which assumed associations between fat consumption, serum cholesterol and mortality to be causal and entirely reversible and predicted an increase of three to four months on average life expectancy, mainly for people over the age 65. Another model used by Grover et al (1992) considered low-risk (5.2 mmol/L serum cholesterol) and high-risk (7.8 mmol/L serum cholesterol) individuals. They estimated that a reduction in serum cholesterol would increase the average life expectancy of men and women aged 35-65 by between 0.03 (low risk) and 3.16 years (high risk) and it would delay the onset of the symptoms of CHD by between 0.06 to 4.98 years respectively. In other words, average life expectancy of low risk individuals (the majority of the population) would be increased by about 0.03 years (about 11 days) after age 65 and just over four months for a 35 year old man. Actual data from an intervention study support the validity of the models. The Multiple Risk Factor Intervention Trial (MRFIT), a study of men aged 35-57 reported a 50 per cent lower death rate with reduction of serum cholesterol – from 1.4 to 0.7 deaths per 1,000 – a reduction in risk of 0.07 per cent and the Framingham study showed a 0.14 per cent decrease in risk (Robb-Smith 1967).

The attempts to lower cholesterol may even have negative consequences. One set of outcomes appears to be a significant excess of mortality from non-coronary deaths – accidents, suicides and violence (Dunnigan 1993). Muldoon et al. (1990) found that 76 per cent of those receiving dietary or drug intervention for reducing cholesterol were more likely to die from these causes than participants in the control groups although he could give no plausible explanation for this effect. Were they just feeling mean and miserable? Serum cholesterol levels below 4 mmol/l are also associated with increased risk of mortality from other diseases such as cancer, trauma, respiratory and digestive diseases (Jacobs et al. 1992).

In order to reduce serum cholesterol by 10 per cent and reduce ischaemic heart disease by 27 per cent (equivalent to a net gain of 2.5 to 5.0 months in life expectancy), drastic changes in diet are required (Bonneux and Barendregt 1994). These diets are near unpalatable and unacceptable to most eaters (Ramsey et al. 1994). Consequently, the panicked people will resort to any means to reduce the cholesterol without pain. Thus, the FDA (1996) has licensed 'Olestra', which replaces fat, but is not absorbed itself. However, Olestra also stops absorption of the fat soluble vitamins, A, D, E and K, and so these vitamins must be added to the product. In addition, numerous sites on the Internet have warned of the unfortunate side-effects of stomach cramps and 'anal leakage', however, these problems appeared only in early trials, and not the product approved.

It must be realised that animal products provide important valuable nutrients. Dairy products provide calcium for bone development and maintenance, while red meats provide essential iron. Eggs provide high quality nutrition at a reasonable price and are easy to cook and digest, especially for the elderly. Justifiably, McNamara (1986) has queried if it makes sense for these valuable foods to be excluded, or even limited, from everyone's diet on the off-chance that this may reduce blood cholesterol and the incidence of heart disease.

Fat and children

While the epidemiological support may be weak, it could be argued that there is no harm in dietary restraint and erring on the side of caution. However, disturbing evidence is emerging that adult dietary advice is being applied uncritically to children with potentially disastrous consequences. Naturally, parents want to do the best for their children and so they believe that an adult, low fat diet is the ideal. They do not realise that dietary fat is a vital nutrient. It supplies essential fatty acids which are especially important for proper growth in children. It carries fat-soluble vitamins – A, D, E, and K – and aids their absorption. They help the body to use carbohydrate and protein more efficiently. Stores of fat are not only required for energy but insulate the body and support and cushion organs. Most importantly, cholesterol is part of every cell. It covers the fast transmission nerves in the body and so enables efficient brain function. Cholesterol is the essential building block of many of our hormones, including the sex hormones (Weatherall et al. 1989).

Astonishingly, many so called authoritative bodies in the USA now recommend the restriction of dietary fat for all children over the age of 2 years (American Academy of Pediatrics 1992; American Dietetic Association 1997; Fisher, Van Horn and McGill 1997), on the basis of the recommendations for adults which, as we have already seen, are anything but conclusive. The consequence is that recent studies are showing that restricting fat consumption in children is inhibiting growth, delaying puberty and there are now a growing number of malnourished babies in the USA (Mirkin 1997; Canadian Paediatric Society 1994; Schafer 1995). In the UK, there is evidence that underconsumption of fat leads to anaemia, stunted growth, learning difficulties, diabetes or heart disease (Stordy and Wright 1996).

Parental misconceptions due to the dietary guidelines were pointed out by Fima Lifshitz and Omer Tarim of the Maimonides Medical Centre, Brooklyn in 1996:

> In a recent study, it was demonstrated that mothers of children with non-organic failure to thrive had higher levels of dietary restraint. Despite their child's low weight, 50 per cent of mothers were restricting their child's intake of sweet foods, and a further 30 per cent were restricting foods they considered fattening or unhealthy. These data show that parental misconceptions are quite common and play an important role in the aetiology of non-organic failure to thrive, a common paediatric problem (Lifshitz and Tarim 1996).

Ahrens (1985) states there is 'no evidence that optimum growth and good health will be maintained in growing children on such a diet' and why start at 2 years when fibrous plaques manifest themselves at 15-19 years. If 80 per cent of heart disease occurs in people over 65 and 50 per cent after 75, and the process of atherosclerosis starts young, it must indeed be very slow (Robb-Smith 1967). In any event, cholesterol levels in children are a poor predictor of high levels in young adulthood and even poorer at predicting CHD later in life (Newman et al. 1990). A recent AHA report points out children and teenagers have wildly fluctuating blood cholesterol levels (Labarthe et al. 1997).

Even if a screening and intervention programme was successful, 100-200 boys found with the highest level of blood cholesterol would have to follow the diet for 50 years to prevent one premature death (before age 65) from CHD. For girls the figure would be 300-600 girls, and if the downward trend in CHD deaths continues, these numbers would be even higher (Newman et al. 1990). It could be argued that even one premature life is worth saving, but is important to remember that intervention to lower serum cholesterol levels may result not only in malnutrition but more importantly, as we have seen, an increase in accidents, violence or suicide – an iatrogenic tragedy (Newman et al. 1990).

Some voices of reason

Fortunately, eminent voices are expressing their dissent and questioning the wisdom of applying the dietary guidelines across populations, especially to women, children and the elderly.

After reviewing the evidence exhaustively, Atrens has emphasised the inadequacy of a diet–heart disease association:

> The image of dietary fat congealing in arteries to form a lethal sludge occupies a prominent place in the public psyche. The view that a high fat diet is atherogenic and that reducing fat intake and serum cholesterol levels is a viable means of combating heart disease persists, in spite of the fact that every main tenet of this belief structure has been found to lack consistency and, in some cases, any support (Atrens 1994).

Thus, he has concluded that there is lack of logic for the dietary recommendations:

> The most basic precept of medicine, dating back to the works of Hippocrates, is *primum non nocere* [foremost cause no harm]. This precept appears to have been forgotten in the illogic that generated and sustains the demonisation of dietary fat and cholesterol ... Until there are sound reasons for dietary change and a general cholesterol lowering, urgent messages to adopt these courses of action must be considered as irrational and mischievous ... It is unacceptable that the zeal and pronouncements of consensus committees and national panels continue to substitute for logic and data. In their advocacy of unpleasant, ineffective and likely dangerous medicine, public health authorities and the medical profession promote ill health, undermine their own credibility and inhibit the discovery of genuine therapies for pressing health problems (Atrens 1994).

The levels set in the recommendations and the problems of comparing various populations has been questioned by Walter C Willett, Professor of Epidemiology and Nutrition at Harvard:

> the recommendation to eat a 30 per cent fat diet to protect against heart disease is not based on facts. Research has not proved that there is an ideal amount of fat to be aimed for in the diet. The link between high-fat diets and cancer was tenuous at best. Studies within populations have repeatedly failed to find a relationship between fat intake and cancer incidence. And comparisons of populations in different countries may neglect other factors contributing to low cancer rates. In rural China, where the population eats little fat and has a low breast cancer rate, people also have limited amounts of food in general and the average age of menarche is 18. The later the onset of menstruation, the lower the cancer rates (Willett 1995).

The reasoning behind the dietary guidelines was questioned by Dr Petr Skrabanek, late Fellow of Trinity College and Ireland's Royal College of Physicians:

> There is no doubt that the Spaniards, the French, the Italians, and the Greeks enjoy their cuisine, their drinks and *l'amour*. But the engineers of

our diet do not mean this when they talk about the 'Mediterranean diet', they just mean olive oil and greens. The simplistic reasoning behind this idea is based on the observation that in Mediterranean countries the mortality from coronary heart disease is lower than in Britain or the United States. Yet people in the Mediterranean region do not on average live longer than the British; they simply die of something else. The life expectancy at birth for English men in 1988 was 73 years, the same as for French and Italian men (Skrabanek 1995).

Whilst Donald McNamara, Professor of Nutrition and Food Science at the University of Arizona at Tucson pointed out in 1986:

> Mankind now experiences the longest life expectancy ever known; in many ways due to advances in medicine and to better overall nutrition, both quality and quantity. It seems curious that the diet we now enjoy, which has in part provided this extension of the lifespan, is accused of being the culprit in the development of a number of diseases associated with ageing. Today's snake oil salesmen now write popular books on nutrition and play on man's fear of mortality by promising much more than is reasonable, let alone proven (McNamara 1986).

The costs incurred by governments and quasi-governmental agencies were questioned by McCormick and Skrabanek (1988):

> Review of the present experimental evidence that we can prevent much coronary heart disease provides no data to justify the time, energy, and money which are being devoted to this crusade. The cost of drugs and the burden of their side-effects, particularly in the treatment of slightly raised blood pressure and blood cholesterol are by no means negligible. Finally, the false promises of benefit may induce, in the short term, morbid preoccupation with death rather than the enjoyment of life and, in the longer term, disillusion and the loss of professional credibility. The time-scales of intervention trials may be too short to show any effect, but to base population strategies on unproven hypotheses seems unreasonable.

The problem of advocating the dietary guidelines for children was highlighted by A. E. Harper, Professor of Biochemistry and Nutrition at the University of Wisconsin in 1996 in the following quotation:

> The objective of dietary guidance is to provide advice about healthy diets that is reliable, effective and safe (will do no harm). The purpose of dietary guidance for children is to ensure, above all, that they will achieve their genetic potential for growth and development. The primary need for children to achieve these goals is a diet that provides adequate quantities of essential nutrients and energy sources. Impaired develop-

> ment and malnutrition are encountered mainly in areas where diets are low in fat and are most commonly associated with inadequate energy intakes. Neither the safety nor the efficacy of adopting the uniform dietary guidelines for adults as guidelines for children before they have reached maturity has been demonstrated. It is proposed on the basis of extrapolation from studies mainly of sedentary, middle-aged males (Harper 1996).

Similarly, concern has been voiced by Lifshitz and Tarim that children are not little adults:

> Recommendations to enforce severe dietary restrictions on children should be based on data demonstrating that these diets are healthy and, to date, these data are lacking. It has to be kept in mind that hypercholesterolaemia in childhood can only be a risk factor for CHD, and, at present, there is no evidence that early intervention reduces the CHD incidence (Lifshitz and Tarim 1996).

As for the call to screen all children for cholesterol levels, Thomas B Newman and colleagues from the University of California at San Francisco in 1990 stated:

> The first two decades of life are a time when many other things are more important than cholesterol – like growth, surviving adolescence, and enjoying ice cream. The potential benefits of cholesterol screening in children do not justify the risks of interfering with these more important things (Newman et al. 1990).

Conclusions

From this review, it is very clear that the dietary guidelines are varied and confusing. Thus, it is not surprising that the ordinary person is bewildered and cynical about dietary advice.

Of course, excessive consumption of any one dietary constituent is likely to be unhealthy, but safe limits cannot be prescribed for the individual. Obviously, there are medical conditions which require rigid dietary restraint, but for most people there are wide ranges of tolerance.

Of special concern is the unthinking application of adult dietary guidelines to children, who need the fats to develop physically and the joy of chocolates and chips to grow mentally.

Altogether, we should be thinking of food as fun and not food as fuel. Thus, we should reverse Molière's lines from 'Il faut manger pour vivre et non pas vivre pour manger' to 'Il faut vivre pour manger et non pas manger pour vivre' – One should live to eat and not eat to live.

Acknowledgements

The research was supported by a research fellowship to Eve Sweeney from ARISE (Associates for Research Into the Science of Enjoyment). Information on ARISE can be found on the www.arise.org site.

We would like to thank the following for their assistance in supplying information on dietary guidelines:- Dr Wulf Becker, Sweden [wube@msmail.slv.se]; Professor Dr Jan Pokorny, Department of Food Science, Prague Institute of Chemical Technology, Czech Republic; Dr Ctibor Perlin, Institute of Agriculture and Food Information, Prague, Czech Republic; Dr Jiri Ruprich, National Institute of Public Health in Prague, Czech Republic; Dr Karen Hulshof, TNO Nutrition and Food Research Institute, The Netherlands; Katarzyna Kozlowska, MSc, Department of Human Nutrition, Warsaw Agricultural University, Poland; Conseil National de la Nutrition, Belgium; Conselho Nacional de Alimentacao E Nutricao (CNAN), Portugal; DGE, Frankfurt, Germany; Istituto Nazionale della Nutritzione, Italy; Institut FranVais pour la Nutrition (ifn), Paris.

References

Ahrens, E. H. Jr (1985). The diet–heart question in 1985: Has it really been settled? *Lancet*, **2,** 1085-1087.

Ahrens, E. H. (1979). Dietary fats and coronary heart disease: unfinished business. *Lancet*, 1345-1348.

AIM (1996). Worldwide recommendations on sensible drinking. *Alcohol in Moderation Digest*, **5,** (2) 11.

Alderman, M. H., Madhavan, S., Cohen, H., Cealey, J. E. and Laragh, J. H. (1995). Low urinary sodium is associated with greater risk of myocardial infarction among treated hypertensive men. *Hypertension*, **25**, 1144-1152.

American Academy of Pediatrics (1992). Statement on cholesterol (RE9258). *Pediatrics*, **90**, 469-473

American Dietetic Association (1997). *The food guide pyramid... your guide to food variety.* [http://www.eatright.org/fgp.html]

American Heart Association (1997). Home, health and family, heart and stroke A-Z guide. *FAT – AHA scientific position.* [http://www.amhrt.org.hs96/fat.html]

American Heart Association (1997). Home, health and family, heart and stroke A-Z guide. *Alcohol.* [http://www.amhrt.org/heartg/alcohol.html]

Anderson, P. (1996). Less is better. *Addiction*, **91**, 25-26.

Atrens, D. M. (1994). The questionable wisdom of a low-fat diet and cholesterol reduction. *Social Science and Medicine*, **39**, (3), 433-447.

Bonneux, L. and Barendregt, J. J. (1994). There's more to heart disease than cholesterol. *British Medical Journal*, **308**, 1038.

British Medical Association (1995). Alcohol: guidelines on sensible drinking. In *Addiction*, (1996), **91**, (1), 25-33.

Browner, W. S., Westenhouse, J. and Tice, J. A. (1991). What if Americans ate less fat? A quantitative estimate of the effect on mortality. *The Journal of the American Medical Association*, **265**, (24), 3285-3291.

Chew, R. (1995.) OHE Compendium of Health Statistics (Ninth Edition). London: Office of Health Economics.

Canadian Paediatric Society (1994). Report of the joint CPS/NHW working group on dietary fat and children. [http://www.cps.ca/english/statements/N/n9401.htm]

Cappuccio, F. P., Markandu, N. D., Carney, C., Sagnella, G. A. and MacGregor, G. A. (1997). Double-blind randomised trial of modest salt restriction in older people. *Lancet*, **350**, 850-854.

Davey-Smith, G., Neaton, J. D., Wentworth, D., Stamler, R. and Stamler, J. (1996). Zip codes and premature death. *Am J Public Health*, **86**, 486-496.

Department of Health (1994). Nutritional Aspects of Cardiovascular Disease: Report of the Committee on Medical Aspects of Food Policy. London: Her Majesty's Stationary Office.

Dunnigan, M. G. (1993). The problem with cholesterol. No light at the end of the tunnel? *British Medical Journal*, **306**, 1355-1356.

Fisher, E. A., Van Horn, L. and McGill, H. C. (1997). Nutrition and children. A statement for healthcare professionals from the Nutrition Committee, American Heart Association. *Circulation*, **95**, 2332-2333.

Food and Drug Administration (1996). Taking the fat out of food. A reprint from FDA Consumer Magazine by Paula Kurtzweil. [http://www.pueblo.gsa.gov/cic_txt/fd&nut/fatout.txt]

Food and Drug Administration (1994). A consumer's guide to fats by Eleanor Mayfield. [http://222.pueblo.gsa.gov/cic_txt/fd&nut/fatguide.txt]

Fuchs, C. S., Stampfer, M. J., Colditz, G. A., Giovannucci, E. L., Manso, J. E., Kawachi, I., Hunter, D. J., Hankinson, S. E., Hennekens, C. H., Rosner, B., Speizer, F. E. and Willett W. C. (1995). Alcohol consumption and mortality among women. *The New England Journal of Medicine*, **332**, (19), 1245-1250.

Goldstein, M. R. (1990). Daddy likes ice cream too. *American Journal of Med*, **88**, 666.

Grover, S. A., Abrahamowicz M., Joseph L., Brewer C., Coupal L. and Suissa S. (1992). The benefits of treating hyperlipidaemia to prevent coronary heart disease. Estimating changes in life expectancy and morbidity. *The Journal of the American Medical Association*, **267,** (6), 816-822.

Harper, A. E. (1996). Dietary guidelines in perspective. *Journal of Nutrition*, **126,** 1042S-1048S

Harper, A. E. (1983). Diet and heart disease – a critical evaluation. in *Dietary fats and health* eds. Perkins, E. G. and Visek, W. J., American Oil Chemists' Society, USA.

Hopkins, P. N. and Williams, R. R. (1981). A survey of 246 suggested coronary risk factors. *Atherosclerosis*, **40**, 1-52.

Jacobs, D., Blackburn, H., Higgins, M., Reed, D., Iso, H., McMillan, G. et al. (1992). Report of the conference on low blood cholesterol: mortality associations. *Circulation*, **86,** 1046-1060 .

Jousilahti, P., Vartiainen, E., Tuomilehto, J. and Puska, P. (1996). Twenty-year dynamics of serum cholesterol levels in the middle-aged population of eastern Finland. *Annals of Internal Medicine*, **125,** 713-722.

Keys, A. (ed.) (1970). Coronary heart disease in seven countries. *Circulation*, **41**, (suppl 4), I 1-211.

Labarthe, D. R., Nichaman, M. Z., Harrist, R. B., Grunbaum, J. A. and Dai, S. (1997). Development of cardiovascular risk factors from ages 8 to 18 in project heartbeat! *Circulation*, **95**, 2636-2642.

Le Fanu, J. (1986). *A Diet of Reason* Anderson D (ed.), Social Affairs Unit, London.

Lichtenstein, A. H. (1997). Trans fatty acids, plasma lipid levels, and risk of developing cardiovascular disease : a statement for healthcare professionals from the American Heart Association. *Circulation*, **95,** 2588-2590.

Lifshitz, F. and Moses, N. (1989). Growth failure. A complication of dietary treatment of hypercholesterolaemia. *Am J Dis Child,* **43,** 537-542.

Lifshitz, F. and Tarim, O. (1996). Considerations about dietary fat restrictions for children. *Journal of Nutrition*, **126,** 1031S-1040S.

McCormick, J. (1996). Health scares are bad for your health. In *Pleasure and Quality of Life* D. M. Warburton and N. Sherwood (eds.). Chichester, John Wiley and Sons Ltd.

McCormick, J. and Skrabanek, P. (1988). Coronary heart disease is not preventable by population interventions. *The Lancet,* **2,** 839-841.

McMichael, J. (1979). Fats and atheroma: an inquest. *British Medical Journal*, **1**, 173-175.

McNamara, D. J. (1986). *A Diet of Reason*. Anderson D (ed.), Social Affairs Unit, London.

Mann, G. V. (1994). Metabolic consequences of dietary trans fatty acids. *The Lancet,* **343**, 1268-1271.

Mezzetti, M. and Franchesi, S., (1998). Population attributable risk for breast cancer: diet, nutrition and physical exercise. *Journal of the National Cancer Institute*, **90**, 389-394.

Midgley, J. P., Matthew, A. G., Greenwood, C. M. T. and Logan, A. G. (1996). Effect of reduced dietary sodium on blood pressure. A meta-analysis of randomised controlled trials. *Journal of the American Medical Association*, **275**, 1590-1597.

Mirkin, G. (1997). Growth of Vegetarian Children. [http://www.wdn.com/mirkin/6803.html]

Muldoon, M. F., Manuck, S. B. and Matthews, K. A. (1990). Lowering cholesterol concentrations and mortality: a quantitative review of primary prevention trials. *British Medical Journal,* **301,** 309-314.

National Cholesterol Education Programme. Highlights of the report of the expert panel on blood cholesterol levels in children and adolescents. The population approach.
[http://www.medaccess.com/h_child/diet/bld_ch103.htm]

Newman, T. B., Browner, W. S. and Hulley, S. B. (1990). The case against childhood cholesterol screening. *Journal of the American Medical Association*, **264**, (23), 3039-3043.

Peele, S. and Grant, M. (eds.) (1999). *Permission for Pleasure*. Philadelphia, PA: Taylor and Francis

Ramsey, L. E., Yeo, W. W. and Jackson, P. R. (1991). Dietary reduction of serum cholesterol concentration: time to think again. *British Medical Journal,* **303**, 953-957.

Ramsey, L. E., Yeo, W. W. and Jackson, P. R. (1994). Effective diets are unpalatable. *British Medical Journal*, **308**, 1038-1039.

Readers Digest (1999). Food and health – Eggs. [http://www.readersdigest.co.uk/CTHE-3YVKAN.htm]

Reid, V., Mulcahey, R. (1987). Nutrient intakes and dietary compliance in cardiac patients: 6-year follow-up. *Hum Nutr Appl Nutr*, 41A, 311-318.

Reiser, R. (1984). A commentary on the rationale of the diet-heart statement of the American Heart Association. *The American Journal of Clinical Nutrition,* **40,** 654-658.

Robb-Smith, A. H. T. (1967). *The Enigma of Coronary Heart Disease.* London, Lloyd-Luke.

Schafer, E (1995). National network for child care: adult diets don't work for babies.
[http://www.exnet.iastate.edu/Pages/nncc/Nutrition/adult.diet.html]

Secretary of State for Health (1992). Health of the Nation. London: Her Majesty's Stationary Office.

Secretary of State for Health (1998). Our Healthier Nation. London: Her Majesty's Stationary Office.

Shaper, A. G., Pocock, S. J., Walker, M., Cohen, N. M., Wale, C. J. and Thomson, A. G.

(1981). The British Regional Heart Study: cardiovascular risk factors in middle-aged men in 24 towns. *British Medical Journal,* **283,** 179-186.

Skrabanek, P. (1995). Fat Heads *National Review,* **43,** 30-32.

Stordy, J. and Wright, C. (1996). Low-fat food stunts children *The Sunday Telegraph* 12/5/96.

Surgeon General's Report on Nutrition and Health (1988). Washington D C, US Department of Health and Human Services.

Taylor, W., Pass, T., Shepard, D. and Komaroff, A. (1987). Cholesterol reduction and life expectancy: a model incorporating multiple risk factors. *Ann Intern Med,* **106,** 605-614.

Weatherall, D. J, Ledingham, J. G. G. and Warrell, D. A. (Eds.) (1989). Oxford Textbook of Medicine (Second Edition) Oxford: Oxford University Press.

Willett, W. C. (1995). Amid inconclusive Health Studies, Some Experts Advise Less Advice. Kolata Gina (1995) *The New York Times, May 10 1995* in CHANCE News 4.08.
[http://freeabel.geom.umn.edu/docs.../chance_news_4.08.html]

World Health Organisation (1993). European Alcohol Action Plan. Copenhagen: World Health Organisation.

8 The perversity of agricultural subsidies

Linda Whetstone

Summary

The original aims of Europe's Common Agricultural Policy were to guarantee food security at stable and reasonable prices by maximising production and protecting agriculture. This was largely a response to fears of food shortages and convoy blockades of the kind that had occurred during WWII. Today, not only have these fears receded, the CAP itself has proved an expensive failure. Its subsidies and price supports use half the total European Union budget, while its unintended consequences undermine the viability of farming and hamper environmental improvement. It is recommended that the CAP be reviewed, subsidies to production gradually be reduced and any transfer payments to be fully evaluated and assessed on their aims and effects.

Introduction

> In the United States, the government pays farmers not to grow grain; in the European Communities, farmers are paid high prices even if they produce excessive amounts. In Japan, rice farmers receive three times the world price for their crop; they grow so much that some of it has to be sold as animal feed – at half the world price. In 1985, farmers in the EU received 18 cents a pound for sugar that was then sold on the world markets for 5 cents a pound; at the same time, the EU imported sugar at 18 cents a pound. Milk prices are kept high in nearly every industrial country, and surpluses are the result: Canadian farmers will pay up to eight times the price of a cow for the right to sell that cow's milk at the government's support price. The United States subsidises irrigation and land clearing projects and then pays farmers not to use the land for growing crops (World Bank 1986).

Since this description of chaotic farming subsidisation was written, the details have changed a little, but the substance is still the same and the costs still outrageous.

Transfers or subsidies are payments from one agent in the economy to another. In this instance government takes resources from taxpayers and consumers and gives them to farmers. Agricultural subsidies are amongst the biggest and most universal of all such transfers in existence at the end of the twentieth century. No developed country has managed to avoid them and although one or two have now almost extricated themselves from the process, most are still transferring huge annual sums to their farmers.

Subsidies are granted to any group making a strong enough demand and with enough political clout to persuade government to coerce others into paying. Subsidies are paid from general taxation, so individual payers may not notice a small increase or trade-off against another service. Those who stand to gain will make their arguments as plausible as possible, drawing into the political process any others who could benefit, such as the politicians who would grant the subsidy and the bureaucrats who would administer them. Government rarely, if ever, proposes subsidies except in response to lobby pressure.

Successful lobbyists focus their case on a single issue, making it seem urgent and overwhelming, and will avoid discussing any ramifications of their group being favoured over another. Taken as a single case, the general public may sympathise with the group's demands and agree that the government should indeed do something to help them. However, if the demand for subsidy was fully considered and weighed against claims from another group, sympathy would be divided, and perhaps both would be rejected.

Michael Winter (1996) gives a graphic description of how this process works and how small groups gain power and influence over policy. Once subsidies are made available, agriculture will adapt to the new system, as to a new market. Vested interests will want to get maximum benefit from the subsidy fund; a new job sector will sprout, even to the extent of hiring high-profile lobbying organisations, that will specialise in exploiting and maintaining the system. Such perversity is inevitable under a subsidised system.

The aims of the CAP

Continental Europe traditionally protected its farmers because the agricultural population tended to man their armies, so states had a vested interest in keeping agriculture flourishing. The UK, tending to have paid professional soldiers and relying so heavily on trade for its prosperity, was more interested in keeping trade free. However the experience of blockade during two world wars encouraged the UK government to institute a system of price support for farmers to guarantee food security.

The Treaty of Rome set up the Common Agricultural Policy (CAP) in 1957. Its aims are broadly similar to those of most major systems of government support for agriculture and it has traditionally gobbled up about half of total EU spending.

The original aims were:

1. to increase agricultural productivity by promoting technical progress and ensuring the rational development of agricultural production and the optimum utilisation of the factors of production, in particular labour;
2. thus to ensure a fair standard of living for the agricultural community, in particular by increasing the individual earnings of persons engaged in agriculture;
3. to stabilise markets;
4. to assure the availability of supplies; and
5. to ensure that supplies reach consumers at reasonable prices.

We now consider to what extent these objectives have been met.

Does CAP lead to economic efficiency: the optimum utilisation of factors of production?

The CAP has kept prices artificially high in Europe by a series of measures including tariffs, quotas and government purchase of excess production. In addition there have been subsidies for specific purposes, for example, grants for building, fertiliser and drainage. Later, a payment for not using land (set aside) was introduced, and later still, fixed payments to compensate farmers for a reduction in the level of their price protection.

This has indeed encouraged investment in agriculture with production rising spectacularly. The EU has gone from being an overall importer of agricultural produce to self-sufficiency and then to surpluses, which have been exported with assisted prices, stored or destroyed. Numbers employed in agriculture have fallen from over 21 per cent of the EU's workforce in 1960 to around 5 per cent today, but an increase in output from a reduced labour force does not necessarily result in the optimum allocation of the factors of production.

A report on the subject published by the European Commission in 1994 stated that:

> Economic efficiency requires that capital, labour, land and other inputs are allocated to produce the highest possible level of gross domestic product. As regards agriculture, the main cause of economic inefficiency in developed countries is too many resources being used in that industry relative to other industries. Within agriculture itself, resources

> maybe similarly misallocated. Economic efficiency in general requires the abolition of barriers to external trade and of government subsidies which influence production decisions. Market price support which increases domestic price relative to world market prices, hence retaining more resources in agriculture and discouraging consumption has therefore a cost in terms of economic efficiency (EU, 1994).

Put simply, they are saying that agricultural subsidies have caused inefficiencies in the general economy as well as within agriculture itself.

The CAP has been very expensive, devouring up to 60 per cent of the entire EU budget for many years. It is now reduced to 50 per cent, but its costs are spread further than the tax burden. Consumers have had to pay more for their food in 'fortress Europe' than would be the case without protection of the internal market.

The Organisation for Economic Cooperation and Development (OECD) calculates the overall costs. Table 8.1 shows a steady annual increase in the costs borne by taxpayers until 1997. The subsequent fall may be a blip due to high world cereal prices that year. World prices fell dramatically in 1998 so the costs of supporting grain farmers and the grain mountains will probably grow again.

The costs to consumers have genuinely fallen due to recent reforms, principally the MacSharry Reforms of 1992 and the Uruguay round of the General Agreement on Tariffs and Trade (now the World Trade Organisation – WTO), which ended in 1994. Their aims were to cut agricultural subsidies, making farmers reliant on world prices so that they tailor their production to the demand for their produce, and to stop dumping excess agricultural produce outside the EU, which destroys markets of traditional third country suppliers.

Price support was replaced with compensation payments, although only partly in the case of cereals. Payments for cereal

Table 8.1. Total transfers associated with agricultural policies (Billion US dollars)

Year	*1986-88*	*92-94*	*95*	*96p*	*97e*
From taxpayers	38.8	55.4	62.6	67.8	62.8
From consumers	76.3	77.8	73.1	51.7	48.6
Total	115.1	132.8	135.7	119.5	111.4

p = projected
e = estimated

farmers averaged about £100 an acre for cereals in England in 1998 with farmers paid for every acre planted plus the market price for their product when they sell it. Additionally, cereal growers are required to set aside (leave fallow) a certain percentage of their farms, for which they were compensated at about the same rate.

Beef producers got about £85 a head in 1998 on all beef cattle up to a total of 90 cows per farmer, payable twice during each animal's lifetime. Suckler cow payments were about £112 a head and headage payments for sheep, about £12, up to a limit of 500 animals.

Farmers working in certain designated production areas are entitled to compensatory payments designed to de-couple support for farm incomes from agricultural production in the belief that if the farmer received the headage payment and the market price he would react to falling prices by reducing production.

Theory suggests that if prices fall, the producer would pocket the headage payment and reduce production, but if prices are so low that he will go out of business then he almost certainly will use the headage payment to avert that possibility. A £10 headage payment on a ewe would enable him to sell that ewe's lamb for £10 less than in the previous year if his first priority was to remain a sheep farmer. So headage payments are not proving as effective as was originally intended in reducing output.

As with milk quotas they are saleable but the costs involved are a deterrent to potential new entrants or for expansion plans of the more go ahead, whilst they encourage older or less efficient farmers to stay. The value of suckler cow quota was about £140 per head in 1998, which does not entirely equate with the howls of protest from farmers about their economic plight and need for further subsidies.

> Production controls distort the natural development of the pattern of production, encourage inefficient producers and increase the overall level of costs in the industry. Flexibility to respond in light of changing conditions is severely reduced. The efficient will be held back and new entrants will find it increasingly difficult to get onto the farming ladder. In the meantime, inefficient producers with quotas are protected (CLA 1994).

When the sheepmeat regime was introduced in 1982, the payments were welcomed with open arms. But the outcome of paying farmers for either having or selling sheep has been an increase in the number of sheep, oversupplying the market and inevitably depressing the market price. This has led to further calls for financial assistance, so that subsidies have effectively created their own demand, leaving the market in disarray.

A further unfortunate effect of subsidies has been the dramatic rise of farmland prices since 1957. Various studies show that agricultural subsidies have caused farmland prices in the UK to increase by about

50 per cent above what they would have been in the absence of the subsidies (Howarth 1993).

Rents have also been higher, severely deterring those wanting to enter farming. This has inevitably accelerated the recent intensification of agriculture resulting in fewer, larger farms. So subsidies have increased agricultural output by inadvertently causing an intensification of agricultural production, but they have not produced the optimum use of the factors of production and if they were intended to keep labour on the land, they have not succeeded here either. Subsidies have also benefited agricultural suppliers, as a small number of big-spending, mechanised operations are far easier to service than disparate, 'one man and his dog' customers.

Discussions about the social objectives of the CAP give us some leads as to how the developing debate on subsidies for the environment may proceed. The objectives were not specific, they were emotional and undefined. They made assumptions which were probably not true and they were always likely to be conflicting but they sounded pretty good.

We have seen that numbers employed in agriculture in the UK have fallen significantly but, 'quite what is meant as the policy goal for living standards is by no means well articulated in the EU or in most industrialised countries, though the prevention of poverty is a major implied objective' (Hill, 1996).

Very few economists would now accept that subsidies improve economic efficiency. The theoretical case against them is overwhelming: they distort factor prices, leading to inefficient use of resources; they encourage rent-seeking behaviour, undermining even the benefits to those for whom they are intended. Research into the economic effects of the CAP suggests that in practice such subsidies lead to utterly inappropriate use of factors of production, distorting not only the market for food but all related markets and producing harmful externalities, such as phosphate run-off, to boot. Consider the impact on manufacturers of food and drink products. This is the single largest manufacturing industry in the UK with gross output of £55 billion, employing some half a million people (11.9 per cent of total manufacturing employment) and contributing over £6.3 billion worth of value added exports to the UK balance of trade. Agricultural commodities are the most important raw material used by member manufacturers, who purchase over 70 per cent of UK agricultural output.

> The availability, quality and price of agricultural raw materials are crucial to the food and drink industry's own efficiency, its ability to satisfy consumer demands for a wide choice of foodstuffs and its international competitiveness. The industry has therefore a major interest in the current operation and the future direction of the Common Agricultural Policy. Its prime objective ... is to see progress towards a market orientated CAP, in all sectors, encouraging an efficient and viable supply base to provide manufacturers with adequate quantities of agricultural ingredients. The

current CAP does not address the longer-term needs of either the agricultural or food manufacturing sectors (FDF, 1998).

The new round of GATT negotiations is likely to lead to limits on subsidised exports, coupled with increased imports. These risk leading to further production controls which, in turn, make EU farmers even less competitive in world terms. 'For the food industry, the CAP is a direct impediment to competitiveness. By forcing up raw material price and artificially distorting the origin and level of ingredient supplies it has a profound impact on productivity and efficiency' (Ridley, 1995).

Does CAP ensure a fair standard of living for farmers?

Assessing the individual earnings of those engaged in agriculture is a surprisingly difficult thing to do and comparing it to the earnings of those employed elsewhere is equally fraught with problems. Firstly, farmers live in their businesses. This has considerable tax advantages as they are able to offset many costs which others would have to meet from taxed income. It would be a foolish farmer who did not reduce his visible income to a minimum, as that also reduces his tax burden.

Secondly, it is not in the interest of the farming community for its average income to become completely clear because if no one knows exactly what it is then it cannot be compared to the living standards of others. 'The inference is often made that agricultural households are typically in a disadvantaged position and that governments must somehow act to narrow the gap between farmers and non-farmers' (Hill, 1996).

Another interesting, and often avoided, aspect of agricultural subsidies is why those involved in agriculture should be entitled to a standard above that available to everyone else. If their incomes fall below what is considered an unacceptable level why should they not be supported simply via means tested direct-income payments. Why should farmers get preferential treatment to miners or steelworkers or small shopkeepers for instance? And why should farmers in the UK get support with no account whatever taken of their assets whilst the rest of the population must have less than £8,000 in assets before social security benefits kick in?

> When assessing the economic position of farmers, it is necessary to go beyond the income from current farming activities and to consider the possibility of capital gains or losses and also their net worth. This is particularly the case when attempting to use the concept of a poverty line to distinguish between the poor and the non-poor. Conventional indicators largely depend on money incomes, but this is clearly unsatisfactory with

the agricultural population, where low incomes are frequently found combined with substantial wealth (Hill, 1996).

Farm incomes in the EU do not compare unfavourably with the income of other groups and in some countries they compare very favourably. They are higher on larger farms and, when averaged over three years, the number of low income cases falls.

> Agricultural households have average disposable incomes per household that are typically higher than the all-household average. The relative position is eroded or reversed when income per household member or per consumer unit is examined. Non-farm income transforms the income situation among small farms, so that on average their total incomes are frequently satisfactory (Hill, 1996).

There is an argument that farmers need support because of the time scale involved in farming. That the production cycle is long and without certainty of returns no one would undertake the necessary risks. The risks are however broadly the same in many other industries. A furniture maker, large or small, must invest in training, premises, machinery and man power without the knowledge that he will have a definite market for whatever he chooses to produce at a price that covers his costs. If he misjudges his market or his market changes then it is equally expensive and miserable for him to give up his business and lay off his workers as it is for the farmer to stop production.

Furthermore, the farmer has available to him two mechanisms for reducing the risks he faces: insurance and futures markets. Insurance is available against a broad range of risks, including unfavourable rainfall and other natural disasters. Futures markets enable farmers to sell their products at known prices before they have even started to produce them.

Does CAP lead to market stability?

The third aim of CAP activity is to stabilise markets, although which markets are not specified. In real terms farm gate prices and world market prices have fallen since the inception of the CAP, reflecting increases in productivity brought about by technological development. However, the decline in world prices has been greater than that of EU farm gate prices with greater short term fluctuations in the former.

Pricing policies within the EU elicited increased production which gradually squeezed out imports from the rest of the world as the EU became ever more self-sufficient in an increasing number of commodities. For example the EU 6 were 86 per cent sufficient in cereal production in 1970/71 and by 1985/86 the EU 12 were 118 per cent self-sufficient. With oilseeds the figures were 27 and 138 per cent, beef and veal 93 to 109 per cent and fresh fruit 76 to 94 per cent. Overall the nine EU countries converted themselves from net importers of 20 million tons of wheat a year to net exporters of 10 million tons between 1965 and 1983.

Supplies continued to expand, turning self-sufficiency into surpluses which could not be sold outside the EU because they had been produced at well above the prices pertaining in world markets. This was when the infamous butter mountains and wine lakes were created. The reaction of the then EEC was to dispose of surpluses with the aid of export subsidies, effectively paying foreign importers the difference between EEC prices and world prices, with the obvious detrimental effect on traditional suppliers.

Export subsidies are a euphemism for 'dumping' which destroys the markets of third country suppliers and also the market for domestic producers within the importing countries.

> The CAP, together with the agricultural policies of some other developed countries, has driven down world prices and distorted the relative prices between different agricultural commodities and of agricultural commodities in relation to industrial products. These policies have also destabilised international commodity markets, amplifying fluctuations in prices and supplies to the detriment of third countries (NCC, 1998).

> The Common Agricultural Policy has had adverse effects on the agricultural sectors and economies of other agricultural exporting countries. This is because it has: reduced the volume of agricultural products that they can sell in the European Community; reduced world market prices; reduced the volumes that other countries can sell on world markets and destabilised international market prices (World Bank, 1986).

These effects should be familiar to anyone who has studied other attempts by central planners to stabilise markets. Ultimately, the attempt to replace private market transactions with decisions by non-omniscient 'experts' is doomed to failure. Of course, price fluctuations and unforeseen events pose problems for agricultural producers but, as noted above, these can be solved very well using private insurance and futures markets.

Does CAP assure the availability of food supplies?

Guaranteeing food supplies would be a high priority for any country. The argument goes that we must always be able to provide enough food in crises, whether they be created by war or weather or some other major international event. It continues that paying a higher price in perpetuity is worth it for this peace of mind.

No one ever asks the central question, which is whether having a self-sufficient or nearly self-sufficient level of production all the time is more or less likely to produce enough to feed us in any of the above situations. Future wars are less likely to be long drawn out affairs and more likely to last a minimal amount of time with devastating effect. In this situation the fact that we had been self-supporting would be unlikely to help.

If a war is to last several years then having a high level of production at the moment may make it less likely that we could continue it during a blockade situation. During two World Wars the UK was able to replace some of its traditional imports by ploughing up permanent pasture which produced reasonable crops for several years without a high level of inputs. Currently, we have a higher acreage under the plough and maintain a higher level of production from every acre by the increased use of fertilisers, pesticides and big machinery.

According to the Fertiliser Manufacturers Association, fertiliser use has risen from 860,000 tonnes in 1952-55 to 2,354,000 tonnes in 1996-97 and although the majority of this can be produced in the UK, potash is a critical ingredient of one type and that is only available from abroad.

With regard to agricultural machinery (excluding tractors) the UK imports about 60 to 70 per cent at present and the trend is upwards. Admittedly, in a blockade agricultural production could be maintained for some time, even without any supply of potash or new machinery or replacement parts, nevertheless our intensive agricultural activities are more vulnerable than is immediately apparent.

Another and more likely scenario is a Chernobyl closer to home or particularly severe weather conditions that specifically affected European production, so the most sensible long run solution to security of supply is to trade with a wide variety of food producing countries on the basis of their ability to produce at the lowest price. If the USSR could manage to purchase a record quantity of imports despite the US grain embargo in 1980 there is little chance that the EU would ever starve.

Clearly, then, CAP has increased agricultural production in the EU. However, it is highly questionable as to whether it has increased the security of supply. Indeed, to the extent that it has led to depletion of soil nutrients and to the extent that it has distorted world prices, making procurement of products from overseas more difficult, CAP could be said to have had the opposite effect.

Does CAP ensure that supplies reach consumers at reasonable prices?

The final aim of the CAP in the original Treaty of Rome is that supplies reach consumers at reasonable prices. People may dispute the meaning of reasonable in this context but there is almost unanimous agreement that there is nothing reasonable, under any definition of the word, about the price that consumers have had to pay for the Common Agricultural Policy. The most surprising thing has been their apparent willingness to go on paying such inflated prices for so long.

The quality and variety of food available to EU consumers has been adversely affected by the distortionary pricing policies of the CAP. Food prices have been very much higher, which is hardly surprising as the CAP's main activities have been to increase farm gate prices. Quotas and tariffs on imports deny consumers access to lower world prices. In 1988 the National Consumer Council wrote that 'During the 1980s agricultural commodities have been, on average, 70 per cent more expensive in the Community than on the world market' (NCC 1988, BAE, 1985).

One point of agreement on healthy eating is that an adequate daily intake of fruit and vegetables is important and yet the EU has a policy of keeping up prices of certain fruit and vegetables by 'quality controls'. This is a euphemism for market support because it involves destroying smaller or less than perfect fruit and vegetables. It keeps prices higher and this can only have the effect of reducing the intake of the lower income groups.

Subsidies to tobacco production also seem ironic from the perspective of a Commission concerned to promote healthy lifestyles. Most of the reasons for agricultural subsidies hardly apply to tobacco production! 'Rational development of agricultural production' and 'certainty of supply at reasonable prices' sound rather ludicrous while many governments tax tobacco products heavily in order to discourage consumption. Yet the EU subsidises the production of tobacco – by over £600 million in 1996 (Bate, 1997). For many years it has been one of the most heavily subsidised crops with support per hectare in 1991 being 23 times greater than that for cereals and even after capping reforms in 1992, the EU taxpayer still coughs up about £2 million per day for tobacco subsidies.

As with other subsidies, they are paid regardless of quality, so Europeans generally grow a high tar tobacco which is banned in the EU, so much of Europe's subsidised production is dumped on markets in Eastern Europe and Africa at a fraction of the price paid to the farmers, courtesy of European tax payers (Harris 1997).

The tobacco story may seem particularly outrageous but exactly the same forces are involved with all other subsidies. A focused group of beneficiaries, including those involved in the transfer process who also benefit, fight for the continuance of the subsidies. These include the bureaucrats who handle the subsidies, those who find and arrange the dumped sales, those who transport the product, those who buy it and sell it to its eventual market. They all make a living from the existence of the subsidy and so have every interest in perpetuating the arguments in favour. Conversely, those who pay the bills are dispersed and only pay a small amount for each subsidised product, so they lack the knowledge, time and inclination to fight each subsidy one by one. Farmers literally get hooked on the subsidies and even if they do not like the political uncertainties inherent in getting much of their income via government, they may fear they have been lured so far from a commercial path that their only option is to fight for the continuance of subsidies.

The OECD produces figures to indicate the costs of the CAP to consumers and taxpayers.

Table 8.2. European Union: consumer subsidy equivalents

These are an indicator of the total value of transfers to consumers (or from consumers where the % is negative) resulting from agricultural policies in a given year (OECD, 1998, p. 76).

	1986-88	1992-94	1995	1996p	1997e
Wheat	–50	–32	–8	11	–2
Maize	–47	–37	–26	–2	– 10
Rice	–58	–53	–47	–30	–22
Sugar	–69	–56	–42	–47	–50
Beef & Veal	–47	–52	–52	–50	–48
Pipmeat	–28	–22	– 13	2	–3
Poultry	–33	–30	–35	–24	–20
Sheepmeat	–64	–39	–36	–22	9
Eggs	–19	–8	–10	–5	1

p = projected
e = estimated

Table 8.3. European Union: producer subsidy equivalents

These are indicators of the total value of transfers to agriculture resulting from agricultural policies in a given year as a % of the total value of production (valued at domestic producer prices)

	1986-88	1992-94	1995	1996p	1997e
Wheat	56	52	44	28	36
Maize	55	52	45	24	33
Rice	66	59	51	34	27
Oilseeds	68	58	54	41	48
Sugar	73	61	48	53	54
Milk	64	62	60	57	54
Beef & veal	51	58	61	64	60
Pigmeat	6	8	11	10	9
Poultry	24	25	33	32	26
Sheepmeat	72	62	64	51	42
Eggs	5	0	7	10	2

p = projected
e = estimated

Since the MacSharry reforms, the burden falling on consumers has dropped while that falling on taxpayers has risen. This shift in burden of support from consumer to taxpayer is an improvement in that the higher the proportion of the cost borne by consumers the more regressive the policy. Low income consumers spend a higher than average proportion of their income on food. Generally though, it is the big farmers (who tend to be the richest ones) who have benefited more from the CAP than small farmers, so the subsidies have genuinely redistributed money from poor consumers to rich farmers which is certainly a perverse outcome.

In 1996 the average transfer to farmers in the EU from the rest of the population amounted to about £12,000 for every full-time farmer equivalent. That is to say that on average every full time person working in agriculture in the EU, including hired employees and unpaid family workers, received a transfer of £12,000 from the rest of us. The estimate for 1997 had fallen to about £10,500 but is likely to rise again (OECD, 1998).

In 1994 the Commission said that about 20 per cent of farmers (the biggest) received around 80 per cent of the CAP budget. Since the CAP accounts for half the EU budget, then 20 per cent of farmers accounted for 40 per cent of the entire EU budget. We have shown that these transfers impinge most heavily on the poorest – a grotesque result of agricultural subsidies, as is the prospect of a few wealthy individuals being handed cheques of up to a quarter of a million pounds every year simply because they are farmers.

One benefit of the change to direct payments however is that we can see what farmers are getting much more clearly than when they received their support through market price support.

The environmental consequences of the CAP

We have seen that EU agricultural subsidies have failed to achieve some of their aims. In addition, they have produced perverse consequences, many of which relate to the environment and have been caused by the intensification of agriculture encouraged by the higher prices.

Undoubtedly, technical innovation over the years since the inception of the CAP would have led to an intensification of production. However, it is impossible to say how much intensification should be attributed to technical progress and how much to increased prices. Subsidies are blamed for environmental damage but the extent of this depends on the extent to which subsidies have caused the intensification.

Higher output relies on increased inputs such as fertiliser, machinery, pesticides and use of antibiotics in feed and so on. Other controversies are nitrates leaching into water and the effect of routinely adding antibiotics to animal feeds, both topics being covered elsewhere in this volume (Bate, L'hirondel).

Capital grants for farm buildings have undoubtedly led to an increase in modern farm buildings above that which would have occurred in their absence. Ironically, many metres of hedge, grubbed in the name of increasing productivity, may now be replaced with the help of grants; farmers who were paid to pull hedges out may now be paid to put them back.

> Acres of marginal land have been drained or ploughed up, prompting fears about soil erosion. In the UK uplands, agriculture and nature conservation are in conflict. Substantial losses have occurred of semi-natural vegetation and wildlife habitat, notably moorland and rough grassland, much of it attributable to agricultural intensification (especially cultivation and reseeding). The scale and pace of the losses appear to be unprecedented: it has increased markedly in the last decade (Smith, 1985).

Habitat changes and loss of biodiversity have undoubtedly affected many species. The loss of three million skylarks since the 1970s is a loss indeed, whether from increased use of insecticides as suggested by the British Trust for Ornithology, or from the reduction in winter stubble brought about the replacement of spring sowing by autumn cultivation. Similarly the Royal Forestry Society is able to document a serious decline in the numbers of some butterflies because of habitat changes, caused by the ending of coppicing and disappearance of downland (Ridley 1995).

With the intensification of livestock production so the risk of water pollution from silage effluent and slurry pollution has increased. Straw yards have been replaced by cubicle housing, feeding hay by feeding silage and stocking rates have increased because of improved grassland management. As stocking rates increase so does the amount of slurry but the area on which to spread the increased supply of slurry does not. In Holland, intensification of livestock production has produced a crisis in manure management.

The nationalised and subsidised UK forestry industry produced howls of dismay when acres of marginal land was covered in conifers. Monoculture replaced moors and other biodiverse habitats and eventually the outcry and the threat of privatisation persuaded the forestry industry to rethink its planting strategy with an eye on environmental, social and leisure considerations.

Some of the other environmental 'crimes' said to be perpetrated by the CAP are set out in a document produced in 1998 by an alliance of UK NGOs working on CAP reform (NGO Alliance 1998).

- Between 1968 and 1990, UK yields of wheat increased from 3.9 to 7.3 tonnes per hectare and the European Commission estimates that between 1995 and 2005, cereal yields will increase from 5.3 to 6.3 tonnes per hectare.

- In the same periods, milk yields increased from 3722 to 5372 litres per cow and will increase from 5000 to 6000 litres per cow.
- Between 1975 and 1995 the percentage of crops treated with herbicides doubled.
- Fungicide use increased 6 times and insecticides increased from a negligible area sprayed in 1970 to almost 100 per cent of the cropped area in 1990.
- Between 30 to 80 per cent of nitrogen applied to crops is lost to water, the atmosphere or the soil and is not taken up by the crops.
- Far less support is given to vegetables than arable, for instance 0.2 per cent of total subsidies went to horticulture compared to 43 per cent for arable in the UK.
- Since the 1930s the area of ancient woodland has decreased by 45 per cent.
- Between 1978 and 1987, 46 per cent of Dorset heathlands were lost to agriculture.
- Between 1987 and 1995, 60 per cent of Berkshire's wildlife-rich grasslands were lost to agriculture.
- Between 1990 and 1993, 48,000km of hedges were lost in England to neglect.
- Between 1971 and 1995 the populations of farmland skylarks have declined by 61 per cent, lapwings by 63 per cent, song thrush by 70 per cent (NGO Alliance 1998).

Some will not think that all of the above are bad consequences. Most will debate to what extent the CAP was solely responsible for such changes but no one will doubt that the CAP has had some environmentally perverse consequences.

The earliest EU policy reaction to environmental damage came in 1975 with the introduction of the Less Favoured Areas Directive, which made payments to farmers in these areas to help conserve them. Unfortunately, the scheme backfired as farmers increased livestock numbers and in order to feed them, replaced the traditional heather moorland with more productive grassland.

During the 1980s capital grants for ploughing up moorland, grubbing up hedges and draining were ended and new grants were brought in to rectify the damage caused by those that had just been rescinded. Environmentally Sensitive Areas were established and farmers were paid for not doing things that might have damaged the environment. There have certainly been occasions under this legislation where farmers are paid for not doing things that they probably would not have done anyway. Generally, they are paid for not doing things that they would not have done in the absence of subsidies.

Set-aside was brought in as a remedy for over-production caused by agricultural subsidies and it involves paying farmers higher prices to increase production on one part of their holdings whilst simultaneously paying them not to produce on another. It has encouraged them to set aside the most difficult acres and to intensify production on the others, so that the reduction in farmed acreage is not matched by an equal reduction in production.

The acres that a farmer has set aside may have been useful in terms of providing habitat for wildlife but they have also led to an explosion of weeds. Witness the great unkempt areas of poisonous yellow ragwort that cover much set-aside land in summer, despite the legal requirement to cut it twice a year. Ironically, cutting set-aside in spring often destroys the nests of those birds who had taken the opportunity it offered them.

A more rational approach to CAP reform

In 1987 the Single European Act amended the Treaty of Rome and was the first occasion when the protection of the environment became a component part of the community's other policies.

Since then 'the environment' has become the refuge of subsidy seekers, who realise that their previous reasons for such transfers are discredited, in alliance with a great number of people with a heartfelt concern for the environment. General emotional desires are difficult to define clearly or implement rationally, but the onus is on all who promote such subsidies to set out exactly what it is they require and whether there is any more chance of subsidies delivering these objectives than on previous occasions. There is however, a simpler alternative. If the subsidies caused the environmental problems, then a gradual ending of subsidies should reduce the problems. This must be a more sensible first step than creating further subsidies to counteract the possible effects of the earlier and continuing subsidies.

The same argument applies to demands for higher subsidies for 'organic' production to counter-balance the subsidies currently given to increase production. We should end the subsidies that caused the imbalance rather than heaping on more subsidies.

New Zealand is almost the only country to have made a serious attempt to unravel its agricultural support system. Since 1984 the level of subsidies has been reduced to practically nothing.

> The effects of removal of agricultural subsidies, when combined with a general downturn in commodity prices, has had some pronounced effects on farm input use and output mix in the short-term. In particular, the use of phosphate fertiliser, pesticides and the rate of development of marginal lands into pasture have fallen significantly. These changes lessen the likelihood of off-site contamination and of severe land degradation of newly developed land (OECD 1998).

Likewise, the reduction in sheep numbers and the diversification of pastoral farming operations will have environmental benefits:

> Given that the agricultural sectors of most of the OECD countries are much more heavily assisted and more intensive in their use of inputs than in New Zealand, the environmental gains of subsidy reductions and a shift away from production based assistance in these countries are likely to be substantially greater than those experienced in New Zealand (Reynolds et al.1993).

The climate in much of Europe and the grade 1 and 2 land is as good for growing crops as almost anywhere in the world. The biggest market is on the doorstep, so why should Europe's farmers not compete with produce from around the world and give EU consumers the benefit of greater choice at a lower price?

Many in the UK, even in farming, now accept that agricultural subsidies do not have the desired effects and that many farmers would manage perfectly well without them, providing they had sufficient warning of any changes. But what about the less favoured areas, such as the uplands in Scotland and Wales? Should such farmers be kept on their farms so that the landscape 'as we know and love it', is maintained in the way that it has been for generations?

But that pattern of farming existed before subsidies, it was not subsidies that created it and in some instances an effort to maintain it by subsidies has already had perverse effects. Many in the UK state walking in the countryside as one of their prime leisure pursuits, but there are many others who simply cannot afford to get there and they are the ones on whom the current costs falls heaviest. Then there are others who argue that a mixture of agriculture and wilderness has benefits for them and for the environment.

So how is the political process to reconcile these different views? How is it to give due weight to the different preferences? How, indeed, is it to define clear aims, which are essential before any subsidies are given at all – because without clear aims how can anyone assess the potential success of different courses of action? Because it is so difficult to do any of the above, it is easy for the subsidy seekers to argue for the redirection of current agricultural subsidies in the knowledge that their new haven will be a safer, softer option. As yet, they have failed to define how such transfers could be administered. They must do that and they must also demonstrate that any new transfers are not going to have the adverse effects of their predecessors. That will not be so easy to do.

Ancient forests were replaced by hay meadows many years ago before anyone could object to the changes. Hay meadows have been replaced by fields of silage. Other areas have seen the establishment of conifer forests, which have been decried in the name of the environment. These forests are coming to maturity and it is already evident that the felling and replanting of them is going to cause objection. Most humans do not like change, although they enjoy the effects of change. How can this be logically incorporated into the assessment of potential subsidies?

Those looking for environmental subsidies are talking in terms of much greater regulation of the countryside, but will regulation create the beauty which so many seek to preserve and which was created before subsidies themselves? The talk is of targeting the budget at rural development and paying farmers to manage the countryside and the environment. Funds must be diverted towards support for agri-environmental schemes, conservation, diversification and local food initiatives. But there already seems to be a problem with agri-environmental schemes.

> The Less Favoured Area (LFA) policy overhaul signalled in Agenda 2000 (the EU's latest reform document) is to some extent at odds with the mainstream commodity support regime. It would be unfortunate if a re-vamped LFA policy had to compete with the livestock subsidies as is the case already with agri-environmental schemes. It is vital therefore that any reforms to the LFA payments take place alongside tighter environmental conditionality and extensification rules within the mainstream support mechanisms. GB Countryside Agencies have proposed an area based two-tiered structure of Environmental Value Allowances with payments for meeting specific environmental conditions (Winter et al. 1998).

These types of suggestions abound in proposals from the EU itself and from others seeking to continue with the subsidy-go-round, but how could such extraordinarily complicated suggestions be defined sufficiently rigorously for them to be judged? How could they possibly be implemented or policed when existing cross-compliance measures seem incapable of policing? Why should they be capable of achieving their objectives if no previous subsidies have done so? And why should they be preferable to farmers than a much freer market?

The much heralded Agenda 2000 Reforms are set to increase spending from Euro 45.2 billion in 1999 to Euro 51.6 billion in 2006, so there will be no respite for the taxpayer if these come into force and it seems, little benefit for many others either.

Conclusions

Agricultural subsidies in the EU and elsewhere have generally failed to achieve their objectives and have often produced adverse effects, some of which have had damaging environmental consequences.

As the discredited subsidies come under political threat those who currently benefit are joining forces with environmentalists to redirect them to environmental goals. Like the average farmer's income, the environmental effects of farming (positive and negative) are difficult to define, examine and quantify, which is why they present the perfect smokescreen for those fighting to retain the structure of government support for agriculture.

If the earlier mistakes are not to be repeated the onus is on those demanding these transfers from other groups in society to:-

1) Set out clear objectives
2) Detail strategies for their implementation
3) Identify the beneficiaries
4) Quantify the returns
5) Identify those who will pay the costs
6) Quantify the costs

That will be a demanding exercise, and it won't happen at all unless the silent majority insist on it.

References

BAE (1985) *Agricultural policies in the European Community*, Bureau of Agricultural Economics, Policy Monograph no 2.

Bate, R. (1997) The Marlboro Man Can't Make You Smoke, *Wall Street Journal Europe*, 15 January.

CLA (1994) *Focus on the CAP*, CLA Discussion Paper, London: Country Landowners Association, September 23.

EU (1994) EU Agricultural Policy for the 21st Century, *European Economy Number 4*, Reports and studies from the Directorate-General for Economic and Financial Affairs of the European Commission.

FDF (1998) European Commission's Agenda 2000 Proposals. Food and Drink Federation Response to the Legislative Texts of 18 March 1998.

Harris, J. (1997) 'Warning: These subsidies can Damage Your Health,' *The Readers Digest*, August.

Hill, B. (1996) *Farm Incomes, Wealth and Agricultural Policy, Second Edition*, Athenaeum Press Ltd.

Howarth, R. (1993) *The Political Economy of British Agricultural Policy 1045-1990*, Thesis Submitted to the University of Wales, May.

NCC (1988) *Consumers and the Common Agricultural Policy*, A Report by the National Consumer Council, London: HMSO.

NCC (1998) *Consumers and the Common Agricultural Policy*, London: National Consumer Council 1998.

NGO Alliance (1998) *CAP Fact Sheet produced by an Alliance of UK NGOs Working on CAP Reform*.

OECD (1998) Agricultural Policies in OECD Countries, Measurement of Support and Background Information, Paris: OECD.

Reynolds, R., Moore, W., Arthus-Worsop, M et al. (1993) Impact on the Environment of Reduced Agricultural Subsidies: A Case Study of New Zealand, New Zealand MAF Policy Technical Paper 93/12 (Executive summary).

Ridley, M. (1995) *Down to Earth*, London: The Institute of Economic Affairs.

Smith, M. (1985) *Agriculture and Nature Conservation Conflict – The Less Favoured Areas of France and the UK*, Langholm: Arkelton Trust.

Winter, M. (1996) *Rural Politics: Policies for Agriculture, Forestry and the Environment*, Routledge.

Winter, M., Gaskell P. & Short, C. (1998) Upland Landscapes in Britain and the 1992 CAP Reforms, *Landscape Research* **23** (3), Landscape Research Group Ltd.

World Bank (1986) *World Development Report*, Oxford: Oxford University Press.

9 Politics, policy, poisoning and food scares

Barrie Craven and Christine Johnson

Summary

Increasing affluence has allowed the modern reliance on outside food services for provision of food and consumption of processed rather than fresh food. Given the resultant 'moral hazard' an increase in food poisoning was and remains inevitable. A four-fold increase in recorded food poisoning incidents since the early 1990s has been accompanied by dozens of new food regulations designed to reduce risks from food. Regulation is not achieving its desired effects in protecting consumers but is imposing significant costs upon both those whom the regulators are trying to protect; the producers and the consumers. These regulatory costs fall disproportionately on small companies. The increase in costs to the health services, the family and industry as a result of the increasing incidence of food poisoning is small by comparison with other causes of morbidity and mortality. The much larger costs to the nation resulting from back pain and cancer do not receive the same media treatment given to food poisoning because they are so common.

In dealing with food poisoning a balance has to be struck between placing responsibility on the individual for their own actions and protecting the individual from the mistakes, incompetence or deceit of others; there is now too much emphasis placed on attempting to protect the individual.

Background: food scares and food poisoning

> Cod liver oil, the foul-tasting vitamin supplement forced on generations of children, has been found to contain traces of dangerous industrial chemicals including some linked with cancer, a report said yesterday ... Sainsbury's said last night that it was considering putting a warning on bottled cod liver oil that it should not be

given to children under five or adults of low body weight after discussions with the Department of Health and MAFF

Daily Telegraph, 20 January 1998

Food scares arise when suddenly some food or food processes are asserted to contain a new and unexpected risk to the public health. Some of these alleged risks are treated with ridicule and contempt by the public but, others cause massive switches in consumer preferences and official policy. Sometimes the reaction is a mixture of the two. The alleged threat of contracting new variant Creutzfeldt-Jakob disease (nvCJD) after eating T-bone steaks raised the prospect of a law to ban the sale of beef on the bone. Consumers reacted by rushing to load their freezers with the product before it was removed from the butchers' shelves.

There is superficial evidence that the incidence of food poisoning is increasing. There were 53,731 notifications of food poisoning in 1997; nearly a six fold increase between 1982 and 1997 (Figure 9.1).

Many of these will be false positives, which may be due in part to a greater awareness among doctors following circulation of the definition in 1992 (CDR 1996). The considerable increase in laboratory reports of confirmed cases of *salmonellæ* (12,322 cases in 1982 to about 32,169 cases in 1997) and *campylobacter* (about 12,000 in 1982 to over 40,000 in 1995) suggests that incidence of food poisoning is increasing (Figure 9.2). The Public Health Laboratory Service (PHLS) considers that the trends are genuine because there is no evidence that clinicians have changed their criteria for submitting specimens for examination, nor that laboratories have changed in significant ways their reporting practices.

Deaths from the main forms of food poisoning rose from 187 in 1986 to 356 in 1986 (Table 9.1). Included in these figures are deaths from *lis-*

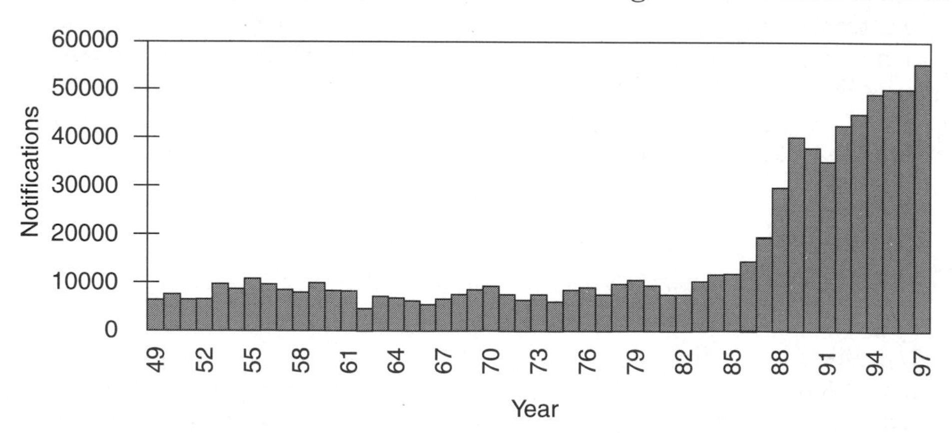

Figure 9.1. Food poisoning notifications 1949–1997 (Source: Communicable Diseases Surveillance Centre)

Table 9.1. Deaths from main causes of food poisoning, Engand and Wales: 1986 – 1997

	1986	*1987*	*1988*	*1989*	*1990*	*1991*	*1992*	*1993*	*1994*	*1995*	*1996*	*1997*
Cholera	0	0	0	0	0	0	1	0	0	0	0	0
Typhoid fever (Salmonellæ)	0	2	0	0	2	0	1	0	0	0	0	0
Paratyphoid fever (Salmonellæ)	0	0	0	0	0	0	0	0	0	0	0	0
Other salmonellæ infections	40	52	58	61	68	62	59	35	39	36	48	43
Shigellosis	3	3	2	1	0	3	3	0	3	1	1	0
Other food poisoning (bacteria)	3	3	2	2	1	1	0	2	0	0	1	0
Amoebiasis and other Protozoal intestinal diseases	2	1	0	2	4	2	4	2	6	5	5	2
Intestinal infections due to other organisms	52	26	36	54	43	34	82	79	117	172	223	272
Ill defined intestinal infections	76	75	65	65	69	67	90	76	57	63	61	66
Listeriosis	11	17	11	16	9	6	5	9	11	14	18	14
Total	187	179	174	201	196	175	245	203	233	291	357	397

Source: Office for National Statistics, Mortality Statistics, cause, 1997; 1986-1996 figures supplied from Public Sector Laboratory Service

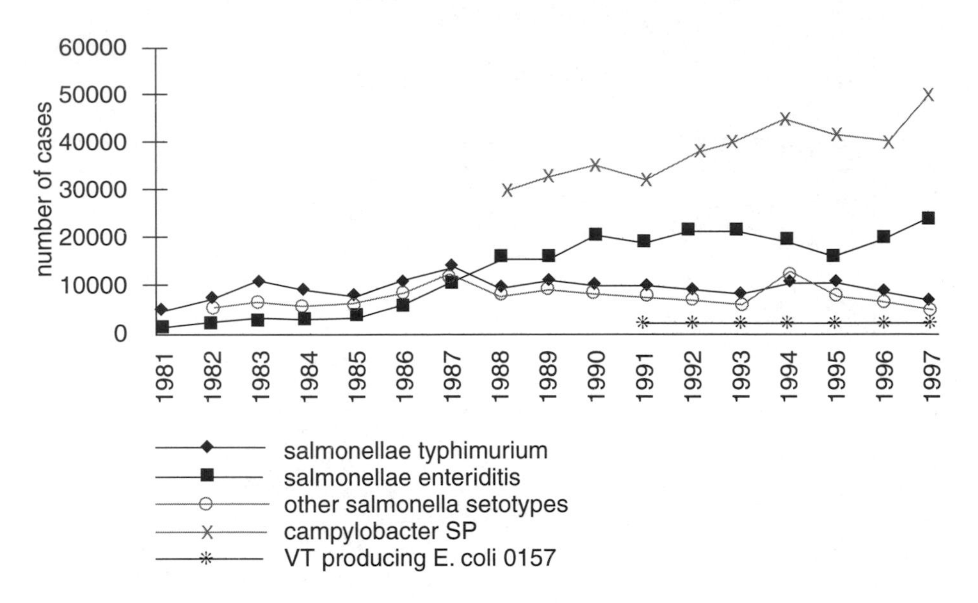

Figure 9.2. Main causes of laboratory confirmed food poisoning: England and Wales 1981–1997 (*Source*: Public Health Laboratory Service)

teriosis (18 in 1996), *salmonella* infections (49 in 1996) and other food poisoning bacteria (1 in 1996). Deaths from *E. coli* and *campylobacter* are not reported specifically in this table, and are classified as deaths from intestinal infections due to other organisms. This category increased rapidly after the definition was introduced in 1992. Of the 357 deaths from food poisoning in 1996, 306 (86 per cent) occurred in people aged over 65 and 255 (71 per cent) in people aged over 75. Only 34 (10 per cent) of all deaths from food poisoning were in people aged between 15 and 64 years.

Spectacular examples of catastrophic poisoning, such as the Toxic Oil Syndrome in Spain, where between 1981 and 1982 about 20,000 people were poisoned and 350 died from food poisoned by the manufacturing process, are extremely rare. Newspaper headlines involving food scares have become more common during the last 15 years. Earlier scares included the 1964 incident in Aberdeen when hundreds of people suffered after eating corned beef and other meats from a local supermarket. But even when, in 1976, more than 1,000 people fell ill from food poisoning in Leeds in the space of three months the story failed to make national headlines (North and Gorman 1990). Two scares stand out in the UK and both proved to be disastrously costly for thousands of small agricultural producers. The first occurred when a misleading statement by a Junior Health Minister, Edwina Currie, in 1988 resulted in the slaughter of more than a million hens, mostly involving small producers, with no impact on salmonellæ poisoning:

"We do warn people now that most of the egg production in this country, sadly, is now infected with *salmonellæ*". The second occurred in 1996 when the then Health Secretary, Stephen Dorrell, made a statement to Parliament officially linking bovine spongiform encephalopathy (BSE) with its human equivalent, CJD. Despite this link being only a theoretical possibility it resulted in a European Union (EU) world ban on all UK beef exports and slaughter of hundreds of thousands of the UK herd. The EU deemed that British beef was safe enough for only the British to eat.

Increasing affluence: hazards in consumption and production

The dangers from inadequate hygiene standards and food have been known about for at least 4,000 years. The book of Leviticus records the laws made by Moses to protect his people against disease. These included the importance of washing hands. The eating of swine, now known to excrete *salmonellæ* and harbour *taenia solium* (tape worm), was forbidden. Other forbidden foods included mice, rats, lizards snakes, vultures, eagles and shellfish (Hobbs and Roberts 1990).

The big advances in the prevention of ill health were made in the 19th century. These included measures to improve the disposal of sewage and the heat treatment of milk by pasteurisation. Pasteurisation is a common method of treating food for preservation or to destroy harmful bacteria. Food borne infection (such as *salmonellæ* and *campylobacter*) can still occur in tuberculin tested raw milk. Chlorination of drinking water began in 1905 in Britain.

At the turn of the century poor people were ill-fed and ill-nourished. Two out of five possible recruits for the Anglo-Boer war in South Africa of 1899 to 1902, who had to be at least five feet tall, were rejected as too feeble. Whilst the government may have been indifferent to the personal health of its citizens "there was cause to worry if national security was to depend on an army of crippled dwarfs" (Seddon 1990). By the time of the start of the First World War in 1914 the recruitment position had not changed. In the depression years of the 1930s food choice was very limited – the working class ate white bread, margarine, sugared tea, potatoes and corned beef. Those exposed to the austerity of this period knew how food keeps and decays. They knew that sugar, salt and vinegar preserve. They knew that raw meat must be treated with extreme care and that unsmoked fish had to be used within a day of purchase. When Britain entered the Second World War in 1939 two out of three called up were recruited. Malnutrition then was more often a consequence of insufficient food. Little has been heard during the past few decades about malnourishment as a consequence of lack of quantity of food. Rather, the issue has become one of nutritional food content and safety of manufactured food. Furthermore, there are other clear indications, such as height and other physical measures, overall that food quality is far better than it has ever been.

During the industrial age of the 19th century, it was common for people to witness death, particularly of the young. Stillborn deaths were much more frequent. Illness and fatal accidents were endemic and life expectancy was short (see Table 9.2).

Table 9.2 shows how life expectancy has increased steadily over the 20th century. Increased life expectancy of this magnitude has overwhelmingly been caused by improved living standards, while the contribution of socialised medicine has been marginal. Longer life expectancy has been accompanied by improvements in general health. Increasing affluence has enabled millions of households to afford to eat protein-rich food regularly, as well as enjoy warm accommodation and good sanitation facilities. Growing affluence led the economist, J. K. Galbraith, writing perspicaciously in 1973, to speculate on future lifestyles predicting the "contracting out of consumption from the household to the independent entrepreneur. The cooking of food by the housewife shifting to the restaurant, the food that is still consumed at home is otherwise pre-cooked or pre-prepared".

Increased affluence has fed demand for manufactured food – from canned produce and biscuits which were novel to the Victorians, to the complete, chilled gourmet meals from supermarkets shelves today. But mass production of food and food products bring, in turn, the need for scrupulous hygiene standards and the scope for wide scale poisoning incidents, as shown by the prominent example of factory farming of chickens and eggs where, despite regulatory controls, in the UK, *salmonellæ* is ubiquitous in poultry, as mentioned above. In another example North (1993) showed how even large slaughterhouses, like the best equipped operating theatres, are unable to eliminate cross contamination, and how regulatory attempts to do so imposed such absurdly strict requirements that meat quality fell in consequence.

Affluence also allowed the market for deep freezers to grow and mature where cooked food could be stored for long periods before use. In 1981, 49 per cent of British households possessed a deep freezer; by 1996 this had risen to 89 per cent. Affluence permitted most households to own a microwave oven which facilitated quick defrosting of precooked frozen food. From no recorded statistics in 1981, by 1996, 70 per cent of British households possessed a microwave oven (Social Trends 1997). Food could be cooked in minutes which otherwise would have required hours of traditional oven cooking. Both these undoubtedly useful labour saving innovations, when used improperly, can provide conditions where food-borne pathogens flourish.

In addition, increasing affluence has separated, the principal from the agent, as alluded to by Galbraith above. When more people eat communally in restaurants, the eater of the food, the principal, is separated from the preparer, the agent. This puts the onus on the agent to adopt the, presumably high, standards of the principal. Because of the obvious scope, and possible incentive, for the agent to adopt lower standards of cleanliness than the principal, there is a problem of 'moral hazard' resulting in an increased risk of poisoning.

On the other hand, a higher standard of living is accompanied by higher expectations, and a greater propensity to complain if disappointed or inconvenienced. This could mean that, although the level of food poisoning incidents may remain constant, the numbers reported may not, and this may explain some of the recorded increase.

Other retailing trends also present potential hazards. Increased affluence has allowed increased car ownership. This has encouraged the growth of large out of (or edge of) town hypermarkets (Burke and Shackleton 1996). Weekly buying of food at the hypermarket has replaced frequent shopping for food at the local butcher, baker or grocer (North and Gorman 1990). Food is being kept longer in the home before consumption. Increasing affluence has stimulated a huge growth in sales of fresh chilled meals and more than a thirty per cent decline in home cooked meals since 1980. So concerned are supermarkets to avoid selling any food that subsequently poisons, they have replaced the 'sell by' date with a 'use by' date. At one Tesco store £3,500 per week of fresh food is destroyed because the 'use by' date has passed. There are about 1,000 stores owned by Sainsbury's, Tesco and Waitrose (*Daily Telegraph* 1998). Sainsbury's finds 2.5 tons of fresh food per week which has passed the 'use by' date, from 10 London stores, safe enough to pass on to good causes. These actions are not required by law but are the self interested responses of a company with a good reputation to defend and promote.

Home cooked meals have also been replaced by the purchase of already pre-cooked hot food. Hot food retailing began in Britain with fish and chip shops. Competition came first from Chinese restaurants and later 'take aways' (to go) and, secondly, from Indian restaurants. The fast food and restaurant chain concept, such as the American burger and pizza chains, are a relatively new phenomenon in the UK. All these examples of the growth in food manufacturing and processing are potentially able to increase the number of food poisoning cases.

Protecting public health and eliminating risk

In their efforts to protect the public health, regulatory bodies too often attempt the impossible task of eliminating risk. Perhaps the most ludicrous example of this in the UK is the illegality of the sale of beef on the bone despite the risks it poses being unquantifiably small. The result is over-regulation; that is to say regulations that impose great costs and hence actually increase the level of risk faced by the general public. By contrast, economists and others argue that a balance has to be struck between placing responsibility on the individual for its own actions and protecting the individual from the mistakes, incompetence or deceit of others.

In consumer law the concepts of *caveat emptor* (let the buyer beware) and *caveat venditor* (let the seller beware) vie for supremacy according to circumstances. Where the knowledge of the seller is assumed to be considerably greater than that of the buyer – for example when a dealer sells a used car to a private buyer – the principle of *caveat venditor* prevails: the seller owes a duty to the buyer to take care over the representations he makes with regard to the characteristics of the product. Where the knowledge of the buyer is likely to be as good as that of the seller – as for example when a dealer buys a car from a private seller – the principle of *caveat emptor* more nearly prevails. In contract law, the result is that when an expert makes a representation it is more likely to be considered a term of the contract than when a non-expert makes such a representation, especially if the purchaser is an expert; compare *Dick Bentley Productions Ltd v Harold Smith (Motors) Ltd [1965]* with *Oscar Chess Ltd v Williams [1957]*. An economic explanation for this apportionment of responsibility is that the costs of passing on incorrect information fall on the party who is able to acquire the correct information at least cost, creating an economic incentive to provide accurate information. So, for example, the vendors of food products will have incentives to ensure that the claims made about the food they sell are accurate because they will be expected to have more knowledge about the veracity of those claims than will the buyers.

Whilst it may be desirable to ensure that claims made by vendors of food products are accurate, this is not the same as arguing that the products they sell should carry no risks. All it means is that *if* vendors make claims about the risks entailed by their products, then these should be true insofar as they can be ascertained. A consequence of this is that manufacturers of less risky products will be able to advertise those lower risks to consumers. If consumers then elicit a strong demand for lower-risk products, the proportion of low-risk products on the market is likely to increase and may even crowd-out high-risk products. The question remains whether it is desirable or even possible to have regulations that eliminate the remaining high-risk products. In our view, the objective should be not to eliminate high-risk products (such as those that may contain poisonous bacteria) but to establish an environment that results in an acceptable (optimal) level of risks from

Table 9.2. Expectation of life by gender at birth, UK, 1901–1996

	1901	*1931*	*1961*	*1991*	*1992*	*1996*
Males	45.5	57.7	67.8	73.2	73.6	74.4
Females	49	61.6	73.6	78.7	79	79.7

Table 9.3. Death by selected causes (England and Wales) 1996

All Causes		560 135
Malignant neoplasms		139 459
Digestive organs and peritoneum	38 226	
Trachea, bronchus and lung	32 273	
Breast	12 179	
Ischaemic heart disease		129 047
Cerebrovascular disease		55 021
Pneumonia	54 137	
Diabetes	5994	
Suicide		3445
Motor vehicle accidents		3134
All other accidents	8368	
Accidental falls		3637
of which		
Fall on or from stairs in home	558	
Slipping, tripping or stumbling	144	
Fall from chair or bed	83	
Fall into hole in ground	4	
Accidental poisoning		1089
of which		
by medical drugs, medicaments and biologicals	937	
by alcohol	152	
byutility gas in the home	70	
Choking on food		262
Meningococcal infection (meningitis		245
Accidental drowning	216	
Struck accidentally by falling object		47
Clothes catching fire		45
Choking on objects other than food		43
In the bathtub		35
All food popisoning other than E. coli, listeriosis and salmonellae (persons aged 15–64)		**22**
Injury caused by animals		12
Plastic bag		9
***Listeriosis* (persons aged 15–64)**		**6**
***Salmonellae* (persons aged 15–64)**		**6**
Hornets, wasps and bees		5
Other venomous arthropods	1	
In sports		5
Lightning		4
Dog bite		3
Fireworks		3
E. coli (persons aged 15–64		0

Source: Office for National Statistics

Table 9.4. Risk of an individual dying in any one year from various causes

Smoking 10 cigarettes per day	1 in	200
All natural causes age 40	1 in	850
Any kind of violence or poisoning	1 in	3300
Influenza	1 in	5000
Accident on the road	1 in	8000
Accident at home	1 in	26 000
Accident at work	1 in	43 500
Radiation working in radiation industry	1 in	57 000
Homicide	1 in	100 000
Poisoning from salmonellae in poultry meat	1 in	5 000 000
Hit by lightning	1 in	10 000 000
Release of radiation from nuclear power station	1 in	10 000 000
Eating beef on the bone	1 in	1 000 000 000

Source: BMA (1987), Maitland (1998)

such products (implying an optimal level of food poisoning). In spite of the apparent economic desirability of this situation, it does not seem to be popular either with the public or with the regulators. At present the public seem to expect their politicians to disinfect the nation's kitchens; we now consider why this is the case.

Fear and death

Public concerns over food safety fall predominantly into three categories:

1. Contamination by infectious agents
2. Contamination by chemicals
3. Nutritional concerns

Nutritional concerns include fat, salt, sugar, cholesterol, and junk food (with concomitant overeating/obesity and poor nutrition, which are generally recognised as major contributors to degenerative diseases such as heart disease and diabetes). However, the public does not exhibit a great deal of fear regarding these issues and they do not tend to escalate into major food scares. By contrast, when viewing Tables 9.1 and 9.3, which list the numbers of deaths from selected causes, it is apparent that deaths ascribed to contamination by infectious agents account for a minuscule number of the total, and deaths from contamination by chemicals are not recorded at all. Yet the public experiences a great deal of fear and apprehension over possible microbial and chemical contaminants and many disastrous food scares have resulted.

The fact is that food poisoning threatens only the elderly and young and those with compromised immune systems. There are more deaths every year from people drowning in their bathtub than of food poisoning in those aged between 15 and 64. As with AIDS, the general public is not at risk (Craven et al. 1994).

While lethality of the risk is an important factor in how the public assesses its severity, it is not the actual lethality that counts, but rather the perceived lethality. Deaths from nuclear power plants have been very few, and the actual risk very small indeed (Table 9.4), but perceived risk of death is high because the public thinks in terms of the potential for huge numbers of deaths if something goes wrong.

Another example is the food scare involving BSE, a disease that is tenuously linked to nvCJD, an extremely rare condition that has accounted for only 35 deaths in total (February 1999) in the UK. The fatality of the disease, combined with media speculation about impending huge epidemics of nvCJD, led to a very high perceived risk among the public. Fear of nvCJD was also no doubt influenced by the tendency of people to over-estimate deaths from rare diseases while deaths from common diseases are underestimated (Slovic et al. 1981).

Experts versus the public

It has been well established that experts differ tremendously from the public in the way they view and assess risk (Slovic et al. 1981, Fife-Schaw and Rowe 1996). For example, on a list of 30 hazards, experts rated nuclear power generation at the lower end of the scale (less risky), believing it to be approximately as dangerous as food colouring and home appliances, and less dangerous than food preservatives, riding bicycles, and swimming. On the other hand, the public rated it at the number one position (most dangerous of all) (Slovic et al. 1981). Experts are least uncomfortable with a hazard that has the potential to kill many people, but which has a very low probability of happening (such as massive nuclear power radiation or meltdown), whereas the public feels more comfortable with a hazard that affects fewer people, such as a road accident, but which has a much higher likelihood of occurring. This accounts for the fact that what experts communicate about the low risks involved in various food scares is often almost totally disregarded by the public. An example of this dynamic was a major food scare in the 1980s that involved daminozide (Alar), a ripening agent used on apples. Experts considered this agent to be harmless, but the safety of its breakdown products was under investigation (Lee 1989). The Natural Resources Defense Council, a consumer advocacy group, announced that Alar in apples created a 45 in a million cancer risk, and predicted that 6,000 American school children could get cancer from Alar residues. This created a media stam-

Table 9.5. Public risk perception

Less perceived risk		*More perceived risk*
Voluntary	vs.	Coerced
Natural	vs.	Industrial
Familiar	vs.	Not familiar
Not memorable	vs.	Memorable
Not dreaded	vs.	Dreaded
Chronic	vs.	Catastrophic
Knowable	vs.	Unknowable
Individually controlled	vs.	Controlled by others
Fair	vs.	Unfair
Morally irrelevant	vs.	Morally relevant
Trustworthy sources	vs.	Untrustworthy sources
Responsive process	vs.	Unresponsive process

pede and public hysteria, and apples were banned from school lunchrooms across the nation and removed from supermarket shelves. Eventually, the dust cleared after the increased cancer risk from Alar had been calculated at 0.025 per cent (considered to be trivial) and apples were restored to school menus. However, "No danger from Alar in apples," doesn't make a newsworthy headline, so damage from media coverage of the scare was not reversed, but left a lingering psychological impact. Ultimately, trust in foods was damaged and losses to the apple industry were estimated at over $100 million (Lee 1989). Ten years on, many of the characteristics of the media's treatment of the Alar scare correspond to the hysterical coverage of genetically-modified food.

The outrage component

Experts may often feel frustrated that the public ignores the facts and bases its opinions on what seem like illogical thought processes. However, the public's method of risk assessment involves many complex cognitive processes which need to be understood and addressed in order to communicate risk in a way that people will understand and accept. Sandman (1994), America's foremost expert on risk communication, has developed the following formula to explain this phenomenon:

RISK = HAZARD + OUTRAGE

According to Sandman, traditional risk assessment defines risk as magnitude (how bad the problem could be) times probability (how likely it is to happen): "However, experts tend to focus on this definition, (let's call it hazard), and so underestimate actual risk, because they ignore outrage. The public tends to focus instead on outrage and pay less attention to hazard" (Sandman 1994). Most public policy disputes over food safety are concerned with whether a given risk is *acceptable*. That value judgement depends on outrage factors more than on the size of the risk (Groth 1991). In order to effectively communicate a risk to the public, it is necessary to focus on the outrage. Sandman has identified 12 components of outrage and how they affect perceived risk.

Once again, using nuclear power plants as an example, the outrage components fall almost exclusively into the right column, creating a high degree of outrage, and concomitantly a high level of fear and a low level of acceptance. Food hazards involving chemicals and microbial contamination also score high on the outrage scale. Other analysts (Slovic et al. 1981, Fife-Schaw and Rowe 1996) have pinpointed the three most important categories as being: first, dread, secondly, the unknown and thirdly, the number of people exposed.

The 'dread' component is of particular importance. Perceived risk can be predicted almost perfectly from firstly, how much a hazard is dreaded and secondly, how lethal the hazard is considered to be (Slovic et al. 1981). 'Dread' includes such things as greater public concern, more serious effects on future generations, threats of widespread disastrous consequences, and potential to become more serious (Sparks and Shepherd 1994). According to Slovic, the higher the dread rating of an activity, "the higher its perceived risk, the more people want its risk reduced, and the more they want to see strict regulation employed to achieve the desired reduction in risk."

All of the above explains why nutritional concerns don't generate widespread public fear and huge food scares, whereas microbial and chemical contaminants do. Food hazards such as alcohol, sugar, fat, nutritional deficiencies and so forth adversely affect large numbers of people, yet they rate very low on the 'dread' scale, and are so very familiar that they rate very low on the 'unknown' scale as well (Sparks and Shepherd 1994). Conversely, chemical contamination rates very high on all three scales (making it no surprise that Alar in apples generated such a massive food scare). Contamination by infectious agents rates high on the dread scale but low on the unknown scale (Sparks and Shepherd 1994). No matter what the actual risk, hazards which rate high in outrage are most likely to become scares.

Consider the source

Sandman did a content analysis study on 248 environmental risk stories from New Jersey's 26 daily newspapers (Sandman 1994) and found that 68 per cent of the paragraphs contained no risk information at all, another 15 per cent dealt with whether the potentially risky substance was present or absent, and only 17 per cent of the paragraphs dealt with whether the substance was risky or not. A panel consisting of an environmental reporter, an activist, an industry spokesperson, and a technical expert was asked to assess these stories further. Although members of the panel disagreed about almost everything else, they strongly agreed that "environmental risk information was scanty in these stories. Technical content was especially lacking. What risk information was provided came mostly in the form of opinions, not evidence." Government is by far the most common source of environmental risk information used by the media. In the New Jersey study, government officials accounted for 57 per cent of all source attributions. On network television, government represented 29 per cent of the on-air sources, but when only one source was used, it rose to 72 per cent (Sandman 1994). According to Sandman, "Reporters typically start with a government official, the swing vote. If the government says 'dangerous,' they look for an industry source or possibly an expert to say 'safe.' If the government says 'safe,' they look for a citizen or possibly an activist to say 'dangerous.'" In view of the above, it is of particular interest to note that government officials are the source which is least trusted by the public. On the hazards of both food poisoning and excessive alcohol use, a committee of medical doctors was rated as highly credible, whereas government was rated as having low credibility (Frewer 1997). Rating for distrust, members of parliament, government ministers, and government ministries were rated among the top four most distrusted sources, along with tabloid newspapers (Frewer 1996). It is not surprising to see how food scares escalate when most information about them comes from the least trusted sources.

If it bleeds, it leads

What explains media attraction to food scares? How does a concern over food safety escalate into a food scare? Obviously, the media often play a crucial role in this dynamic. Which issues the press decides to cover, how it covers them, and how much coverage it gives them, often determines whether the issue becomes a scare. Is it the seriousness of a food hazard (the potential risk) that attracts the media? Not necessarily. Sharon Begley, science editor at *Newsweek* magazine, states that "Risks surrounded by uncertainty and controversy are good contenders for coverage" (Begley 1991). Traditional journalistic criteria, such as timeliness, proximity, prominence, human interest, drama and visual appeal, make a big controversy intrinsically newsworthy even if it's not a serious health threat (Sandman 1994).

When asked why journalists write about some risks and not about others, Begley stated that "Journalists are paid to engage their audience ... We can't do it by boring them. One way to avoid boring people is with stories that have a strong element of uncertainty. Reporters crave uncertainty because it is generally equated with drama ... Most of what science 'knows' about risks such as pesticides or hormones in food or food irradiation is uncertain and highly debatable." Begley also says that journalists "seek out the contrarian," that they are attracted to "subjects, discussions, and controversies that go against conventional wisdom, that will evoke in our readers a sense of surprise." Unsuspected risks in food make good stories because they oppose the conventional wisdom (that food in first world countries is safe). It's really not news that smoking still causes cancer or that driving without seat belts is still a foolish thing to do, but new food hazards always have potential for a good story.

The media play into what Slovic (1981) calls "availability bias." This means that people will judge an event as more likely to occur (and thus be more apprehensive) if it is easy to imagine or recall. Even though frequently occurring events are easier to imagine or recall, the availability bias will often come into play regarding events that don't occur frequently, or perhaps don't occur at all. As Slovic points out, "Discussing a low-probability hazard may increase its imaginability and hence its perceived riskiness, regardless of what the evidence indicates" (Slovic 1981). Availability is similar to Sandman's item on the outrage scale "memorable v. not memorable." Since the media tend to cover events that are more dramatic (even though the actual risk may be low), and since they cover these events more often, media coverage increases both availability and memorability. In an attempt to ascertain how much the media influences risk perception, two groups of people were asked to rate items on a list of some 90 hazards according to how risky they believed each hazard to be (Kone and Mullet 1994). One group consisted of Burkina Faso intellectuals, the other of French students. France and Burkina Faso are markedly different in terms of geography, economics, politics, and ethnic background, but quite similar as to media coverage. There are also extreme differences in the real risks that exist in either country (many risks one might be exposed to in third world countries do not occur in first world countries). Even so, the media in Burkina Faso give coverage to hazards that may not even be present there (but would be of concern to citizens of France). The results of the study were that the Burkina Faso inhabitants had "approximately the same preoccupations as the French respondents and to the same degree." This indicates that reality plays a secondary role to the media's representation of reality.

But the decisive factor is that food hazards involving chemicals and microbial contamination inspire outrage. "Outrage is newsier than mere risk. Outrage is active, risk is a number that just sits there" says Begley (1991). "The media are in the outrage business. Most of the coverage isn't about the risk. It's about blame, fear, anger, and other non-technical issues–about outrage, not hazard ... In their focus on outrage rather than hazard, journalists are at one with their audience" (Sandman 1994). The BSE scare, for example, ran almost its entire course without any scientific proof that there was anything to be scared of. It is reasonable to conclude that after 1996 the outrage component of public risk perception of BSE/nvCJD would have increased substantially.

This, of course, attracted more media attention, which in turn increased public apprehension, which in turn increased pressure on government to 'do something.' This something took the form of a grossly exaggerated regulatory response.

Food law and the regulatory framework

Principal Acts of Parliament:

The Food and Drugs Act 1938.
The Defence (Sale of Food) Regulations 1943.
The Food and Drugs Act 1955.
The Food Act 1984.
Food Safety Act 1990.

In an excellent summary Thompson (1996) described the regulatory changes which began in 1938. It was then that food poisoning became notifiable under the Food and Drugs Act 1938. This Act enabled ministers to implement regulations to control the composition and labelling of foods. It also enabled Local Authorities to make bye-laws on the sanitary conditions of food for human consumption. The food shortages resulting from the Second World War gave rise to food substitutes; chicory in coffee for example, and developing technology enabled the possibility that substitutes could be passed as the real thing. The Defence (Sale of Food) Regulations 1943 strengthened the law on labelling and food content. The Food and Drugs Act 1955, had two main objectives: to safeguard health and prevent deception and fraud. By this Act, Ministers were empowered to make regulations "in the interests of public health or otherwise for the protection of public health" (Thompson 1996). As we have seen, by the 1980s, food manufacturing became more complex, sophisticated and technological. Economic growth had generated markets for pre-packed and pre-cooked foods, the deep freezer and the microwave oven (see earlier section) and it was felt that legislation required updating. As a consequence the Food Act 1984, which replaced the Food and Drugs Act 1955, was introduced to deal with these developments. The objective of the Food Act 1984 was to ensure consumers receive a pure and wholesome food supply. The Act prohibits food from being sold which:

1. Has been made injurious by addition or abstraction of substances or by treatment.
2. Is not of the nature, substance or quality of food of a different kind, or adulterated or inferior.
3. Has false or misleading labelling or advertisement.
4. Is unfit (decomposed or contaminated).

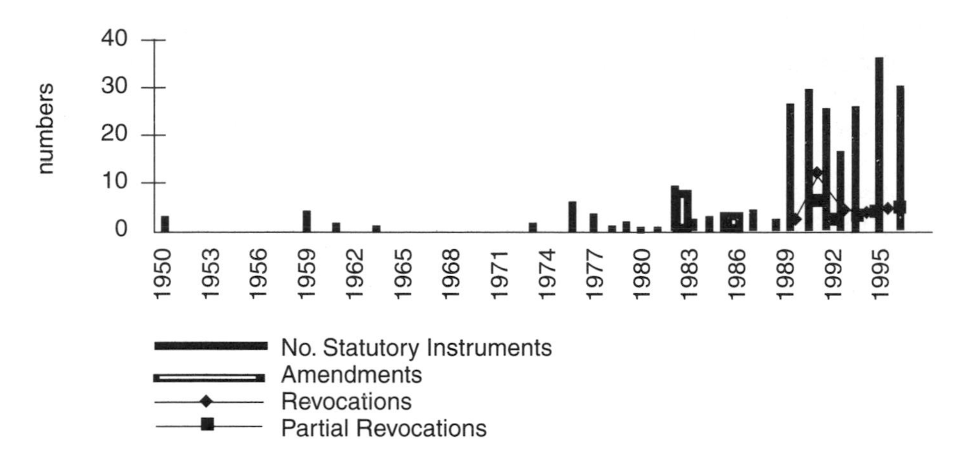

Figure 9.3 Number of statutory instruments (Food) (Regulations), amendments and revocations 1950–1966 (*Source*: Halsbury's Statutory Instruments (1996))

Enforcement allowed premises to be closed within three days and remain closed pending the outcome of legal proceedings under the Act. Despite this Act, incidents of food poisoning continued to increase in the 1980s (Figure 9.3). The Government believed these incidents were preventable and, a White Paper, *Food Safety – Protecting the Consumer*, (CM 732) was published in 1989. The White Paper had several main objectives. These included ensuring that new methods of production and distribution were safe. A second objective was to prevent food being mislabelled or misleadingly presented. A third objective was to ensure that European Community directives could be implemented in UK law, and a fourth to set standards which spelt out consumer rights. Other changes in the law were proposed too. These included reinforcing penalties against law breakers and measures to strengthen controls to prevent unfit food reaching the consumer. Responsibility for enforcement was given to Local Authorities (Scott 1990) with power to charge for inspections.

Despite the objective of increasing regulatory powers the "White Paper portrayed a legislative system which was working extremely well" (Thompson 1996). Scott (1990) stated that the promotion of the new *Food Safety Act* based on the findings of the 1989 White Paper was little more than political expediency. The new Act reasserted the old law with some changes. The most conspicuous change was the addition of the word 'safety' to the title.

Although the 1980s showed a rise in the number of food regulations by comparison with the previous 30 years, this rise was small by comparison with the unprecedented increase in the number of food regulations (Figure 9.4) following the *Food Safety Act* (1990). It is not in doubt that there are substantial costs to the tax-payer and industry of attempting to regulate against food poisoning illness. There are also costs to the health sector from food poisoning. General practitioners treat cases in the community, hospitals treat at accident and emergency units and there are follow up costs. Whether the regulatory approach which does not tackle the main cause of the increase in food poisoning (and largely preventable by adopting good hygiene standards in the handling and heating of food) is cost effective, is a moot point. It is of little surprise that the stronger regulatory regime has not been accompanied by a fall in food poisoning. In short, the UK is awash with regulations designed to improve food safety. All the requirements of these regulations are sensible, but do they require such immense legal status and force?

Theory and politics of regulation

Food poisoning imposes costs on employers, the health services, the family and society in taking care of those ill. It was estimated that 10,000 working days were lost in the survey reported by Roberts and Sockett (1994). Whilst this appears to be a huge cost, it pales into insignificance by comparison with the 116,000,000 working days lost through back pain (NPBA). Roberts and Sockett estimated the production loss to industry from sickness related absence from work to be a maximum of £334m. By comparison, the estimated cost to industry from back pain was £5,100 million (NPBA). The response to the consequences of food poisoning, as we have seen, has been a deluge of regulatory statues. In context regulation is perceived as necessary for requiring or controlling a process or treatment used in the preparation of food intended for human consumption (Schefferle 1990). Regulation may also be intended to generate consumer information (labelling, contents, processes used). It is rarely denied that regulation is other than well intentioned in democratic countries and designed to protect public and private interests (Tower 1993, Veljanovski 1991). The reality is often quite different from these good intentions.

The causes of food scares

Food scares and consumer problems with food have been shown to be based largely on: consumer ignorance about the risks involved, lack of information on product content, misleading information on product content and lack of adequate standards of consumer hygiene. Some argue that these are the result of 'market failures'; imperfections caused by the fact that in the real world information is always asymmetrically distributed and the idealised vision of the market as a system in general equilibrium does not reflect this reality. However, it is equally plausible that they are the result of regulatory failures; inhibitions in the market caused by state intervention. The issue is one of relative importance. Are the greatest food problems (and hence potentially greatest scope for successful problem solving) associated with lack of information and inadequate hygiene standards, 'bad' food products or are they, perhaps, the consequences of regulatory action?

There are two main views as to how to approach the idea of 'market failure'. The first (and minority view) is that in any dynamic society, where technology is constantly changing and information is asymmetrically distributed, a minimally regulated competitive process produces the best results. This market process view derives from the work of Carl Menger and his intellectual followers, including Ludwig von Mises, and Friedrich Hayek. For a more modern account of the approach, see Langlois (1986). This theory says that 'market failures' are best corrected by the actions of market participants utilising both the market itself and its supporting institutions, including the courts. The alternative and still dominant orthodoxy differs in that it advocates government intervention to correct for these 'market failures', through state supply, fiscal measures or regulation. It is generally understood in a theoretical sense that regulatory interventions must not be so severe as to impose costs which outweigh benefits or which would drive businesses abroad. In practice cost-benefit approaches to food safety regulation are seldom adopted or even attempted. In reality it appears that regulators often appear to fail to take into account that although there may be benefits from regulation there are also costs. These costs can be so great that they make the situation worse ('regulatory failure'). The costs arise not only from the regulatory authority in its day to day business of regulating but also in compliance costs imposed on the regulated (Veljanovski 1991). A good example occurred when the British Agricultural Minister, after stating on 3 December 1997 that "our beef is the subject of the most rigorous safeguards of any beef, anywhere in the world," announced that the sale of beef on the bone was to be banned. The estimated cost of removing the bone would push the price for prime sirloin up from £4.50 to £6.50 per pound. This was added to an estimated £70 per animal in compliance costs of existing government regulations on beef.

Principal versus agent and asymmetric information

Regulation is predicated on the belief that 'experts' or 'the government' have superior knowledge to the consumer. In practice politicians when making decisions on complex issues have to rely on a public sector bureaucrat for a supply of filtered information. The bureaucrat in turn has to acquire summarised and filtered information from 'experts'. In attempting to act to improve the public health Government decision makers are severely limited in their ability to process all relevant information affecting a complex issue; they must make decisions based on a limited set of information which imposes boundaries on the rationality of their choices (Williamson, 1975). In response to this bounded rationality, decision makers acquire only such information as they consider necessary to satisfy themselves that they can make a reasonable attempt at an informed decision – they 'satisfice'. Decision makers are therefore unlikely to be aware of the subtleties associated with the regulatory functions they are recommending and funding. The process of regulation is thus one where Governments request and receive expert advice and often, after consultation with civil servants, make policy decisions. The process assumes that experts are neutral and objective and that experts and policy makers act rationally. Governments use the process too to legitimise their actions and thus can avoid blame when errors and failures occur.

The public and the private interest can, unfortunately, easily conflict especially where there are extreme information asymmetries between different market actors. A good example of such asymmetry can be found in health care markets where the producer (physician) typically has much greater information than the consumer (patient). Other examples include the markets for financial insurance services, and for electrical and motor vehicle repairs. In such cases consumers (encouraged by anti-business media reports) may suspect that producers are exploiting them. Such suspicions enhance calls (often made by the media themselves) for 'something to be done'.

More pertinent in the context of this chapter is the likelihood of consumer ignorance (and a lack of readily accessible reliable and accurate information) concerning an alleged risk from consumption of a particular food product – and this is strikingly the case with genetically modified foods. When such alleged risks arise (and are given tabloid coverage) a government spokesman typically feels obliged to give reassurance. Information asymmetries between government departments and their expert advisers generate suspicions in the public's mind of hidden agendas, especially given that politicians appear to be amongst the least trusted people in society. This arises when there is doubt about the probity of government advice and information where government departments represent conflicting interests. This was a real problem in the UK following the BSE scare in 1996 when it was recognised that the government ministry, MAFF, despite reassurances to the contrary, in effect was charged with both furthering producer interests and protecting the consumer.

To address the conflict of interest, in 1998 the Government proposed the establishment of a new self-financing regulatory agency (SEFRA see later section), The Food Standards Agency (FSA). The proposal was met with almost universal criticism. One problem is the chosen funding mechanism. Nearly 50 per cent of the annual cost of FSA is to be funded through an industry wide flat-rate levy of about £90 per annum (although some estimates put it as high as £600) which will be collected from the approximately 490,000 food establishments registered with local authorities. Thus a mobile hot dog stand will pay the same levy as the largest supermarket. However, some organisations, such as kiosks selling crisps and soft drinks and bakers' delivery vans, will be exempt. Food manufacturers and processors are also exempt. Local authorities object to the requirement that their inspectors should collect a new and unpopular 'poll tax'. The main effect of the agency will be to deter thousands of small shops from selling sandwiches. If the FSA is vital to public health – if there is an external benefit – then it should be financed out of general taxation.

Another problem is that 'experts' may not be reliable advisers when problems occur which are new, unpredictable and either unprojectable, such as with nvCJD, or projectable with wide confidence limits as with AIDS in the early stages. Indeed in the case of AIDS, reliance by governments on imprecise information from medical experts was the main explanation for serious resource misallocation and misuse (Craven, Stewart and Taghavi 1994, Craven and Stewart 1997). The erroneous view of many experts that most cases of *salmonellæ* poisoning were caused by *salmonellæ* infected eggs is another example. 'Experts' have typically accumulated specialised knowledge about a particular scientific field (such as biochemistry, immunology or epidemiology), at the cost of not acquiring specialised economic knowledge, so their advice should not be the unique basis of public policy.

Regulatory capture

Whilst regulation may be intended to protect the public from exploitation by suppliers with superior information, it can also be used to protect or enhance the interests of the regulated. In health care, for example, professional qualification is a form of public regulation designed to impose minimum standards for treatment but also a form of private regulation because it enhances the economic rents (incomes) of the regulated: health physicians (Stigler 1971). This can lead to what is known as regulatory 'capture', where the regulated may eventually control or 'capture' the regulators (Kay 1988). A survey by Cannon (cited in Scott 1990) found that of 370 people sitting on Government advisory committees involving food only a hundred advisors were independent in having no links to the Government or the food industry. Of the rest 133 worked or had worked for the food industry; 65 were funded by, or were advisors to, the food industry and 156 had links to the British Nutrition Foundation, which is funded by the industry. These circumstances are propitious for regulatory 'capture'.

Empire building

In practice the efficacy of regulation is also likely to be compromised for several other reasons (Posner 1974). Regulators may fail to use their power over the regulated, or may use it in unnecessary or harmful ways. In the field of public health, North (1992) demonstrated how legislation imposed unnecessary costs on the regulated. Regulators, if beneficiaries of political patronage (*Financial Times* 1994), may not be the most appropriate people to appoint. More important is the recognition that regulators cannot be presumed to be without self interest. They have an interest in maintaining their jobs and enlarging their budget by taking advantage of their ability to disguise costs or their inability to cost activities. This will result in the service being supplied at greater than minimum (productive efficiency) cost (Niskanen 1971, 1973). There is also an incentive for regulators to overstate the importance of their work and to ensure that they do not remove the need for regulation. If the problem they are attempting to control (say food scares) worsens, there is likely to be a call for greater funding and more regulation. If the problem diminishes it is because of the success of the regulation; funding must continue. Food safety regulation should be like a good steak; lean and simple.

There is a first class example of this behaviour in the response of the Department of Health (DoH) in explaining the failure of AIDS to spread to the heterosexual population. There is strong evidence that individuals are not adopting safe sexual practices. There has been a continued increase in visits (547,437 in 1984, 732,000 in 1994 in England and Wales) to clinics for treatment of sexually transmitted diseases and rising figures of more than 160,000 abortions per year in England and Wales. In 1996 there were 8,000 pregnancies to girls under the age of 16; the highest for 10 years. Despite this the DoH consistently asserts that health education expenditures have succeeded in persuading people to adopt safe sexual practices. The DoH then further asserts that this is the reason why AIDS has not spread heterosexually!

Nonetheless, it is unwise to assume a homogenous set of utility maximising bureaucrats (Downs 1957 1967). Some, for example, will seek to conserve power, prestige and budget. Others will seek to maximise these attributes to gain promotion. Some will combine loyalty, altruism and self interest. Altruistic bureaucrats may seek to act in the public interest perhaps out of loyalty to the state (the classical civil servant), subject to conserving power and influence. Taken together, it is certain that the cost of activities supplied by the bureaucrat will be excessive. Bureaucrats have much managerial discretion over what activities they choose to perform because they need not respond directly to the customer or tax-payer. Sometimes this will result in satisfycing behaviour where the tasks performed suit the bureaucrat rather than the public health. At other times they will result in empire building. The public caricature of the lazy bureaucrat is misplaced. Staff will often be employed to work long hours, at a fast pace perhaps doing a particular job well, even though that may mean searching for lost memos. The essential point is that often, activities will be performed which are of little or no value to the customer or taxpayer. Given the additional problem of output measurement, annual budget allocations are likely to be based on incrementalism, where the budget increases by a small amount each financial year rather than on the size, nature and seriousness of the problem to be addressed (Wildavsky 1973 1975). In all cases, there is a supply of regulatory activity which is excessive and inefficient in both accounting and economic terms.

In recent years a new type of bureau, the SEFRA (Self Financing Regulatory Agency), and bureaucrat has emerged. The name was coined by Booker and North (1994). The SEFRA is an agency which can charge those it is regulating for the costs of the regulation. Examples include Her Majesty's Inspector of Pollution, the Environment Agency, the Medicines Control Agency, and the Veterinary Medicines Directorate. The SEFRA arises from an ideological belief that regulatory agencies must be self financing. Recall that the need for regulation arises from the demands of the consumer for protection from exploitation from information asymmetries in the market. It must follow therefore that the consumer is willing to pay, in higher taxes or higher prices, for the costs of regulation. In giving SEFRA officials the authority to charge producers for their services and the power to close producers unable or unwilling to pay or comply, a powerful force for generating rents from unnecessarily restrictive regulatory activity results. The current fees, for example, authorised under the Dairy Products (Hygiene) (Charges) Regulations 1995 SI 1995/1122 are £94 for a general dairy farm visit and £63 for a sampling visit. The claim that the costs for inspection of licensed slaughterhouses required under European Union Directive 93/119/EC are unduly onerous and possibly illegal (Thompson 1996) was tested in *Woodspring District Council v Bakers of Nailsea Ltd* (unreported). The charging was found not to be illegal but the judge did consider there were arguable grounds for challenging the Directive and Regulations which required some inspections to be carried out by veterinarians rather than by cheaper meat inspectors. When severity of regulatory controls differs between countries, the producers in countries with strongly enforced regulations will be at a competitive disadvantage by comparison with producers in countries with liberal regimes. There is evidence that this is becoming a real problem in England and Wales. Booker and North (1994) have documented dozens of cases where European Union Directives have been translated into UK law with much more onerous regulatory force than required by the Directives.

Defusing a food scare

Given the goals of the press, it is not surprising that alarming content about risk is more common than reassuring or intermediate content (Sandman 1994). However, it is sometimes difficult to determine what will be construed as alarming or reassuring. Industry tends to construe mildly alarming stories as being highly alarming. Experts consider test sample results showing low levels of contamination to be reassuring; however, many lay persons focus more on the presence of the contaminant than on its concentration and find the same data exceedingly alarming (Sandman 1994). Sandman reports that "Explicit statements by official sources minimising the risk ('the levels are low,' 'it hasn't spread,' 'don't worry') are often considered offensive, incredible and therefore alarming by citizen readers. One-sidedly reassuring risk information is likely to strike readers as incredible and therefore produce a boomerang effect." Once a scare has emerged, reassurances from government and scientists that the risk is actually quite small will be ineffective in subduing panic and de-escalating the scare. The outrage must be dealt with in an effective manner.

The science of risk communication is still relatively new, though valid and effective precepts have been clearly defined. Mitigating a hazard itself does not mitigate the outrage about the hazard. To defuse a scare, outrage must be addressed, that is, the public's particular concerns must be addressed and dealt with in a way responsive to their emotional needs regarding the issue. Sandman, who serves as a consultant to industry, gives the following example of good risk communication:

> About six months after the Exxon Valdez oil spill, a ship carrying BP oil ran aground and spilled at Huntington Beach, California. The BP CEO flew to the spill, and had obviously planned his risk communication carefully. When he was asked, "Whose fault was this spill?" you could see he wanted to say "Look, it was a contract ship, with a contract crew. They spilled our oil!" But instead he said: "My lawyers say this was not our fault, but I feel as if it were our fault, and we will deal with it as if it were our fault." Six months after the spill, they polled the residents, and BP had a higher approval rating than before the spill.

This 'responsive process' of communication has four facets (Snow 1997):

1. Openness and disclosure;
2. Acknowledgement of wrongdoing;
3. Courtesy (even if public response is angry and impolite);
4. Compassion (recognising and addressing people's fears and apprehensions).

According to Sandman, "Risk communication comprises two facets: 'scaring people' and 'calming people down,' or alerting and reassuring people. There are moderate hazards that people are apathetic about or minor hazards that people are outraged about." The goal of risk communication is to create a level of outrage appropriate to the level of hazard (Snow 1997).

Since the public responds more to outrage than to hazard, "Risk managers must work to make serious hazards more outrageous, and modest hazards less outrageous." Stoking the outrage has been a successful strategy in increasing public concerns about the serious hazards of drunk driving for example (Sandman 1987). Conversely, any strategy that attempts to decrease public concern about minimal hazard must decrease the outrage. Sandman states, "When people are treated with fairness and honesty and respect for their right to make their own decisions, they are a lot less likely to overestimate small hazards. At that point risk communication can help explain the hazard. But when people are not treated with fairness and honesty and respect for their right to make their own decisions, there is little risk communication can do to keep them from raising hell–regardless of the extent of the hazard."

Any agency responsible for conveying risk information to the public should consider using the services of professional risk communicators. Adams and Sachs (1991) suggest that "'active listening' may be a more important component of effective communication than 'telling.' People are more open to hearing facts, or data, if they feel they have been listened to" (Adams and Sachs 1991). They also describe community outreach programmes instituted by the US Department of Agriculture Food Safety and Inspection Services, which were designed to interface with the public in such a way as to obtain "advice, comments, and recommendations from consumers, industry, public-interest groups, our labour union, and our inspectors." To prevent panic in the face of a potential food scare, sources of information must be trusted. The first source of information is always the government: Government agencies, public health departments, government spokespersons and so on. Unfortunately, as discussed above, they are the least trusted. Neither is industry trusted to tell the truth about hazards they are responsible for. On the other hand, university scientists and medical groups are the most trusted sources. If you seek to reassure, then use them to convey the message. A message is more believable when it comes from a totally unexpected source. If government or industry says something is harmless, it doesn't ring true for the public, since no one expects them to tell the truth. On the other hand, since industry is expected to down-play or cover up hazards, if they say it's dangerous, they will be trusted on that issue. And, since it's the job of an environment group or a consumer group to sound the alarm about hazards, if they say there's no danger, they'll be trusted as well. The source of the message is often more important than the content of the message.

Genetically modified foods.

Since 1983, maize, oil seed rape, tomatoes and Soya have all been genetically modified and are currently being grown on about 8m hectares in the USA. These food products have been declared safe by the European Union.

In February 1999, Dr Arpad Pusztai, in an interview for a networked peak time television programme in Britain, claimed that rats suffered damaged health when fed with genetically modified potato containing a snowdrop gene which coded for an insect resistance substance (a lectin). In August 1998 Dr Pusztai had told journalists that rats suffered ill health when fed potato genetically engineered to produce Concavalin A, an insecticide naturally produced by the South American jackbean plant. Dr Pusztai was suspended from the Rowett Institute when, two days later, one of his research assistants, Dr Eva Gelencser, contradicted his claim by asserting that the experiment had been carried out with rats fed potatoes to which Concavalin had been added. An investigation by a panel which included the two scientists did not confirm whether the Concavalin A experiment was carried out. The investigation cleared Dr Pusztai of any misconduct but was not convinced by his research conclusions. Dr Pusztai's research has not yet been published after being assessed by independent peer review.

From such a confusing beginning, a huge new food scare arose. Genetically modified foods fit well on the outrage scale being industrial, not familiar, dreaded, unknowable, from untrustworthy sources, morally relevant, controlled by others and so on. All the usual suspects and tactics were in evidence; suppressed Government reports (*Daily Telegraph* 16/2/99), calls for moratoria on planting and importing of genetically modified foods (*Financial Times* 18/2/99), conflicts of interest between advising scientists and connections with biotech companies (*The Times* 18/2/99), pressure group distortion (Daily Telegraph 16/2/99), government captured by big business (*Financial Times* 19/2/99) and disputes amongst experts (*Guardian* 23/2/99). Within days genetically modified foods were labelled 'Frankenstein foods'. The *Daily Mirror* headlined Prime Minister Blair as "The Prime Monster", sub headlined with "Fury as Blair says 'I eat Frankenstein food and it's safe'" The Daily Express led with the headline "Human Genes in GM Food." A cartoon showed supermarket shoppers wearing hermetically sealed suits whilst pushing the trolley. Another showed giant potatoes running though towns like King Kong. Then a government report expressed fears that the food chain supporting insects, animals and birds could be threatened. The Local Government Association recommended a five year ban on genetically modified food being served in schools, hospitals and old peoples' homes. In the midst of all this confusion the consumer is hard pressed to find any credible scientific evidence that genetically modified foods are either safe or unsafe. Few, if any, lessons from our foregoing analysis of how to deal with food scares appear to have been learned.

Conclusions

Governments and regulatory authorities have become over-protective. It is time to redress the balance by placing more responsibility on individuals for the consequences of their own actions. Mistakes made by regulators are not learning experiences, as they would be for an individual, but causes of recrimination and resentment. Perhaps the following label on a USA food product may be more effective than all the regulatory requirements together!

Safe handling Instructions

This product was prepared from inspected and passed meat or poultry. Some food products contain bacteria that could cause illness if the product is mishandled or cooked improperly. For your protection follow these safe handling instructions.

Keep refrigerated or frozen. Thaw in refrigerator or microwave.

Keep raw meat and poultry separated from other foods. Wash working surfaces (including cutting boards) utensils and hands after touching raw meat.

Cook thoroughly.

Keep hot foods hot.

Refrigerate leftovers immediately or discard.

Figure 9.4 USA food warning label

References

Adams, C. E., Sachs, S. (1991). Government's Role in Communicating Food Safety Information to the Public, *Food Technology*, May, pp. 254-255.

Axelrad, J. (1998). An auto-immune Response Causes Transmissible Spongiform Encephalopathies, *Medical Hypotheses*, 50, 259 – 264.

Begley, S. (1991). The Contrarian Press, *Food Technology*, May 245-246.

Booker, C. and North, R (1994). *The Mad Officials*, London, Constable.

Bowbrick, J.(1977). The Case Against Compulsory Minimum Standards, *Journal of Agricultural Economics*, 28,113.

British Medical Association (BMA) Guide, (1987). *Living With Risk*, J Wiley and Sons.

Bruce, M. E., Will, R. G., Ironside, J. W., et al. (1997). Transmissions to Mice Indicate that 'New Variant' CJD is Caused by the BSE Agent, *Nature*, 389, 498-501.

Burke, T. and Shackleton, J. R. *Trouble in Store*, Institute of Economic Affairs, Hobart Paper No. 130, London.

Cannon, P. (1987) *The Politics of Food*, London: Century, p. 314.

Communicable Disease Report (1996), Food Poisoning: notifications, laboratory reports, and outbreaks – where do the statistics come from and what do they mean?, 6, Review No. 7, 21 June.

Craven, B. M. and Stewart, G. T, (1997). 'Public Policy and Public Health: coping with potential medical disaster' in: *What Risk?* Ed. R Bate, Butterworth-Heinemann Oxford.

Craven, B. M., Stewart, G. T. and Taghavi M, (1994), Amateurs Confronting Politicians: A case Study of AIDS in England, *Journal of Public Policy*, 13,4:305 – 325.

Daily Telegraph (1998). 'Stand Up to the Use-by Bullies' 7 March.

Downs, A. (1957). *An Economic Theory of Democracy*, Harper and Row.

Downs, A. (1967). *Inside Bureaucracy*, Little, Brown and Co. Boston.

Fife-Schaw, C. and Rowe, G. (1996). Public Perceptions of Everyday Food Hazards: A psychometric study, *Risk Analysis*, 16, 487-500.

Financial Times (1994), 'Through Gas and High Water', 30 May.

Frewer, L. J., Howard, C., Hedderley, D., et al. (1996). What Determines Trust in Information about Food-related Risks? Underlying psychological constructs, *Risk Analysis*, 16, 473-486.

Frewer, L. J., Howard, C., Hedderley, D., et al. (1997). The Elaboration Likelihood Model and Communication about Food Risks, *Risk Analysis*, 17, 759-770.

Galbraith, J. K. (1973). *Economics and the Public Purpose* Houghton Mifflin.

Groth, E. (1991). Communicating With Consumers about Food Safety and Risk Issues, *Food Technology*, May, 248-253.

Halsbury's Statutory Instruments, (1996). Vol. 8 1996 Re-issue, London, Butterworths.

Harveys, I., (1991). Infectious Disease Notification. A neglected legal requirement, *Health Trends*, 23, 73 – 74.

Hill, A. F., Desbruslais, M., Joiner, S., et al. (1997). The Same Prion Strain Causes nvCJD and BSE, *Nature*, 389, 448-450.

Hobbs, B. C. and Roberts, D., (1990). *Poisoning and Food Hygiene*, Fifth Edition, London.

Kay, J. (1988). The Economics of Regulation *in* Seldon, A. (ed.), *Financial Regulation or Over-Regulation?,* Institute of Economic Affairs, pp 17 – 31.

Kone, D., Mullet, E. (1994). Societal Risk Perception and Media Coverage, *Risk Analysis*, 14, 21-24.

Lave, L. (1987). Health and Safety Risk Analyses: Information for better decisions, *Science*, 236, 291-295.

Lee, K. (1989). Food neophobia: major causes and treatments, *Food Technology*, December,. 62-73.

Maitland, A., (1998). 'One in a Billion Risk' from Beef on the Bone, *Financial Times*, 10 February 1998.

Morabia, A., Porta, M. (1998). Ethics of Ignorance: lessons from the epidemiological assessment of the bovine spongiform encephalopathy ("mad cow disease") epidemic, *Perspectives in Biology and Medicine*, Winter 1998, 259-266.

NBPA (National Back Pain Association), Statistical sources: Department of Social Security, Department of Health, Health and Safety Executive, Clinical Standards Advisory Group, OPCS, RCN.

Niskanen, W. (1971). *Bureaucracy and Representative Government*, Aldine, Chicago.

Niskanen, W. (1973). *Bureaucracy: Servant or Master?*, Hobart Paperback 5, Institute of Economic Affairs, London.

North, R. (1992). *Death by Regulation: the butchery of the British meat industry*, Institute of Economic Affairs Health and Welfare Unit, Health Series, No. 12.

North, R. and Gorman, T. (1990). *Chickengate: An Independent Investigation of the Salmonellæ in Eggs Scare'* Institute of Economic Affairs Health and Welfare Unit paper No. 10. London.

Posner, R. A. (1974). Theories of Economic Regulation, *Bell Journal of Economics and Management Science*, 5, 335 – 358.

Roberts, J. A. and Sockett, P. N. (1994). 'The Socio-economic Impact of Human *Salmonellæ enteritidis* Infection, *International Journal of Food Microbiology*, 21, 117 – 129.

Sandman, P. M. (1994). Mass Media and Environmental Risk: Seven principles, RISK: *Health, Safety, and Environment*, Summer, 151-260.

Sandman, P. M. (1987). Risk Communication: facing public outrage, *EPA Journal*, November, 21-22.

Schefferle, H. (1990). 'Legislation' *in* Hobbs, B. C. and Roberts, D., *Poisoning and Food Hygiene*, Fifth Edition, London.

Scott, C. (1990). Continuity and Change in British Food Law, *Modern Law Review*, 53,785.

Seddon, Q., (1990). *Spoiled for Choice: Food Scares Unscrambled*, Evergreen Publishing, Essex.

Skegg, D. C. (1997). Epidemic or False Alarm?, *Nature*, 385, 200.

Slovic, P., Fischhoff, B., Lichtenstein, S. (1981). Perceived Risk: psychological factors and social implications, *Procedures of the Royal Soc*. London., 376, 17-34.

Snow, E. (1997). Risk Communication: notes from a class by Dr. Peter Sandman. Found at web site *http://www.owt.com/ users/snowtao/risk.html.*

Social Trends (1997). Central Statistical Office, *HMSO*, London.

Sparks, P., and Shepherd, R. (1994). Public Perceptions of the Potential Hazards Associated with Food Production and Food Consumption: an empirical study, *Risk Analysis*, 14, 799-806.

Stigler, G. J. (1971). The Theory of Economic Regulation, *Bell Journal of Economics and Management Science*, 2:3 – 21.

Thompson. K. (1996). *The Law of Food and Drink*, Shaw and Sons, Crayford, Kent.

Tower, G. (1993.) A Public Accountability Model of Accounting Regulation, *British Accounting Review*, 25,1: 61 – 85.

Veljanovski, C. (1991). *Regulators and the Market*, Institute of Economic Affairs.

Voss, S. (1992). How Much do Doctors Know About the Notification of Infectious Diseases? *British Medical Journal*, 304,726 – 727.

Wildavsky, A. (1973). Does Planning Work? *The Public Interest*, No. 33.

Wildavsky, A. (1975). *Budgeting: A Comparative Theory of Budgetary Processes*, Little, Brown and Co. Boston.

Williamson, O. E. (1975). *Markets and Hierarchies: Analysis and Anti-Trust Implications*, Free Press, New York

10 Five famine fallacies

Peter Bowbrick

Summary

There is a set of views about the causes and cures of famine, which, although radical, is very popular with governments and policymakers. However, these views are tragically misguided, poorly-reasoned and unsupported by evidence. When practice is based on these views, food crises become famines, and famines are worsened. The five most common myths and fallacies are examined in this chapter; these are:

1. It is possible for a proportion of the population to eat so much more food that the great majority of the population is hit by a famine.
2. A substantial heterogeneous population group will significantly increase its food consumption in time of famine, despite famine prices reducing purchasing power for food for all groups; despite rationing, and despite of the absence of food to buy.
3. We can ignore transport constraints as a factor affecting food distribution.
4. Speculation by rational speculators can create a famine.
5. If the Government overestimates the degree of 'shortage', the famine will not be tackled effectively, and many more people will die.

It is shown that these errors permeate Amartya Sen's Nobel Prize winning work on famine, and that the facts he quotes to support them are contradicted by his sources.

Introduction

When famine threatens politicians, administrators, the general public and even economists cling to fallacies and myths on the cause and cure of famine. These fallacies and myths are no less powerful because they have no base in fact or economic theory. If decision makers let these fallacies and myths influence their actions, they will exacerbate the crisis and millions more may die.

To illustrate the tragic process of misinterpretation and mismanagement, we turn to the works on famine by Amartya Sen, who was recently awarded a Nobel Prize in Economics. Acting on Sen's theory of the causes and cures of famine will cause, not cure, famines.

Theoretical conflict: the cause

The mainstream view is the one which seems most like common sense to the man in the street, and it is the maverick's view that is celebrated by the academicians; but in this case, the man in the street is right.

Sen believes that most famines are caused by a sudden redistribution of available food, rather than by a sudden decline in food. He argues that a famine may occur when there is actually plenty of food, but where one part of the population suddenly starts eating a lot more than usual. This reduces the supply available to the rest of the population, and pushes up food prices, so the poor starve. If this is the cause of famine, it is formally possible to control the famine by reversing the change in distribution. This can be done by seizing traders' stocks, imposing rationing, paying food as wages, free food and soup kitchens. Crucially, this means that most famines can be dealt with as an internal problem; that they can be cured without emergency imports of food. In some cases, imports may be necessary – to persuade speculators and hoarders to release their stocks for instance – but the quantity imported would be well below the amount that traditional economists would import to avert a famine. Sen's explanation means that early warning systems designed to identify crop failures are largely redundant, as crop failures do not cause famines. Indeed, he is extremely critical of the Bengal government for trying to determine the amount of the food deficit in the middle of a famine (Sen 1977 p53). It must be said that Sen is anything but consistent, and it is possible to find sentences scattered through his many works which contradict the main points he makes and keeps emphasising.

Sen creates a straw man to oppose his view, the Food Availability Decline (FAD) doctrine, which says that all famines are caused by a fall in food availability so food imports are essential to cure famines; changes in purchasing power are irrelevant and no public distribution system is necessary. This view has never existed.

Mainstream economists have always believed that nearly all famines are caused by a sudden decline in food availability, due to flood, drought or the export of grain, for instance, but that sometimes an invasion or influx of refugees increases demand, with a similar effect. In these cases it is usually necessary to bring more food into the country, but this is not sufficient.

Contrary to what Sen states, the mainstream has always recognised the key importance of the ability to purchase food – indeed his examples of starvation caused by inability to purchase are drawn from mainstream analyses. Detailed analysis of the impact of famine on occupational groups in Bengal may be seen in Hunter (1873), Sir Bartle Frere (1874) and Mahalanobis, Mukkerjee, and Ghosh, (1946). Both Hunter and Frere refer to evidence in previous reports and Blue Books. (See also Adam Smith 1776, ii p26). When crops fail, many people have no money to buy food, especially subsistence farmers, landless labourers and people in related occupations. Others have money, but not enough to buy food at famine prices. Accordingly, rationing, free food handouts and food for work have long been standard parts of the famine control package, and set out most clearly in policy documents such as the *Bengal Famine Code* (1897). In a famine, a few people such as traders may be better off, but generally everyone, particularly the poor, eats less, in contrast to Sen's view that a large proportion of the population eats much more.

To the mainstream economist, the high price of food is the result of the famine, not the cause. The high price determines that the poor die rather than the rich. High prices may mean that more people die, by worsening distribution of available food. Paradoxically, high prices could also mean that fewer die, for if the total quantity of available food is so low that to share it out equally would mean that nobody had enough to survive, concentrating supplies would keep some alive.

The possibility that a famine could be caused by a switch in internal demand has always been recognised, but as a theoretical curiosity. The one thing that the proponents of the two views agree on is that using the wrong theoretical approach (the other view) causes or aggravates famine.

Conflicting theories: the cure

If Sen is right, then famines can be averted by preventing sudden redistribution of food supplies, or rapidly cured by reversing any sudden redistribution. This means that crises can be handled with little or no additional food imports, which is of the greatest importance as it can take a long time to arrange imports of food aid, and transport constraints are serious. Sometimes food imports may be required to persuade speculators and hoarders to release supplies, but the quantities are far less than the mainstream economists would demand, because their views on the total food available vary greatly. It is implied that mainstream economists make poor use of what food is available and, by waiting until imports arrive before they take action, they are ineffectual.

Sen presents detailed analyses, notably of the Bengal Famine of 1943, to show that the mainstream approach, by misdiagnosing the cause of the famine, resulted in the Government's doing nothing to ameliorate the famine, and possibly worsening it. Implicit in the view that nearly all famines are caused by redistribution, is the view that early warning systems are redundant and that organisations like Oxfam are misconceived.

The mainstream view is that it is extremely unlikely that a famine will be caused by a sudden redistribution of food. However, it is almost certain that some politicians will agree with Sen, and talk of hoarding and speculation, particularly by minority ethnic groups. To say that a famine is caused by a sudden redistribution of food is to deny any fall in food availability; to claim that there is more than enough to keep the population alive until the next harvest, so long as the change in distribution is corrected.

If there is a minor drop in food availability it may be theoretically possible to ensure the survival of the whole population by seizing stocks and imposing strict rationing combined with free food. This is only likely to be successful on a large scale where much of the population is very well fed, there is an efficient and honest bureaucracy and there is the political will. It worked in wartime Britain, and it worked for isolated pockets of famine in Bengal in 1928, 1936 and 1941. Most Third World countries do not have sufficient overfed people to make it a realistic option for a widespread shortage. It may be used in the short run to keep people alive until imports arrive.

If there is a Third Degree Shortage, there is insufficient food to keep everyone alive until the next harvest. If there is no intervention many of the poor will die. If there is rationing and free distribution, the food will run out, and everyone will die. Imports are necessary. In a national famine, several districts will have Third Degree Shortages.

Mainstream economists are well aware of the weaknesses of food supply data, particularly in the Third World, and are aware that one cannot be confident of the degree of shortage. They are aware of the horrendous consequences of dealing with a food crisis as though it was a Sen-type 'no shortage' situation. They are aware that an inadequate response to a Third Degree Shortage could cause a population extinction. Rather than put millions of lives at risk by acting on a belief in a theoretical curiosity, they demand imports for all famine situations.

Sen says 'no matter how a famine is caused, methods of breaking it call for a large supply of food in the public distribution system. This applies not only to organising rationing and control, but also to undertaking work programmes and other methods of increasing purchasing power'. This is also the mainstream view. The difference is in the quantity of food. If a famine is caused by a shift in distribution, as Sen argues, this food can be obtained by buying grain on the open market or seizing private or public stocks. Mainstream economists point out that if a government starts buying grain during a shortage, the fall in supply will cause panic buying and enormous price rises, with the result that little grain is purchased (Malthus 1800, Bengal Famine Code 1897). This policy will have made things worse by its attempt to organise distribution and will still require that food is imported, but in much greater quantities than those envisaged by Sen.

Conflicting interpretation: the evidence

Sen's theory has been extremely influential because he presents a large quantity of evidence to show that some famines, particularly the Bengal Famine of 1943, were caused by such a redistribution, and to show that the Government failed to take any effective action because they were obsessed by the mainstream view of the cause of famines. He is scathing about those people who argue that a reduced food supply is the usual cause, and particularly about those who believe that it was a major cause of the Bengal famine of 1943 (Sen 1976, 1977, 1979, 1980a, 1981a,b and 1984), which he calls 'possibly the biggest famine in the last hundred years' (Sen 1977 p33). However, Sen's statement is contradicted by his sources. Even if we accept his estimate of the death toll, 3 million dead, Masefield (1963 pp12-14) who he cites on the history, mentions half a dozen famines where the death toll was higher – India 1876/7, 5 million; China 1876/7, 9-13 million; Russia 1920-1, 'millions'; China Hunan 1929, 2 million; Russia 1932/3 3-10 million. There had also been the Sahel famine and the Great Leap Forward in the years before Sen wrote. Accordingly, this chapter examines the evidence he presents on the Bengal 1943 famine. As it is the focus of Sen's attention, it is the one for which there is most evidence and for which the evidence is most convincing; it is also one of the very few famines for which the necessary preconditions for his theory to work are present – imports and exports were limited by the war, there were no food security stocks, and there was a wartime boom. Fundamental criticisms of some of his other examples are to be found in Basu (1984), Kumar (1990, p184). Dyson concludes 'while they are far from being complete explanations, FADS were probably involved in all five of Sen's famines' (Dyson 1996). This chapter concentrates on a few of Sen's key errors and misstatements of fact, although these and many others are documented in great detail elsewhere (Bowbrick 1986a, 1986b, 1999).

Sen's primary source is the report of the Famine Inquiry Commission (1945a,b) which was damning about the role of the Bengal and Indian Governments to the extent that it is widely believed that they tried to suppress it by publishing only a few copies. (See Wavell in Moon, 1973 pp 36-7; Aykroyd, 1974; Bhatia, 1967). It confirmed the views of contemporary critics such as Dutt (1944), T.C.Ghosh (1944), K.C.Ghosh (1944) and Rajan (1944). The English editor of the Calcutta *Statesman*, who had done most to bring the famine to the notice of officialdom in Calcutta, Delhi, London and Washington (defying censorship and the wrath of the local officials to do so) commented 'The Famine Commission's report is as complete, painstaking and balanced an account of what happened and why, as will ever be achievable.'

(Stevens, 1966). Interestingly, when Pergamon proposed to re-publish the reports, Sen dissuaded them. As a result, few scholars are in a position to check his facts.

In December 1940 Bengal had a much reduced winter rice crop, so there were localised shortages throughout 1941. In December 1941 the rice crop was well above average. War with Japan was declared in December 1941. From March 1942 shipments of rice from Burma were cut off – these had provided the rice supply for Calcutta, as well as rice for many other urban areas in India and Ceylon. On 16th October 1942 a cyclone, accompanied by torrential rains, hit West Bengal, causing wind and rain damage and flooding. Three tidal waves laid waste a strip of land seven miles wide along the coast and three miles wide along river banks. The resulting high tides increased the flooding caused by the rain. Some 4000 square miles were affected. 14,500 people and 190,000 cattle were killed; crops and grain stores were damaged. Fungus disease and root-rot then hit the sodden crops, causing even more severe damage than the flooding itself.

When this poor winter crop was harvested, prices rose rapidly, doubling within a month in country areas. By March 1943 there was hunger throughout Bengal, and from July to November, the famine was in full flood. Between 2 and 4 million people died from starvation and disease. Sen's own estimate of 3 million dead, has been strongly criticised by Dyson and Maharatna (1991) and Dyson (1991). The first is highly critical of Sen's data and the use he made of it, and the second shows that other aspects of Sen's demographic analysis of 1943-44 are also wrong.

The Famine Commission provided a comprehensive analysis of a complex situation, but said that the basic cause of the famine was a sudden decline in food availability (FAD), because of a 30 per cent fall in the rice crop, aggravated by the loss of the Burma rice imports and by the unusually small stocks at the beginning of the year.

Sen says that there was at least 11 per cent more food available in Bengal in 1943 than there had been in 1941 when there was no famine, so the famine could not have been caused by a decline in food availability (For example, Sen, 1977b, pp 42, 53;1980 p 80). Instead, it occurred because of a change in the distribution of existing food supplies arising from wartime conditions, particularly inflation and booming war industries. This meant that some groups of the population received higher incomes and ate more, leaving little for the rest of the population. Others, with less money, could not compete for the scarce food, and starved.

Fallacy 1. It is possible for a proportion of the population to eat so much more food that the great majority of the population is hit by a famine.

How many people?

At one level Sen appears to be arguing that the increased consumption was by 100,000 new war workers and that they and their families ate so much more than in previous years that there was not enough food for the rest of the population. These people made up between ¼ and ½ per cent of Bengal's 61 million population. He states that their extra consumption caused a famine in which 40 million people went hungry, 10–15 million of them suffering very seriously indeed and 3 million died. A few minutes' work with a calculator shows that to achieve this, each of the workers' families would have had to eat 60 to 120 times as much food as in a normal year. 'It would probably be an underestimate to say that two thirds of the total population were affected by the famine' (Department of Anthropology, Calcutta University, quoted by Rajan 1944). An independent estimate was made by Mahalanobis, Mukkerjee and Ghosh (1946, pp 337–400), based on a sample survey of the survivors. They estimated that of the 10.2 million families in the rural population, 1.6 million sold some or all their land or mortgaged it, 1.1 million sold plough cattle, and in 0.7 million the head of the household changed to a lower-status occupation (including 0.26 million becoming destitute). These figures are not mutually exclusive: many families suffered loss of land and cattle, and many became destitute because they had sold all they had. Taking an average family size of 5.4, it seems that perhaps 10–15 million people were affected in these ways. However, many more were affected in ways that would not have been recorded in these statistics. Most went hungry; many were hit by disease; many were impoverished but kept the same occupation; many sold all they had except their land.

> Village labourers and artisans, at a somewhat higher economic level, sold their domestic utensils, ornaments, parts of their dwellings such as doors, windows and corrugated iron sheets, trade implements, clothes and domestic animals if they had any – sold indeed anything on which money could be raised – to more fortunate neighbours (Famine Commission p 67).

Elsewhere, Sen presents different figures, suggesting that the people in Calcutta – 10 per cent of Bengal's population – ate so much more food in 1943 than in 1942 or 1944 that they caused this major famine. Again, a few minutes' work with a calculator will show that they would have had to eat more than three times the normal amount of food to create this shortage. This is an *average* covering war workers, as well as government and local government employees, artisans, traders, the unemployed and the destitute. Sen accepts that many traders and artisans were particularly likely to be unable to buy food, as people who could not afford food certainly could not buy their wares. This means an *average* consumption of 6kg a day of cooked rice as well as the accompanying curry, so some people would have been eating twice this amount. To put this into perspective, at the time Sen was writing, the United States consumed only 81 per cent more calories per head than Bangladesh, but half the amount of cereals – it is not possible to eat this many calories in bulk foodstuffs like rice. (FAO World Food Survey 1977).

Sen produces a third set of figures:

> In a poor community take the poorest section, say, the bottom 20 per cent of the population and double the income of half that group, keeping the money income of the rest unchanged. In the short run prices of food will now rise sharply, since the lucky half of the poorest group will now fill their part-filled bellies. While this might affect the food consumption of other groups as well, the group that will be pushed towards starvation will be the remaining half of the poorest community which will face higher prices with unchanged money income. Something of this nature happened in the economy of Bengal in 1943.

This has been a hugely influential statement. Again, a little work with a calculator shows that even if it was true, the lucky half would have eaten very little more, only 1.8 per cent of total. (See Famine Commission 1945 p204, for the consumption figures in five surveys in the years preceding the famine). This would have had no noticeable effect, if, as Sen says, there was plenty of food available.

But it cannot possibly be true. It implies that 10 per cent of the 61 million population doubled their income. A maximum of one million of Calcutta's population was in the very poor category according to consumption surveys, so five million of them must have moved in from the countryside, to make Calcutta the biggest city in the world for this year only. The people must have moved to Calcutta in the months preceding the famine, and then have suddenly disappeared the moment the next main crop was harvested. This did not happen – wartime rationing means that we have accurate figures on Calcutta's population. It appears to have increased in size by 300,000 to 500,000 from 1939. Sen's figures are more than twelve times the true figures.

Nor was there any way in which the extra 1.8 per cent could be taken away from the poorest part of the population. The effect would have been spread over the market as a whole. In Bengal we know that 66 per cent of the population went hungry, not just the poorest 10 per cent, as Sen states.

Sen has misstated the number of the very poor who increased their income by a factor of more than 12 and has misstated the number of people who went hungry by a factor of 6.6, giving a total misstatement by a factor of 80.

Elsewhere I have shown that these results are of general application (Bowbrick, 1999b). It is extremely difficult to imagine a situation where boom famines can exist except in some science fiction scenario.

Fallacy 2. A substantial heterogeneous population group will significantly increase its food consumption in time of famine, despite famine prices reducing purchasing power for food for all groups; despite rationing, and despite of the absence of food to buy.

Did the people of Calcutta eat more?

Fundamental to Sen's theory is his assertion that, as a result of wartime inflation and a boom, the people of Calcutta ate at least three times more in 1943 than in the previous or following years. The facts are different, as might be expected during a famine: they ate less than usual. Calcutta's net imports of rice in the first quarter of 1943 were 43 per cent of normal consumption, in the second quarter 94 per cent, in the third quarter 104 per cent and in the fourth quarter 124 per cent, with an annual average of 91 per cent (See Famine Commission pp219-33. The figures are analysed in Table 1 of Bowbrick 1999). Sen's own sources give the evidence in great detail. Because of wartime controls on transport within Bengal and between Bengal and the rest of India, and because of wartime rationing, there are excellent figures on the quantities consumed.

Sen's sources also give details of the measures taken by the authorities to prevent wage increases from driving up prices. For example, war workers were given their food in kind, rather than being given wage increases. Rationing and special distribution systems also limited the amount available on the open market.

Sen argued that the inflation and wartime boom should have led to a massive increase in food consumption by war workers in Calcutta, but clearly it did not. Nor did it lead to a famine in any of the other war years. Nor did it lead to famines in all the other 120 countries in the world affected by the inflation and booms of the Second World War – famines occurred only where there was a sudden decline in food availability. Nor has inflation or hyper-inflation usually caused famines before or since.

Since the theory conflicts with all observed facts, it should be dropped. It does food policy analysts no credit to use broad generalisations to analyse the impact of inflation on the many different communities within a population, least of all when many of them are not in the money economy and when farmers only earn money at harvest. Nor is it helpful to insist that x happened when the facts show that y happened, merely because a theory indicated that x should have happened.

Fallacy 3. We can ignore transport constraints, even though in reality these often prevent relief reaching the areas most affected, as Sen's theory requires that at least three times this amount of food leaves the area at the beginning, to create the famine.

Movements of rice from the countryside

Also fundamental to Sen's theory is his assertion that large quantities of rice moved from the countryside to Calcutta, leaving countryfolk without enough food to survive. More than two million tonnes would have had to be moved to cause the famine.

Again, this is totally incorrect, as is shown in Sen's own sources (See Famine Commission pp 219-33). As in previous years, Calcutta got all its grain from outside Bengal. There were some small movements from the countryside immediately after the harvest, never more than 15000 tons net per month, and the aggregate amount never amounted to more than 2½ days' supply for the countryside. By the time the famine had bitten, this amount and more (in rice, millet and wheat) had been shipped back to the countryside. If Sen had been right about the cause of the famine, this extra supply shipped to the countryside would have stopped the famine. In fact it had no impact.

Sen also gives as a causal hypothesis the removal of surplus stocks from country areas threatened with a Japanese invasion in 1942. In fact, this happened in a surplus year, nine months before the famine started to bite. The amount was 40,000 tons, one and a half days' supply for the country areas. It is absurd to suggest that this small movement, well before the famine, caused the famine.

In a completely inconsistent argument, Sen considers the boat denial policy to have been a cause of the famine, (Sen 1984 p 4611, 1980b p 619) however he (1976 p 1279) also expresses the opposite view. In May 1942, orders were issued for the removal of boats capable of carrying more than ten passengers from the coastal areas of Bengal, in order to deny them to the Japanese if they invaded. This slightly reduced the quantity of food (fish) available, and to this extent it was a cause of famine. However, since it also prevented the movement of grain from the starving countryside to Calcutta, it should, under Sen's argument, have prevented the famine.

Fallacy 4. Speculation by rational speculators can create a famine.

Did food availability decline?

Sen's theory would have had no credibility whatsoever without his statement that 1943 food availability was 'at least 11 per cent higher than in 1941 when there was nothing remotely like a famine.'

His statement is contradicted by the statistics and evidence presented in his source documents, both the main report of the Famine Inquiry Commission and the dissenting report of one of its members, Professor

Hussein, who argued that the Commission had understated the fall in availability. I have shown elsewhere in considerable detail that, even if one accepts his production figures, his analysis is not correct, and the figures show a decline in food availability of 10 per cent (Bowbrick 1986a,b, 1999a). An analysis was carried out by O. Goswami (1990) of the Indian Statistical Institute, who dissected Sen's analysis at length and in great detail using the same figures as Sen and came to an almost identical conclusion to my own, without, apparently, having seen mine.

Conservative estimates?

Sen is so emphatic that his estimates are both conservative and reliable that he may be understood to believe there was at least 20 per cent more food available in 1943 than 1941, not just 11 per cent. He talks of presenting 'the results of a food supply calculation, taking into account local production and trade, choosing – wherever the data permit – an assumption as unfavourable to 1943 as possible'. Elsewhere he says: 'This is most certainly an overestimate for 1941 vis-à-vis 1943, but this is an acceptable bias as it favours the thesis we are rejecting'. 'To bias the figures as much as possible against 1943...' (Sen 1977, p 40). He may also be interpreted as claiming a much greater accuracy for them than is justified, because he frequently quotes different secondary sources as giving much the same estimate of total production or import needs (see, for example, Sen 1977, pp 53-4). Since these secondary sources are all based on the same official production estimates, no added confidence is given. His scathing comments on those who consider that the famine was caused by shortages emphasise the impression that he is totally confident of his figures.

In fact, the figures he gives are not in any sense conservative. His 'conservative' adjustments consist of making a slight allowance for unrecorded wheat imports, an alteration of a fraction of 1 per cent of the total. Again, he makes much of choosing a 1 per cent population growth rate instead of 0.46 per cent, which makes a difference of 1 per cent when he uses it for comparing 1941 with 1943. These 'conservative' adjustments do not make any noticeable improvement to the accuracy of the aggregate figures he uses.

To state that such estimates are conservative is a misstatement of perhaps 30 per cent. This, in itself, is enough to cast doubt on all the evidence he presents in favour of his thesis.

Unreliable statistics

If the reliability of the statistics he uses is examined, his claim to produce conservative and reliable food supply figures is seen to be even more false. It can be shown that his figures are not right within +/-3000 per cent. They can give no support whatsoever to his argument.

His production figures came from the Famine Commission, which is at pains to show how unreliable they are. They are based on a crop forecast, with subjective estimates of the areas planted and the probable yield. The cyclone meant that area harvested and actual yield were very different.

A junior officer would guess at area and probable yield for a 400 square mile area 'from personal inspection and by questioning other local officers and cultivators'. His guess would be modified by the guesses of each level of senior officer above him as the figures were aggregated (Famine Commission, 1945b, pp 44-5). In 1942, the estimates would have been particularly bad because parts of Bengal, notably those hit by the cyclone, were on the verge of an insurrection, and the army was busy burning villages and so on. There were political pressures to modify the estimates.

As the cyclone, tidal waves, flooding and disease which caused so much damage had never occurred before, there was no experience to guide anyone. Previous famines had been due to drought in other areas of Bengal.

Desai (1953) compares the official estimates of agricultural surveys with the results of scientific surveys carried out by Mahalanobis. He shows that the discrepancies are large, with survey estimates being between 47 and 153 per cent of the official estimate. The discrepancies also vary from year to year, with the sample estimate of the jute crop being 2.6 per cent above the official estimate in 1941, and 52 per cent above it in 1946. (With jute, where exports provided a check, the sample proved correct.) Since there was no sample survey of the rice crop until after the famine, we do not know how inaccurate the 1942 forecast was. Another form of bias arises from subjective eye estimates of the prospective yields. Mahalanobis (1946a) and Allen (1959) found that, even with scientific sampling methods on a mature crop, enumerators tended to select the best fields and the best areas of damaged fields This gives a large upward bias when much of the crop is damaged, as in 1942. The bias would be much worse with untrained observers using subjective methods.

Even in well-organised studies using scientific sampling, enumeration was often so bad as to invalidate the study (Mahalanobis 1946a).

There is also a more serious form of bias – the scale of incompetence and corruption was so vast that virtually every administrator and politician had cause to want evidence suppressed or altered. There is some indication that pressure was brought on statisticians to do this (Elphinstone commenting on Mahalanobis, 1946, p 374; Aykroyd 1974; Bhatia 1967).

We must conclude from this that the statistics are so bad that one cannot confidently say that the true total production lay within 50 per cent of the official estimate in any one year. However, Sen did not rely on the estimates for total production, but on the difference between two estimates for total production. He relied on his assessment that production for 1943 was at least 11 per cent higher than that for 1941. Since both figures could be 50 per cent out, and the bias is not random, it is quite possible that the 1941 crop could be three times that of the 1943 crop (again if one relies entirely on crop figures). This means that the margin of error of his statement is of the order of 3000 per cent. Sen's production figures are so unreliable that they can give no support to his thesis, a thesis which relies entirely on these figures. His claim to have made conservative, reliable estimates is false.

Non-statistical evidence

Sen makes a great deal of the fact that 'the rice crop in Bengal was recognised to be indifferent rather than exceptionally bad' (Quoting Document No 265, p 357 in Mansergh, 1971). This quotation is incorrect. In fact, the document he quotes stated that there was *both* cyclone damage in certain areas *and* an indifferent crop in Bengal generally. The combined effect was seen as being exceptionally serious. This is a clear misrepresentation of the source document.

Speculation

Sen also offers speculation as a causal hypothesis (Sen 1977 pp 50-51; 1981a pp 75-78), but he fails to understand the distinction between speculation within a year, speculation between years, and speculation between countries.

There was speculation and prices rose, but as Adam Smith points out

> If, by not raising the price high enough, the speculator discourages the consumption so little that the supply of the season is likely to fall short of the consumption of the season, he not only loses a part of the profit which he might otherwise have made, but he exposes the people to suffer before the end of the season, instead of the hardships of a dearth, the dreadful horrors of a famine.

Much of the blame for price rises falls on the Government policy of trying to buy large quantities of grain on the open market 'without limit of price' when there was very little available.

For speculation to cause a famine, it must prevent food from being eaten, not just raise the price. The speculators must destroy some of their stocks, export them, or store them into subsequent periods. There is no suggestion in any of Sen's sources that this happened, much less to the extent needed to cause a famine, nor does Sen himself claim it. On the contrary, there is evidence that speculators imported grain into Bengal, and that even house-to-house searches, for example, in June 1943, failed to find substantial stocks being carried over. Transport was very scarce in wartime and so tightly controlled that it could not handle even the very small quantity of relief grain available in the first half of 1943. Supplies were again tight in 1944 and 1945, despite a good crop in December 1943, which indicates that stocks from 1943 were not carried over to these years. There is no suggestion that any speculator was irrational enough to sell rice at low 1944 prices rather than the famine prices of 1944.

Clearly, speculation was not a cause of the famine, but the suspicion of speculation did have a major effect on this famine. The Bengal Government's policy was to identify speculative stocks with a view to seizing them, and to release rice on the market to 'break the Calcutta market' so that speculators would release stocks. This was ineffectual, and it diverted the Government's attention from the only action which would have cured the famine – imports.

I have shown elsewhere (Bowbrick, 1999c) that it is possible to model speculation causing a famine only in situations akin to the garrison of a beleaguered fort, and even then only if the speculators act irrationally. 'The popular fear of engrossing and forestalling may be compared to the popular terrors and suspicions of witchcraft' (Adam Smith ii p24).

Hoarding

Sen states that there were panic purchases between December 1942 and March 1943, but also up to November 1943. He also talks of panic hoarding from March to November 1943 as being a cause of the famine.

Private hoarding could only have been a cause of inflation if it removed food from general supply during the famine year. If the amount of private stocks remained the same as in previous years, they would have had no effect and if they declined during the famine, they would have mitigated the disaster. It is probable that there was substantial hoarding by private households when war broke out, and stocks may have increased when Japan joined the war and again when Burmese supplies were cut off. It is difficult to argue that there could have been further increases in private stocks during the famine year, given the high price of grain – four to twenty times the normal price, and given that those who could afford to hoard already had large stocks.

How many people could afford to hoard? In Bengal, only 20 per cent of the population were well fed even in a normal year: most of the salary earners lived from hand to mouth, and most cultivators borrowed against next year's crop to buy food. A very few might have bought three or four months' supply, but this would have run out in the middle of the famine.

Sen fails to understand the distinction between holding stocks and increasing stocks, and he presents no evidence that stocks were increased.

Fallacy 5. If the Government overestimates the degree of shortage, the famine will not be tackled effectively, and many more people will die.

What the Bengal Government thought and did

Sen's whole argument tends towards his conclusion that 'the failure to anticipate the Bengal famine ... and indeed the inability even to recognise it when it came, can be traced largely to the government's overriding concern with aggregate food availability statistics' (Sen 1984 p 477). He suggests that if the problem had been analysed using his approach, the famine would not have occurred.

This is untrue in all respects. In fact, the Bengal Government used the same assessment of food availability as Sen. They worked to the same theory of famine causation and cure advocated by Sen, and they acted according to his theory. Between two and four million people died as a result.

Sen's sources are agreed that the Bengal Government made much the same assessment as Sen of food availability until the point in July or August, when the famine was reaching its peak, that they started to realise that there really was a major shortage. Even then, they believed that there was at worst a First-Degree Shortage. Surprisingly, in support of his claim that the Bengal Government was obsessed by the FAD approach, Sen gives two pages of evidence showing just the opposite: that the Bengal Government was firmly convinced that there was adequate food available, and that the hunger was due to changes in distribution (1977 pp 53-4, 1981 pp 80-82).

The Government also concurred with Sen in believing that lack of purchasing power, rather than lack of food, caused starvation. They believed that price control was necessary under wartime inflation to prevent certain groups getting more than their fair share. They believed in public relief schemes; that a large supply of food had to be distributed through the public distribution system. They believed that some degree of rationing was desirable and that speculation and hoarding were major causes of the famine. They attempted therefore to provide the population, and particularly in Calcutta, with the purchasing power necessary to obtain the food. They instituted public relief measures and intervened on the market. Had they been right in their assessment of food supplies and on the cause of the famine, they might have been successful. However, because they held the same views as Sen, and acted accordingly, millions of people died.

If they had held the FAD view, as Sen states, their logic would have been as follows: 'There is widespread hunger and even starvation. Under the FAD approach, the only possible reason is a shortage of food. Therefore we must import one and a half million tons immediately.' Whether their analysis was right or wrong, their response would have saved millions of lives.

Sen's explanations are not new. All were discussed in the source documents, notably the Famine Enquiry Commission (1945a & b). They were part of the popular explanation at the time. Furthermore, much of the policy of the Bengal Government at the time was based on them, with disastrous consequences. They were popular during the 1888 Orissa famine and as far back as Adam Smith at least. The inflation explanation for famines was very popular after the Central European hyperinflations of the 1930s, where they did appear to cause deaths. It was this very popularity that encouraged the Bengal Government to adopt it. The explanation appeared in economics textbooks in the 1950s. It even appeared in chess books of the time: a challenger to Lasker for the world championship later died of hunger in Austria during hyperinflation.

Conclusion

This chapter has concentrated on a few of the crucial weaknesses in Sen's thesis, ones related to these five fallacies alone. Sen's thesis is refuted in its entirety by the evidence that the people of Calcutta ate less than usual, not 3 to 100 times as much as usual, as Sen argues. It is refuted in its entirety by the evidence that grain moved from Calcutta to the countryside rather than vice versa. His hoarding and speculation hypotheses are refuted by the facts. The thesis is refuted in its entirety by the evidence that the Bengal Government did have the same diagnosis of the cause of the famine as he did and acted accordingly, with the result that between two and four million people died. A little calculation shows that it is not possible to create a boom famine in the way he describes.

Evidence has also been produced that throughout his work on the Bengal famine, Sen misquoted his sources, misstated the facts in his sources and made factual statements that were contradicted by the facts in his sources. In the fuller analyses of his work (Bowbrick 1986a, 1986b, 1999a) a great deal more evidence of this is produced. It is shown that he treats any evidence that he thinks supports him as perfectly correct and anything that does not as self-evidently wrong – though more generally he ignores the evidence that contradicts his thesis.

Sen has been aware of these errors of fact since 1985 at least, but neither he nor anybody else has even attempted to refute them. It is a matter of the gravest concern that his books and papers depending on these incorrect facts have not been withdrawn. It is not just that it affects the integrity of our science. It affects the lives of millions of people.

These fallacies are memes, virus-like ideas that are accepted by individuals and spread through the population. Like urban myths, which are also powerful memes, they are believed and passed on because people want to believe them, not because there is any factual or theoretical reason to believe them. In any famine they are likely to switch a major part of the response away from effective courses of action, and so to cause death on a large scale.

Economists can avoid such errors in two ways. The first is to recognise that your first model is inevitably wrong or inadequate, and to keep checking its implications against all the facts. If there is a clash, it is the model that must be changed, not the facts rejected. The second is to recognise that many of the facts and statistics available are incorrect, so you must constantly and continuously check them against each other, looking for contradictions, both direct and indirect. The contradictions must then be resolved. Most economists in the field consider this to be fundamental to their work, but it is not often taught as an academic skill.

References

Aykroyd, W.R. (1974). *The conquest of famine*, London, Chatto and Windus.

Allen, George, (1986). 'Famines: the Bowbrick-Sen dispute and some related issues,' *Food Policy*, 11(3) 259-263.

Bengal Administration, (1897). *Bengal Famine Code*, (Revised edition of December 1895) Calcutta.

Basu, D. (1984) 'Food Policy and the analysis of famine' *Indian Journal of Economics* 64 254: 289-301.

Basu, D. (1986).'Sen's analysis of famine: a critique' *The Journal of Development Studies* 22:3 April.

Bhatia, B.M. (1967) *Famines in India*, Asia Publishing House, Bombay.

Blyn, G. (1966). *Agricultural trends in India, 1891-1947: output, availability and production*, Philadelphia, University of Pennsylvania Press.

Bowbrick, P. (1987). 'Rejoinder: an untenable hypothesis on the causes of famine', *Food Policy*. 12(1) 5-9, February.

Bowbrick, P. (1986a.) 'A refutation of Sen's theory of famine', *Food Policy*. 11(2) 105-124.

Bowbrick, P. (1986b.) *A refutation of Professor Sen's theory of famines*. Institute of Agricultural Economics, Oxford.

Bowbrick, P. (1985). 'How Professor Sen's theory can cause famines', *Agricultural Economics Society Conference*. March.

Bowbrick, P.,(1999a) *How Sen's Theory can Cause Famines*, Quality Economics, http://www.prima.net/bowbrick/Famine.htm

Bowbrick, P. (1999b) 'Are Boom Famines Possible?'

Bowbrick, P. (1999c) 'Can Speculation Cause Famines?'

Desai, R.C. (1953). *Standard of living in India and Pakistan, 1931-2 to 1940-41*, Popular Book Depot, Bombay.

Dutt, T.K. (1944). *Hungry Bengal*, Indian Printing Works, Lahore.

Dyson, T. and Maharatna, A. (1991). 'Excess mortality during the Great Bengal Famine: A Re-evaluation' in *The Indian Economic and Social History Review*, Vol. 28, No. 3.

Dyson, T. (1991). 'On the Demography of South Asian Famines, Part II' in *Population Studies*, Vol. 45, No. 2, July.

Dyson, T. (1996). *Population and food: global trends and future prospects*, Routledge,

London and New York.
FAO (1977). *World Food Survey 1977* Rome.
Famine Inquiry Commission (1945a). *Report on Bengal*, New Delhi, Government of India.
Famine Inquiry Commission (1945b). *Final Report*, Madras, Government of India.
Frere, Sir Bartle *On the Impending Bengal Famine* 1874.
Ghosh, K.C. (1944). *Famines in Bengal*, 1170-1943 Calcutta: Indian Associated Publishing
Ghosh, T.K.O. (1944). *The Bengal Tragedy*, Hero Publications, Lahore.
Goswami, O. (1990).'The Bengal Famine of 1943: Re-examining the Data' in *The Indian Economic and Social History Review*, Vol. 27, No. 4.
Government of India (1942). *Report on the marketing of rice in India and Burma*, Government of India Press, Calcutta.
Goswami, O. (1990). 'The Bengal Famine of 1943: Re-examining the Data' in, The Indian Economic and Social History Review, Vol. 27, No. 4.
Hunter, W.W. (1873). *Famine Aspects of Bengal*, Simla.
Kumar, B.G. (1990). 'Ethiopian Famines 1973-1985; a case study' in J Dreze and A. Sen (eds.) *The Political Economy of Hunger 2* Oxford, Clarendon Press.
Mahalanobis, P.C. (1946) 'Recent experiments in statistical sampling in the Indian Statistical Institute *Philosophical Transactions of the Royal Society*, Part iv, pp326-378.
Mahalanobis, P.C., Mukkerjee, R.K., and Ghosh, A. (1946). 'A sample survey of after effects of Bengal famine of 1943.' *Sankhya* 7(4),337-400.
Masefield, G.B., (1963). *Famine: its prevention and relief*, Oxford, OUP.
Mansergh, N. (ed.) (1971). *The transfer of power 1942-7* vol. III, London, HMSO.
Mansergh, N. (ed.) (1973). *The transfer of power 1942-7* vol. IV, London, HMSO.
Moon, P. (ed.) (1973). *Wavell: the Viceroy's journal*, OUP, Oxford.
Palekar, S.A. (1962). *Real wages in India 1939-1950* International Book House, Bombay.
Rajan, N.S.R. (1944). *Famine in retrospect*, Pamda Publications, Bombay.
Sen, A. (1976). 'Famines as failures of exchange entitlements', *Economic and Political Weekly*, Special Number, August
Sen, A. (1977a). 'On weights and measures: informational constraints in social welfare analysis' *Econometrica* XLV 1539-73,
Sen, A. (1977b). 'Starvation and exchange entitlements: a general approach and its application to the Great Bengal Famine', *Cambridge Journal of Economics*, I 33-59.
Sen, A. (1979). 'Issues in the measurement of poverty', *Scandinavian Journal of Economics*, LXXXI 285-307.
Sen, A. (1980a). 'Famine mortality: a study of the Bengal Famine of 1943' in Hobsbawm et al. *Peasants in history: essays in memory of Daniel Thorner*, Calcutta. Oxford University Press.
Sen, A. (1980b). 'Famines', *World Development* 8 (9) 613-21. Sept.
Sen, A. (1981a). *Poverty and Famines* Oxford, Clarendon Press.
Sen, A. (1981b). 'Ingredients of famine analysis: availability and entitlements' *Quarterly Journal of Economics*, August. 433-464.
Sen, A. (1984). *Resources, values and development*, Blackwell,
Sen, A. (1987). 'Reply: famine and Mr Bowbrick', *Food Policy* 12(1) 10-14.
Sen, A. (1986) 'The causes of famine: a reply', *Food Policy* 11(2) 125-132.
Singh, A. (1965). *Sectional price movements in India*, Benares, Benares Hindu University.
Smith, Adam, (1977). *The Wealth of Nations*, Everyman.
Stevens, I. (1966). *Monsoon morning*, London, Ernest Benn.

11 European regulation of genetically modified organisms

Rod Hunter

Introduction and summary

Julian Morris

Rod Hunter's paper on the Regulation of Biotechnology in Europe is an excellent summary of the relevant legislation and he raises a number of very important issues. However, his analysis may be a little opaque for the non-lawyer, so this introduction is an attempt to summarise the main points in layman's terms and to adumbrate what I believe are some conclusions that might be drawn from it. There is an inevitable risk when making such a summary, that important technical niceties, caveats and exceptions will be omitted, so the reader who is interested in understanding the precise legal position is urged to read Rod's chapter in full. Please note that these are my views and not those of Rod Hunter or of Hunton and Williams, nor are they the views of the Institute of Economic Affairs (which has no corporate view), its Directors, Advisors or Trustees.

Summary

In Europe, the development, cultivation, use and distribution of genetically modified organisms (GMOs) are subject to very stringent regulations. The main features of these regulations are now described and their implications discussed.

Development

The first stage in the development of a new GMO, after a gene coding for a particular desired trait has been discovered in a donor organism, is the insertion of that gene into the recipient organism. Once this has been done, the resultant GMO must be assessed to establish whether or not it has the desired trait, whether the gene has been inserted on the desired chromosome, and other considerations. Laboratory experiments involving such resultant GMOs are covered by the Contained Uses Directive (90/219), Article 6(2) of which requires the users of GMOs to 'carry out a prior assessment of the contained uses as regards the risks to human health and the environment that they may incur.' In effect, this means:

1. Assessing the nature of pathogenicity and virulence, infectivity, toxicity and vectors of disease transmission of the donor and recipient organisms; establishing any significant involvement of those organisms in environmental processes such as nitrogen fixation or pH regulation; evaluating any likely interaction of those organisms with, and effects on, other organisms in the environment, including likely competitive or symbiotic properties; and assessing the organism's ability to form survival structures such as spores or sclerotia.
2. Assessing the resultant GMO for its stability in terms of genetic traits; the capability of the inserted vector to transfer it's genes; and the rate and level of expression of genetic material.
3. Assessing the resultant GMO for potential health impacts, which might include, inter alia, toxic or allergenic properties of non-viable organisms and/or their metabolic products.
4. Assessing of the potential impact of the resultant GMO on the environment, which requires consideration of factors affecting the survival, multiplication and dissemination of the GMO in the environment; available techniques for detecting the transfer of the new genetic material to other organisms; known and predicted habitats of the GMO; description of the ecosystems to which the GMO could be accidentally disseminated; anticipated mechanism and result of interaction between the GMO and the organisms which might be exposed in case of release into the environment; known or predicted effects of the GMO on plants and animals such as pathogenicity, infectivity, toxicity, virulence, vector of pathogen, allergenicity, colonisation; known or predicted involvement in biogeochemical processes; and availability of methods for decontamination of the area in case of release to the environment.

In addition, the user of the GMO must report to the authority any new information they acquire regarding the risk posed by the GMO, no matter how dubious the quality and no matter what the source. The authority is then empowered to revisit the conditions of use of the GMO, including the possibilities of suspension or termination. This effectively compels the users of GMOs to report any and all ludicrous and irrefutable claims made by environmentalists, to the authorities, who may then suspend the use of the GMO indefinitely – on the grounds that the claims could not be conclusively refuted.

Finally, the user of the GMO must, where the national authority deems it necessary, have in place emergency procedures for dealing with an accidental release, and in the event of an accident must notify the public authority.

These requirements are in many cases extremely difficult to meet and they slow down the development of new GM crops considerably, with the result that biotech companies have incentives to shift their research and development programmes to countries where the regulations are less oppressive. Whilst it is of course desirable to ensure that dangerous substances are handled with due care it is not clear that most GMOs are in fact dangerous. To read the list of EU regulations, one would think that GMOs are ultra-hazardous air-borne viruses, capable of devastating large swathes of the world's population. But in most cases they present no more hazard than organisms produced by other means, such as cross-fertilisation, which are not subject to these regulations. It is worth remembering as well that the companies producing these GMOs are concerned both for the health of their workers (who would no doubt sue them if they received substantial injuries from their contact with GMOs) and also to keep the nature of the GMOs secret. The reason for secrecy is not that GMO producers are engaged in a conspiracy against the public; quite the opposite: secrecy is important in order to prevent rival companies from obtaining commercially sensitive information. Secrecy enables the company to reap the benefits of their investment and thereby provides incentives to produce products that are beneficial to society and often also to the environment (see Wilson, this volume). But the notification requirements undermine the ability of companies to maintain high levels of secrecy, creating yet further incentives to develop GMOs outside the EU. Nevertheless, it is likely that the incentives to keep the nature of the GMOs secret will ensure that at the development stage companies have stringent procedures for preventing the release of GMOs into the environment.

Cultivation and use

The cultivation and use of GMOs is covered by the Deliberate Release Directive (90/220), which essentially requires that all GMOs not covered by the Contained Uses Directive except those used in food products (which are covered by separate legislation – see below) be subject to notification, vetting, public consultation and reporting procedures.

For GMOs not yet destined for commercial production, these procedures are onerous. For example, notification of the proposed release entails an assessment of the possible inter-actions between the GMO and the local ecosystem. What is a possible interaction is of course open to debate: environmentalists are fond of inventing all manner of possible interactions, regardless of the likelihood of their occurrence. The Directive lacks any criteria for distinguishing between more and less significant risks and so fails to provide appropriate guidelines for evaluating the impact of releasing a GMO. This lack of clarity inevitably lengthens the already bureaucratic procedure for vetting the use of the GMO, which is carried out by the national authority. Where the national authority does consult 'the public' (i.e. interest groups) and the cultivation of a GMO is nevertheless permitted, these interest groups are able to claim that their concerns have not been fully taken into consideration. However, the Directive also makes provisions for 'reporting', whereby any new information concerning the risks posed by the GMO must be submitted to the authorities, regardless of the source and regardless of the reliability of the information. This in principle enables interest groups to submit all manner of implausible new claims, slowing down the regulatory process and enabling them to assert, once the authorities decide that their claims are irrelevant or groundless, that the democratic decision making process is flawed because their claims have not resulted in the GMO being banned.

Crops that are to be grown commercially are subject to all these regulatory hurdles and more. First, they must go through field tests in order to establish the likely impact on the ecosystems in which they are to be grown. This is obviously a more scientific method for establishing the risks presented by growing GM crops, since it results in data on the actual interactions between the GMOs and the surrounding ecosystem. However, 'unknown risks' remain and these cannot be evaluated. This problem applies to all technologies – and indeed to nature itself – but it is important in the context of GMOs because of the extent to which environmental and consumer organisations are able to intervene in the regulatory process governing the use of GMOs by positing during the Public Consultation phase the existence of these unknown, unquantifiable, and often undiscoverable 'risks'.

Second, although the national authority is the first port of call for vetting of the commercial production of GM crops, all applications must be passed on to the national authorities of all member states. If any state objects (based on experience, a highly likely occurrence), the proposal for commercial production is subject to vetting by the European Community, with decisions made by a qualified majority vote in the Council of Ministers.

Third, the Directive contains a 'safeguard clause' which means that even once all these conditions are met, a member state may 'provisionally' restrict the sale of a product containing GMOs if it has 'justifiable reasons' to consider that it 'constitutes a risk to human health or the environment'. Under this clause, the member state notifies the Commission of the use and reason for the application of the safeguard measure. The Commission then circulates its proposal to the Technical Adaptation Committee (TAC), which must act by qualified majority. In the likely event that the TAC does not reach agreement, the Commission sends its proposal to the Council of Ministers, which has 3 months to act by qualified majority. Again, this is very unlikely to happen. The directive states that under such circumstances the Commission then 'shall' adopt its proposal, which it cravenly and unlawfully has failed to do in the Novartis maize case. Under the proposed amendments, which lower the voting rule to a simple majority, this process will be made even easier for member states to gum up.

Are we not men? No we be GMOs

Article 2(2) of the Deliberate Release Directive defines a GMO as 'any organism in which the genetic material has been altered in a way which does not occur naturally or by mating and/or natural recombination.' The term 'organism' is defined in Article 2(1) as 'any biological entity capable of replication or of transferring genetic material.' In addition to plants and animals, this definition covers micro-organisms including parasites, bacteria and viruses, as well as human beings. Strictly speaking, a patient who has undergone gene therapy could constitute a GMO. We await the day when an environmental organisation (sponsored perhaps by a manufacturer of homeopathic medicine) brings an action against a member state for failing to implement the Directive because a gene therapy patient has been allowed home without undergoing the risk assessment and notification procedures specified in the Directive.

Transport

Transport of GMOs is covered by the Transport of Dangerous Goods by Road Directive (94/55) which classifies GMOs as 'infectious substances' and subjects them to placarding, packaging, labelling and other requirements. Again, these onerous restrictions do not seem to be justified by the scientific evidence relating to the impact of GM foods on health and the environment. Indeed, they are likely to be very misleading and will tend to perpetuate an unjustifiably negative public perception of GM foods.

Food production

Food products containing, consisting of, or produced from GMOs are regulated by the Novel Foods Regulation (258/97), Article 3 of which states that such food products must not: (1) 'present a danger to the consumer'; (2) 'mislead the consumer'; or (3) 'differ from food or food ingredients which they are intended to replace to such an extent that their normal consumption would be nutritionally disadvantageous for the consumer'. In order to obtain an authorisation, the company seeking to sell such a food product must submit to the national authority in the member state where the product is first due to go on sale (the rapporteur state) a copy of studies that confirm the product meets these conditions. The company must also confirm that the GMOs in the product comply with the requirements of the Deliberate Release Directive.

After initial consideration by the authority in the rapporteur state, and within three months, the application is passed on to the Commission, which passes it on to the authorities in other member states, which have 60 days to comment or provide 'reasoned objections'. If there are no objections or requests for additional information during the initial assessment, the rapporteur state must inform the applicant that it may place the product on the market. If there is an objection or a request for additional information then the application is subject to a similar procedure to that for deliberate releases. The matter is referred to the Standing Committee for Foodstuffs. The Commission proposes a draft measure, which must be supported by a qualified majority of member states in order to be adopted. Given the weighting of votes and sensitivity of biotechnology issues, deadlock is the usual result. In that case, the proposal is forwarded to the Council of Ministers, which has three months to act on the basis of qualified majority voting. If the Council fails to act in time, the Commission's proposal becomes law.

In addition to these requirements, the applicant must label all foods containing or consisting of GMOs and also those foods derived from GMOs where those foods are no longer equivalent to conventional foods. This label must inform consumers of the 'presence' of GMOs. It is quite possible that many consumers, already primed to fear GMOs by the wild and scientifically dubious accusations of environmental and consumer organisations that have been regurgitated ad infinitum by a scare-mongering media, might well respond to these labels as though they were an indication that the product contains bubonic plague.

Safeguard clause

A novel food or food ingredient that has been properly authorised or notified under the directive may be used anywhere in the Community, subject to conditions specified in the consent. A member state may, however, 'temporarily restrict or suspend the trade in and use of' a novel food or food ingredient if, on the basis of either 'new information' or 'a reassessment of existing information,' the state has 'detailed grounds' for considering that the use of the food or ingredient 'endangers human health or the environment.' It is worth noting that this provision does not authorise member states to restrict products merely because of unpopularity with certain political groups, but rather states must be able to provide detailed reasons concerning how the product 'endangers human health or the environment.' If a member state chooses to avail itself of this safeguard provision, it must immediately inform the Commission and other member states of its decision and reasons. The Commission, working with the Standing Committee for Foodstuffs, is to examine the member state's actions and grounds and take 'appropriate measures.' In the meantime, the member state may maintain its restrictions.

Conclusions

As should be manifest from this summary of Rod Hunter's review of regulation of GMOs, the Community has an elaborate regulatory framework. The regulation imposes extensive environmental risk assessment requirements as a condition for handling GMOs or releasing them into the environment, and requires health-focused risk assessments where workers will be exposed to GMOs, and where GMOs or GMO-derived products will be consumed by humans. Moreover, the prior authorisation procedure for marketing GMO products, and in the food context GMO containing and derived-from products, allows all member states ample opportunity to raise reasoned objections. Further, where a member state can demonstrate that a previously authorised GMO containing or derived-from product presents a risk to human health or the environment, it may legally restrict trade in that product and trigger reconsideration of the authorisation by the Community.

Yet, this comprehensive regulation has not alleviated popular and political concerns. Quite the contrary. And the Commission and member states bear much of the responsibility for this. Given the lack of public understanding of biotechnology, initial popular disquiet over these novel products might be expected. However, European officials have done little to educate the public, and the way Community decision procedures are operated is not calculated to provide confidence. Deliberations of the Commission and national officials in technical committees over whether to authorise the marketing of GMO and GMO-derived products are confidential. Officials typically provide only perfunctory explanations of the grounds for their decisions, and there is no opportunity for interest groups to obtain judicial review of the decisions reached behind closed doors. By dressing the issues up as merely technical and declining to explain their reasoning in detail, officials should not be surprised by public suspicion of the process and anxiety over the products.

Further, some national governments have been lawless in their manipulation of the regulation's procedures. It seems clear, for instance, that Austria's resort to the Deliberate Release Directive's safeguard provision to ban Novartis' properly authorised maize is groundless and unlawful. And the Commission's failure to require Austria to remove its ban also appears illegal. Putting aside the international trade law issue this raises, the very fact that the Community tolerates such unlawfulness calls into question the legitimacy of the Community regulation and in turn invites further abuse.

Finally, the Community's overbroad labelling rules and policies seem sure to cause confusion. Perhaps limited mandatory labelling obligations could be justified, though one might have thought that, if consumers really cared, shrewd businessmen seeking profit would segregate and label voluntarily (the Community and member states already have legislation prohibiting misleading advertising). However, the Community's failure to define workable compliance thresholds and testing methodologies for the labelling rules seems destined to result in a wide variety in enforcement practices across member states (this is already happening), and legal uncertainty for producers, distributors and retailers. Prudent businessmen may well decide to attach to virtually everything that could conceivably contain GMOs, or in the case of foods, be derived from GMOs, a label stating that 'this product may contain or be derived from GMOs.'

Is there a viable alternative to public regulation?

Given that GM crops have not been shown actually to present any serious threat to health or the environment (see, for example, Wilson, this volume), these onerous restrictions do not seem to be justified. Granted, it is desirable to assess the likely risks from introducing new GM crops and to have in place procedures that will prevent unacceptable harm to health and the environment, but it is not necessary to push each new GMO through the absurd regulatory rigmarole described above. A possibly more effective and certainly more efficient mechanism for ensuring that GM crops are produced safely would be to do away with the current regulatory restrictions altogether and simply make the producing companies liable for proven harms created by the resulting organisms. This would stimulate the companies that develop, cultivate and market GMOs to ensure that their risk assessment procedures are appropriate to the particular kinds of harm that might potentially arise. It would also do away with the costly bureaucratic regulatory apparatus by effectively privatising enforcement. (However, standing to take legal action against the producer of a GMO should be restricted to those who are directly affected, either as a result of impacts on their person or property, or as a result of their dependency on an affected person. Environmental and consumer organisations would nevertheless be free to indemnify plaintiffs against the cost of taking an action. In addition, costs should be awarded against the losing party. These requirements would avoid excessive litigation by parties attempting merely to obtain publicity or to extort money from companies that might settle rather than incur large court fees.)

European regulation of genetically modified organisms – Rod Hunter

The European Union regulates environmental, health and safety concerns arising from genetically modified organisms (GMOs) through four 'horizontal' directives (Hunter and Muylle 1999).

- Contained Uses Directive (90/219)
- Deliberate Releases Directive (90/220)
- Biological Agents at Work Directive (90/679) (Bergkamp and Lucas 1990).
- Transport of Dangerous Goods by Road Directive (94/55), under which GMOs are classified as infectious substances and thus subject to the directive's placarding, packaging, labelling and other requirements (O.J. L 319/7 1994).

GMOs are also subject to 'sectoral' legislation, such as the Novel Foods Regulation (258/97) (Buono 1997). This regulation creates a comprehensive, if cumbersome, regulatory framework for assessing and managing risks arising from biotechnology. In addition, European national liability rules, including laws promulgated pursuant to the Council of Europe Convention on Liability for Damages from Dangerous Activities, create a liability and compensation framework for any damages arising from biotechnology. This European regulation of GMOs surely has its flaws – ambiguous drafting of key regulatory terms, heavy administrative burdens, obscure decision-making procedures and opportunities for abuse by national officials. But it would be implausible to suggest that GMOs are under-regulated in Europe.

This Chapter seeks to explain the web of Community regulation of GMOs (Bergkamp 1998). It begins with a discussion of environmental regulation of GMOs – specifically, the regulation of laboratory tests (contained uses), field experiments (deliberate releases) and marketing (deliberate releases) of GMOs. Then follows a brief review of the regulation of biological agents such as GMOs in the work environment; an examination of consumer safety regulation, in particular regulation of so-called 'novel foods' containing or derived from GMOs; and concludes with some observations on the effectiveness of this framework for risk regulation.

Environmental regulation of GMOs

EU environmental regulation of GMOs is set out in the Contained Use Directive and the Deliberate Release Directive. Roughly speaking, the Contained Use Directive is meant to control environmental risks arising from laboratory experiments involving GMOs. The Deliberate Release Directive is intended to control environmental risks arising from field experiments, and from placing on the market of products containing GMOs. Neither of these environmental directives apply to products derived from, but not containing, GMOs (this limitation is logical given that by definition such derived-from products are not capable of transferring genes, and hence do not pose risks unique to GMOs).

Contained use

The Contained Use Directive, which was substantially modified in late 1998, regulates contained uses of genetically modified micro-organisms (GMMs) and sets forth rules for prior risk assessment and conditions of use. Given that the new contained use requirements do not need to be implemented by member states until mid 2000, we discuss the existing requirements before turning to the recent modifications.

Existing rules

Scope

The Contained Use Directive creates notification and authorisation procedures for GMM-installations and GMMs (a number of member states apply these contained use rules to GMOs, as well as GMMs). The directive's scope is not limited to a specific activity – contained use includes 'any operation in which micro-organisms are genetically modified or in which such genetically modified micro-organisms are cultured, stored, used, transported, destroyed or disposed of and for which physical barriers, or a combination of physical barriers together with chemical and/or biological barriers, are used to limit their contact with the general population and the environment'.

Article 2 of the directive defines a modified genetically micro-organism as 'any micro-organism in which the genetic material has been altered in a way which does not occur naturally by mating and/or natural recombination'. A micro-organism is defined as 'any micro-biological entity capable of replication or of transferring genetic material'. Genetic modification, under the directive, is the result of the use of the following techniques:

Recombinant DNA techniques using vector systems;

Techniques involving the direct introduction into a micro-organism of heritable material prepared outside the micro-organism (including micro-injection, macro-injection and micro-encapsulation); and

Cell fusion or hybridisation techniques where live cells with new combinations of heritable genetic material are formed through the fusion of two or more cells (Annex I A, Pt 1).

The directive excludes from its scope organisms obtained through use of certain techniques. First, the following techniques are not considered to result in genetic modification, on condition that they do not involve the use of recombinant-DNA molecules or GMOs:

in vitro fertilisation;

conjugation, transduction, transformation or any other natural process;

polyploidy induction (Annex 1A, Pt 2).

Secondly, the following techniques are deemed to result in genetic modification, but are nevertheless excluded from the directive, on the condition they do not involve use of GMMs as recipient or parental organisms:

Mutagenesis;

Construction and use of somatic animal hybridoma cells;

Cell fusion (including protoplast fusion) of cells from plants which can be produced by traditional breeding methods; and

Self-cloning of non-pathogenic naturally occurring micro-organisms which fulfil the criteria of Group I for recipient micro-organisms (Annex 1B).

The directive creates a couple of limited exclusions where the GMMs are regulated adequately by other legislation. Its notification requirements do not apply to transport of GMMs by road, rail, inland, waterway, sea or air, covered by article 5 of the Contained Use Directive. Such transport activities are governed by other rules, such as The Transport of Dangerous Goods by Road Directive on Transport of Dangerous Goods by Road, which classifies GMOs as 'infectious substances' and subjects them to placarding, packaging, labelling and other requirements. Also, none of the directive's provisions apply to 'the storage, transport, destruction or disposal or use' of GMMs that have been placed on the market 'under Community legislation, which includes a specific risk assessment similar to that provided in this Directive'. However, the directive fails to define 'similar' so leaving ambiguity about which risk assessments under which legislation are adequate.

Risk assessment, classification and safety measures

Article 6(2) requires GMM users to 'carry out a prior assessment of the contained uses as regards the risks to human health and the environment that they may incur.' In conducting such an assessment, the user is to 'take due account' (Art. 6(3) and Annex III) of characteristics of the donor, recipient or parental organisms, characteristics of the modified micro-organism, health considerations and environmental considerations. The user must keep a record of the risk assessment and make it available to the competent authority upon request (Art. 6(4)).

In effect, these requirements are extremely onerous. Accounting for characteristics of organisms means assessing the nature of pathogenicity and virulence, infectivity, toxicity and vectors of disease transmission; significant involvement in environmental processes such as nitrogen fixation or pH regulation; the interaction with, and effects on, other organisms in the environment including likely competitive or symbiotic properties; and ability to form survival structures such as spores or sclerotia.

The modified micro-organism must be assessed for its stability in terms of genetic traits, the frequency of mobilisation of inserted vector and/or genetic transfer capability, and the rate and level of expression of the new genetic material and the method and sensitivity of measurement.

Health considerations include toxic or allergenic effects of non-viable organisms and/or their metabolic products, product hazards and, if the micro-organism is pathogenic to humans who are immunocompetent, also diseases caused and mechanism of pathogenicity including invasiveness and virulence.

Perhaps most time-consuming are the environmental considerations, which must assess factors affecting survival, multiplication and dissemination of the GMM in the environment, available techniques for detection, identification and monitoring of the modified micro-organism, available techniques for detecting transfer of the new genetic material to other organisms, known and predicted habitats of the GMM, description of eco-systems to which the micro-organism could be accidentally disseminated, anticipated mechanism and result of interaction between the GMM and the organisms or micro-organisms which might be exposed in case of release into the environment; known or predicted effects on plants and animals such as pathogenicity, infectivity, toxicity, virulence, vector of pathogen, allergenicity, colonisation, known or predicted involvement in biogeochemical processes, and availability of methods for decontamination of the area in case of release to the environment.

The directive creates a classification scheme for GMMs and activities in order to adapt regulatory procedures and controls to risks. The directive classifies GMMs into Group I (deemed to pose low risks), and Group II (deemed to pose higher risks). To be classified as Group I, for example, requires that the recipient organism contains no adventitious agents, that the vector is free from known harmful sequences, and that the GMM is non-pathogenic (Art. 4(1)). Activities involving GMMs are divided into Type A (small-scale operations used for teaching, research, development, or non-industrial/commercial purposes (Art. 2(d)), and Type B activities (other operations) (Art. 4 and Annex II of Directive 90/219 and see appendix for Commission decision 91/448).

In handling GMMs, users are to follow principles of good microbiological practice and principles of good occupational safety and hygiene (Art. 7 and Annex IV). For Group II GMMs, specific containment measures must be taken to ensure a higher level of safety (Annex IV).

Notifications and vetting

The directive's notification requirements may be triggered by the first use of installations and the first use of GMMs.

Installations. A prior notification to the national authority is required when an installation is to be used for the first time for the contained use of GMMs (Art. 8(1)). This notification must include, in addition to a summary of the risk assessment, (Art. 6(4)) information regarding personnel, their qualifications and training, a description of the installation and the likely scale of the operation (Art. 8 and Annex V). Depending on the classification of the GMMs, either a 'negative vetting' or a 'positive vetting' procedure applies to first-time use of installations. As to operations involving the first time use of a GMM, if (1) the GMMs are classified to Group I and are to be used in Type B operations, or (2) the GMMs belong to Group II and are to be used in Type A installations, the use may proceed after 60 days from submission of the notification, unless the national authority objects (Arts 11(5), 9(2) and 10(1). The contained use of Group II GMMs in Type B operations, on the other hand, may not proceed until formally approved by the national authority. For these perceived higher risk activities, the authority must communicate its decision in writing, at the latest 90 days from the date of notification (Art. 11(4)).

GMMs. Depending on the type of operation and classification of GMMs, the data to be included in notifications is more or less extensive (Arts. 9, 10 and Annex V).

Group I GMM/Type A Installation. This use need not be notified to authorities. Users must simply keep records of work carried out and make the records available to the authority on request (Art. 9(1)).

Group I GMM/Type B Installation. Users must submit a notification containing the following information:

Date of the installation's first use notification;

Parental micro-organisms used and the host-vector systems used;

Source and intended function of the genetic material;

GMM's identity and characteristics;

Purpose of the contained use and expected results;

Culture volumes; and

Risk assessment summary (Annex V, Pt B).

Once the notification has been submitted, the contained use may proceed after 60 days (or earlier with the authority's agreement), in the absence of any indication to the contrary (Art 16(5)(a).

Group II GMM/Type A Installation. Users must submit a notification including the information required for users of Group I, Type B (above), plus:

A description of the installation and methods for handling GMMs;

A description of predominant meteorological conditions and potential danger sources;

A description of protective and supervisory measures to be applied during contained use;

Containment category allocated, as well as waste treatment and safety measures (Annex V, Pt C.).

The contained use may proceed 60 days after notification (or earlier with the authority's agreement), in the absence of any indication to the contrary (Art 11(5)(a)).

Group II GMM/Type B Installation. Users must submit a notification containing:

Information on the GMM;

Information on personnel and training;

Information on the installation;

Information on waste management;

Information on accident prevention and emergency response plans; and

The risk assessment (Art 10(2)).

The directive, Annex V, Pt D, lays down detailed rules on the contents of this information. When the provision of some of the information is not possible for technical reasons, or if it does not appear necessary, the user must state the reasons. The level of detail required by the authorities may vary depending on the nature and the scale of the proposed contained use. The contained use may not proceed without the consent of the authority. That authority must communicate its decision in writing within 90 days of the notification (Art. 11(5)(a)).

The authority may ask the user to provide further information or to modify the conditions of the proposed contained use (Art 11(3)(a)). If the authority requests further information, the proposed contained use may not proceed until the authority has given its approval. The authority may also limit the time for which the contained use is permitted or impose other conditions (Art. 11(3)(b)).

Reporting and modifications

The directive imposes a broad reporting obligation on users. The user must inform the authority as soon as possible and modify the notification if the 'user becomes aware of relevant new information or modifies the contained use in a way which could have significant consequences for the risks posed by the contained use' (Art. 12(1)). A strict reading of the provision might require reporting of *any* information the user learns of, no matter how dubious the quality and no matter what the source. Moreover, the directive provides no guidance on the meaning of significant, thus adding further ambiguity to the scope of the reporting obligation.

The directive also gives authorities broad powers to revisit conditions of use. 'If information subsequently becomes available to the competent authority which could have significant consequences for the risks posed by the contained use,' the authority may require the user to modify the conditions of, or suspend or terminate the contained use (Art. 12(2)).

Emergency planning and accidents

Authorities must ensure that before the start of an operation concerning the contained use of GMMs some special precautions are taken. In particular, where so determined by the national authority, an emergency plan must be drawn up for protection of human health and the environment outside the installation in the event of an accident, and the emergency services must be informed of potential hazards (Art. 14). In the event of an accident, the user must immediately notify the authority and provide the following information: (1) circumstances of the accident; (2) identity and quantities of GMMs released; (3) information necessary to assess the accident's effects on human health and the environment; and (4) emergency measures taken. While article 2(f) defines an accident as 'any incident involving a significant and unintended release of genetically modified micro-organisms in the course of their contained use which could present an immediate or delayed hazard to human health or the environment', no guidance is given on the meaning of 'significant' nor of 'hazard,' thus leaving the definition and the scope of reporting obligation vague. The directive provides for co-ordination between member states and with the Commission regarding the security of the installations for contained use of GMMs. Member states must make available to other member states concerned information on emergency plans for contained use. (Art. 14(b)). In the event of an accident, member states that could be affected must be immediately alerted (Art. 15(2)). Member states must also inform the Commission, as soon as possible, of any accident (Art. 16(1)(b)).

Public consultation

The directive states that, 'Where a Member State considers it appropriate, it may provide that groups or the public shall be consulted on aspects of the proposed contained use'(Art. 13). Thus, the directive itself does not require public consultation procedures for contained uses, but member states are free to impose such requirements.

Revised rules

The revised Contained Use Directive, which is to be implemented by member states by mid-2000, will substantially restructure notification and containment requirements, in order to calibrate better regulatory controls to potential risks. Much of the rest of the contained use legislation, described above, remains essentially the same.

Risk assessment and classification

The Revised Contained Use Directive (Art. 5(1)). requires member states to ensure that 'all appropriate measures are taken to avoid adverse effects on human health and the environment which might arise from the contained use of GMMs'. To this end, the revised directive requires that the user, or 'person responsible for the contained use of GMMs' (Art. 2(e)). carry out 'an assessment of the contained uses as regards the risks to human health and the environment that these contained uses may incur, using a minimum the elements of assessment and the procedure set out in Annex III, sections A and B'. (Art. 5(5) states that the assessment 'shall especially take into account the question of disposal of waste and effluents. Where appropriate, the necessary safety measures shall be implemented in order to protect human health and the environment'. This assessment is to result in the 'final classification' of contained uses into four classes applying the procedure set out in Annex III, which in turn results in assignment of containment levels:

Class 1 activities of no or negligible risk –activities for which Level 1 containment is appropriate.

Class 2 activities of low risk – activities for which Level 2 containment is appropriate.

Class 3 activities of moderate risk – activities for which Level 3 containment is appropriate.

Class 4 activities of high risk – activities for which Level 4 containment is appropriate (Art. 5(3)).

The revised directive states that where there is 'doubt as to which class is appropriate for the proposed contained use, the more stringent protective measures shall be applied unless sufficient evidence, in agreement with the competent authority, justifies the application of less stringent measures' (Art. 5(4)). The user is to keep a 'record of the assessment' and make it available in an 'appropriate form' to the authority as part of a notification or on request (Art. 5(6)). The legislation does not indicate, however, how long the user must keep such records, thus implying an open-ended record-keeping obligation.

The revised directive then defines in detail the containment and protective measures required for each class of activity, specifically, that measures are to be reviewed periodically and immediately if 'there is reason to suspect that the assessment is no longer appropriate judged in the light of new scientific or technical knowledge' (Art. 6(2)).

Notification and vetting

These activity classifications are used to determine notification and approval procedures, with the more potentially risky activities being subjected to more extensive administrative supervision.

Premises first-use. Before premises are used for the first time for contained uses, the user must submit a notification containing:

Name of users, including those responsible for supervision and safety;

Information on training and qualifications of persons responsible for supervision and safety;

Details of any biological committees;

Description of premises;

Description of work that will be done;

Contained uses classes; and

Only for Class 1 contained uses, a risk assessment summary and waste management information (Art. 7, Annex V, Pt A).

Class 1. Following such first-use notifications, subsequent Class 1 contained uses may proceed without further notification. However, the user is required to keep a record of the risk assessment (Art. 8).

Class 2. For first and subsequent Class 2 contained uses in previously notified premises, a notification must be submitted containing:

Date of the premises notification;

Names of persons responsible for supervision and safety and training and information on their qualifications;

Recipient, donor and/or parental micro-organisms and, where applicable, host-vector systems;

Sources and intended functions of genetic materials involved in the modifications;

Identity and characteristics of the GMM;

Purpose of the contained use, including expected results;

Approximate culture volumes;

Description of containment and protective measures, and waste management information, including wastes to be generated, their treatment, final form and destination;

Risk assessment summary; and

Information necessary for the authority to evaluate any emergency response plans (Art. 9(1) and Annex V, Pt B).

If the premises have been the subject of a previous notification to carry out Class 2 or a higher class of contained uses and any associated consent requirements have been satisfied, the Class 2 contained use may proceed immediately following the new notification. Article 9(2) allows the applicant to request a decision on a formal authorisation from the authority, provided the request is made within of 45 days from the notification. If the premises have not been subjected to Class 2 or higher notification, the Class 2 contained use may, in the absence of indication to the contrary from the authority, proceed 45 days after submission of the notification, or earlier with the authority's agreement (Art. 9(3)).

Classes 3 and 4. For first and subsequent Class 3 or 4 contained uses in previously notified premises, a notification must be submitted containing:

Date of premises notification,

Names of those responsible for supervision and safety, and training and qualifications information;

Recipient or parental micro-organisms,

Host-vector systems,

Sources and intended functions of genetic materials involved in modifications,

Identity and characteristics of the GMM,

Culture volumes;

Description of containment and protective measures, including waste management information,

Purpose of contained use and expected results,

Description of installation;

Information on accident prevention and emergency response plans,

Hazards arising from the installation's location,

Preventive measures and procedures and plans for verifying effectiveness of containment measures,

Description of information provided workers,

Information necessary for the authority to evaluate any emergency response plans;

The risk assessment (Art. 10(1) and Annex V, Pt C.)

A Class 3 or higher contained use may not proceed without the authority's prior written consent (Art. 10(2)). The authority is to communicate its decision to the notifier within 45 days of submission of the new notification in the case of premises that have been subject of a previous notification for a class three or higher contained use and where consent requirements have been satisfied for the same or higher class contained use. In other cases, the authority has 90 days from submission of the notification.

The authority may ask the user to provide further information or modify conditions or classes of proposed contained uses. The authority may require that the contained use not begin until the authority has given its approval on the basis of the further information obtained or of the modified contained use conditions. The authority may limit the time for which the contained use is permitted or impose other conditions (Art. 11(3)). For further guidance, article 11(4) states that, for the purpose of calculating the periods referred to in arts. 9 and 10, any period of time during which the competent authority is awaiting any further information which it may have requested from the notifier, or is carrying out a public inquiry or consultation in accordance with art. 13, shall not be taken into account.

Deliberate release

The Deliberate Release Directive regulates releases into the environment of GMOs for experimental and marketing purposes. The directive requires that, at the research and development stage, any product containing or consisting of GMOs be put through field tests in the ecosystems that could be affected by it. The directive establishes procedures and criteria for a case-by-case evaluation of potential risks arising from experimental releases or marketing of GMOs.

There is an important difference between, on the one hand, procedures for experimental releases of GMOs and contained uses of GMMs, and, on the other, the procedure governing marketing of GMOs. Under the Contained Use Directive and, with respect to research releases under the Deliberate Release Directive, consent from the national authority, if required, suffices. By contrast, in the case of marketing of GMOs, the Deliberate Release Directive (Arts. 13(3) and 21) effectively allows any member state to initiate a procedure involving all member states. Moreover, once a product consisting of GMOs is on the market, a member state may still temporarily restrict its sale under article 16(1), if the product is believed to pose a risk to man or the environment. These diverging procedures, of course, reflect the fact that contained uses will probably have effects only in the locality in which that use occurs, while marketing of GMOs could have effects in other jurisdictions as well.

Notifications and consents for deliberate releases have provoked considerable controversy over the past several years. The Commission has proposed a substantial amendment to the Deliberate Release Directive. The Commission's proposed amendment is presently before the European Parliament and Council, and may yet be significantly revised before adoption. Accordingly, this discussion focuses on the existing directive, and discusses the proposed amendment only in brief.

Scope

'GMO' is defined, much like GMM, as 'an organism in which the genetic material has been altered in a way that does not occur naturally by mating and/or natural recombination' (Art. 2(2)). The techniques constituting genetic modification are contained in Annex 1A, part 1, as listed above. The term 'organism' is defined in article 2(1) as 'any biological entity capable of replication or of transferring genetic material.' In addition to plants and animals, this definition covers micro-organisms including parasites, bacteria and viruses, as well as human beings. Strictly speaking, a patient who has undergone gene therapy could constitute a GMO. The legislative intent was presumably not to subject doctors that treat such patients to regulatory and liability rules – presumably doctors may release such patients from hospitals without having to go through a risk assessment and notification procedure. GMOs used in connection with gene therapy, however, would fall under the regulations (van der Meulen 1997). The definition excludes, however, biological entities not or no longer capable of replication or transferring genetic material. As a consequence, the Deliberate Release Directive's notification and vetting procedures do not apply to products derived from but not containing GMOs.

'Deliberate release' is defined as 'any intentional introduction into the environment of a GMO or a combination of GMOs without provisions for containment such as physical barriers together with chemical and/or biological barriers used to limit their contact with the general population and the environment' (Art 2(3)). Placing on the market means 'supplying or making available to third parties'.

Experimental releases

Any person who intends to make a deliberate release of GMOs into the environment must submit a notification under article 5(1) to the competent authority of the member state in which the release will first occur.

Notification

With regard to notifications concerning deliberate releases for 'research and development, or for any other purposes than for placing on the market', the notification must include a technical dossier and a risk assessment (Art. 5(1)-(2)). The scope of these provisions is not directly defined, but the directive states that 'placing on the market' means 'supplying or making available to third parties' (Art. 2(5)). Roughly speaking, any release not meant to result in transfers to third parties should fall within this rubric. However, a transfer to a corporate subsidiary could conceivably constitute the making available to a third party and hence a 'marketing' release. While this threshold language is inadequately defined in the GMO context, the same language is used in EC marking legislation, and in that context the Commission has prepared somewhat useful interpretative guidance. The risk assessment rules have been the subject of criticism. Specifically, the criticism is directed at the lack of substantive criteria, an adequate description of the risks to be considered, and their relative weight. Food and pharmaceutical legislation provides much fuller definitions of the objectives of the risk assessments to be performed for such products, it is said. In addition, pleas are made for a much more differentiated approach commensurate with the specific risks involved in the release at issue. Indeed, not all releases pose the same level of risk and, consequently, merit the same level of administrative oversight. The current directive, however, does not provide for any risk classification nor for any differentiation of administrative requirements according to degree of risks. The proposed amendments, discussed below, would address some of these issues. The technical dossier must include information necessary for 'evaluating the foreseeable risks,' and include in particular information on:

Personnel and training;

The GMOs;

Conditions of release and the receiving environment;

Inter-actions between the GMOs and the environment; and

Monitoring, control, waste treatment and emergency response plans (Art. 5(2)(a) and Annex II).

As for the risk assessment, it must be a 'statement evaluating the impacts and risks posed by the GMO(s) to human health or the environment from the uses envisaged' (Art. 4(2)(b)). The precise contents of the notifications are specified in considerable detail in the recently amended Annex II (Pt A of the annex applies to releases of GMOs other than higher plants; Pt B applies to releases of modified higher plants). Member states may ask the Commission, in certain circumstances listed in article 6(5), to apply simplified notification procedures for GMOs for which sufficient experience has already been obtained. This simplified procedure has been established only for certain modified plants, and used only by certain member states, allows a single notification dossier to be submitted for (1) more than one release of genetically modified plants that have resulted from the same recipient crop species, and (2) a research programme with a single specific recipient plant species over several years and on several different sites (See appendix for commission decisions 93/584 and 93/730).

Data on any prior releases by the notifier of the same GMO, occurring within or outside the Community, must also be submitted under article 5(4). It is interesting to note that the term 'notifier' is defined as 'the *person* making the presentation of the notification documents' (Art. 2(6) emphasis added). This language could be construed to mean that a corporate entity making a notification might not need to report on data resulting from releases by a corporate entity with the same or overlapping shareholders. Multiple releases or a release involving a combination of GMOs may, depending on national law, be covered by one notification (Art. 5(3)). As under Community chemicals legislation, the directive allows the notifier to refer also to data from prior notifications submitted by others, provided that the prior notifiers have consented in writing (Art. 5(4)).

Vetting

As for vetting of research and development and other non-commercial releases, it is handled by the national authority concerned. Other member states receive a summary of the filing and may comment, but have no decision-making role. The authority is to assess the risks posed by the release, and approve or reject the application within 90 days of receipt of the notification. The deliberate release may proceed only after the authority has given its written consent and in accordance with conditions prescribed in the approval (Art. 6(4)).

Public consultation

As with the Contained Use Directive, the Deliberate Release Directive states that member states, where they consider it 'appropriate,' may provide that 'groups or the public ... be consulted on any aspect of the proposed deliberate release' (Art. 7). Some member states (for example, Ireland and the UK) require experimental release notifiers to run advertisements in newspapers indicating the GMOs they intend to release, the purpose of the release and the location. Private parties are then allowed to make representations to authorities on the prudence of allowing the experimental releases.

Reporting

The directive creates an expansive obligation on notifiers to modify their notifications and the conditions of release in light of new information. More specifically, under article 5(6) of the directive, in the event of 'any modification of the deliberate release of GMOs ... which could have consequences with regard to the risks for human health or the environment,' or 'if new information has become available on such risks,' the notifier must 'immediately' (1) revise measures specified in the notification, (2) inform the competent authority in advance of any modification or as soon as the new information is available, and (3) take the measures 'necessary' to protect human health and the environment. A literal reading of this provision would require notifiers to inform officials of each and every evolution of scientific knowledge regarding the risks related to a GMO, no matter how insignificant, and even if the information implies lower risks than previously foreseen. As the obligation refers broadly to 'new information,' without qualitative or possessive qualifiers, arguably *anything* on the subject of the GMOs' risks published by any source, however dubious, would need to be notified. Further, this reporting obligation is temporally open-ended, it would arguably require reporting of new information even well after a release has been terminated.

In addition to this expansive obligation, the directive creates a specific post-release reporting obligation. Article 8 requires the notifier to report on the 'result of the release in respect of any risk to human health or the environment, with particular reference to any kind of product that the notifier intends to notify at a later stage'.

Marketing

Notification

The placing on the market of products containing or consisting of GMOs requires more extensive notification and prior consent procedures. Such consent, which may include conditions, may be granted only if:

> A written consent has previously been given with respect to research and development, or if a risk analysis has been carried out in accor-

dance with the directive;

The product complies with the EC product legislation; and

The product complies with the Deliberate Release Directive's environmental risk assessment rules (Art. 10(1)).

The notification for the marketing of a GMO must be submitted under article 11(1), to the competent authority of the member state where the GMO will first be placed on the market. The notification must include an extended risk assessment reflecting the 'diversity of sites of use of the product' including information on data and results from prior experimental releases – Annex II of the directive provides detailed guidance on the contents of this part of the notification. The notification must also set forth conditions for the placing on the market, including 'specific conditions of use and handling and a proposal for labelling and packaging ...' – Annex III provides detailed guidance on the contents of this part of the notification.

The notifier must also include data on any prior releases by the notifier of the same GMOs, occurring within or outside the Community. Further, 'each new product which, containing or consisting of the same GMO or combination of GMOs, is intended for a different use, shall be notified separately' – so, a separate notification, including a risk assessment, must be made for each new commercial application of a GMO-product. If, however, a notifier believes that the marketing and use of a product poses no risk, he may request an exemption under article 11(2) with respect to certain items of the notification.

Vetting

A marketing release notification triggers a centralised Community vetting procedure, with decision-making by member states on the basis of qualified majority voting. These marketing release procedures do not apply to foods. The marketing of foods containing or derived from GMOs is governed by Regulation 258/97 (see appendix). Within 90 days of receipt of a notification (excluding periods during which the authority is awaiting information requested from the notifier), the national authority may either reject the proposed release or forward a summary of the notification (Decision 92/146) to the Commission with a suggestion for a favourable opinion (Directive 90/220, art. 12(2) and 5). On the notification's receipt, the Commission forwards the notification to other member states (Art. 13(1)). In the (unusual) event that none objects within 60 days of distribution, the originating state may give its written consent to the GMO release (Art. 13(2)).

If, however, a member state objects ('the reasons must be stated'), and no agreement can be reached, the national authority has no decision-making power (Arts. 13(3) and 21). The Commission and a committee of national representatives, the Regulatory Committee, are to rule. If no agreement can be reached between the Commission and the committee acting by qualified majority voting, the Commission may submit a proposal to the Council. The Council must decide within three months by qualified majority whether or not to approve the proposed release (Art. 21). If the Council fails to act within those three months, the Commission is to adopt its proposal regarding the release (Art. 13(4)). Once a decision at the Community level has been made, the originating state issues that decision.

Labelling

As alluded to above, a marketing notification must include a labelling proposal, which must cover information specified in the directive's Annex III, specifically:

- Names of the product and GMOs;
- Name of the manufacturer or distributor and his address in the Community;
- 'Specificity of the product,' exact conditions of use including, when appropriate, the type of environment and/or geographical areas of the Community for which the product is suited; and
- Types of expected use – industry, agriculture and skilled trades, or consumer use by public at large (Art. 11(1) and Annex III).

In its decision approving the marketing release, the authority is to spell out precise labelling requirements (Reg. 258/97).

The labelling of GMO-products under the Deliberate Release Directive has been the subject of controversy. Some member states have objected to marketing of products whose labelling did not indicate that they were genetically modified. In 1997, through the Regulatory Committee procedure, the Commission amended the Annex III labelling requirements (Commission Directive 97/35 in appendix). Under the 1997 rules, any person putting a GMO-product on the market must provide a specific label for the GMO. The label or an accompanying document must indicate that the product contains or consists of GMOs. With regard to GMO-containing products to be placed on the market in mixtures with non-GMO containing products, it is sufficient that the label refers to the possibility of GMOs being present.

Reporting

As under the experimental release procedures, the directive imposes a potentially expansive obligation on marketing release notifiers to modify notifications and conditions of release in the event of new information. More specifically, 'if new information has become available with regard to risks of the product to human health or the environment,' the notifier must 'immediately' (1) revise the measures specified in the notification, (2) inform the competent authority, and (3) take the measures 'necessary' to protect human health and the environment (Art. 11(6)).

Safeguard clause

A GMO-product that has been properly authorised for marketing under the directive may be used anywhere in the Community, subject to conditions specified in the consent (Art. 13(5)). A member state may not prohibit, restrict or impede the deliberate release of the GMO in that product if the consent's conditions are respected (Art. 15). It may, however, 'provisionally' restrict the use or sale of that product on its territory if it has 'justifiable reasons' to consider that the GMO-product 'constitutes a risk to human health or the environment' (Art. 16(1)). If a member state chooses to do so, it must immediately inform the Commission and other member states of its decision and reasons. The Community, following the Regulatory Committee procedure described above, is to take a decision on the matter within three months (Art. 16(2)). This procedure is awkward at best. For example, the technical adaptation committee procedure, where the Council is unable to obtain a qualified majority, necessarily takes more than three months.

Marketing release decisions

The Commission has in many cases sought to authorise marketing releases of products consisting of or containing GMOs (See appendix for Commission Decisions: 93/572; 94/385; 94/505; 96/158; 96/281; 96/424; 97/98; 97/392;97/549; 98/291,2,3,4). In its decisions authorised under the Deliberate Release Directive, the Commission has consistently found that any risks arising from such GMOs were entirely acceptable. In a number of cases, including the gm virus contained in Raboral V-RG (93/572) and the herbicide-resistant tobacco variety ITB 1000 OX (94/395) the Commission concluded that any potential risks for human health and the environment 'are not expected to be significant'. In considering the intra-dermal use of the vaccine Nobi-Porvac Aujesky live (94/505), the Commission found that any potential risks were 'no different from those presented by' its intra-muscular use. Having concluded that the risk of establishment of herbidice-tolerant Swede-Rape Seeds (96/158) was low, the Commission was also satisfied that the risk of transfer 'could be controlled by existing management strategies'. In most decisions, however, the Commission has employed a negative justification to the effect that 'there is no reason to believe that there will be any adverse effects on human health and the environment'.

Some Commission authorisation proposals have been opposed by member states, with the result of the Commission authorising marketing releases over national objections from 13 member states (EPR 1997 in appendix). A Commission Decision of January 23, 1997, authorising marketing of Novartis' genetically modified maize was followed by bans imposed by Austria and Luxembourg (European Report 1997). Under the directive's safeguard provision, the Community has three months to take a decision on the member state's restrictions. However, the Community has well exceeded that deadline. Following opinions of three scientific committees, the Commission proposed decisions requiring Austria and Luxembourg to withdraw their bans on genetically modified maize. The Commission's proposal was referred to the Regulatory Committee in January 1998. The committee, which has to adopt a position by qualified majority to approve the proposal, was unable to reach a decision (European Report 1998). The matter was then referred to the Council, which must also act on a qualified majority basis. The Council also failed to act within its allotted three months expiring in September 1998, leaving the Commission free (or, more precisely, obliged) to adopt its original proposal as law. As of this writing in early April 1999, the Commission has yet to adopt its proposal, even though Austria and Luxembourg are maintaining their bans on the properly authorised maize. Given the Commission's failure to promulgate in timely fashion its decision requiring Austria and Luxembourg to remove their bans, Novartis could bring an action against the Commission before the European Court of Justice for damages (for example, lost profits) arising from the Commission's 'failure to act' (See appendix for EC Treaty art. 175).

Proposed amended directive

In February 1998, the Commission submitted a proposal for amendments to The Deliberate Release Directive (Amendment May 1998). The proposal, prompted by the row over Novartis' maize, was discussed by the European Parliament in February 1999 (Amendment proposal February 1999). The Novartis' maize case is thought to underscore the necessity of modifying the decision-making process. Under the current system, a qualified majority in the Council in favour of or against a Commission proposal for authorisation, is required. If there is no qualified majority in the Council, the final decision-making power rests with the Commission. The proposed amendments would change the decision-making procedure to enable the Council to reject a Commission proposal by simple majority. This would in effect increase the influence of member states in the decision-making process, since the opposition of fewer member states would be required to block a Commission proposal. In short, it will become easier for those national governments opposed to GMOs to prevent the placing on the market of GMO products anywhere in the Community.

The proposed amendments would also provide for a more flexible and tailored administrative procedure for experimental releases of GMOs. There would be a simplified procedure for Category I releases, defined as releases for which there is knowledge of safety for health and the environment. The full procedure would apply to Category II releases, which would include all other releases (Directive 90/220; revising art. 6(2)). As to low risk (Category I) releases, the time limit within which authorities would have to make a decision would be reduced from 90 to 30 days and the file would not need to be circulated to other member states and the Commission (Art. 1(2) creating a new art. 6a). Instead, every year, each member state would submit a list with the releases approved under this simplified procedure.

The procedure for placing products on the market would also be simplified. In contrast to current arrangements, all member states would be involved in the procedure from the beginning. The period for submitting comments or raising objections would be reduced from 60 to 30 days (Art 1(3) creating art. 13). The new rules would also provide a simplified procedure for the renewal of consents and for cases where specific criteria and information requirements on the basis of safety experience have been established (Art. 1(4) (creating a new art. 13a). The proposal further sets forth a procedure for authorisations for multi-state releases (Art. 1(2) (creating a new art. 6c).

To promote greater uniformity and consistency in decision-making, the proposal includes common principles for risk assessment (Art. 1(13) (revising Annex II). These principles deal with:

- Identification of any hazardous characteristics of GMOs;
- Assessment of consequences of the hazard;
- Likelihood of the hazard;
- Estimation of risk posed by the identified hazard;
- Application of management strategies for risks arising from GMOs; and
- Determination of overall risk of adverse effects.
- These principles would apply to all releases, experimental and marketing.

Lastly, the proposal would provide more detailed labelling guidelines. In addition to labelling information currently required, notifiers would have to include in notifications a proposal for 'mandatory labelling' that (1) 'this product contains GMOs,' either on the label or in accompanying documentation, whenever there is evidence of the presence of GMOs in the product' or (2) 'this product may contain GMOs,' where the presence of GMOs in a product cannot be excluded but there is no evidence of any presence of GMOs (Art. 1(13) (revising Annex IV). While these labelling rules might provide a little more certainty for manufacturers and distributors, their potential scope is quite broad. It would seem that manufacturers, distributors and retailers will need to label all products that they are not certain contain no GMOs, with the result of virtually all viable organisms on the market, from bacteria for wastewater treatment to household pets, having to bear the label 'this product may contain GMOs.'

Occupational health and safety regulation

The Biological Agents at Work Directive is aimed at protecting workers against risks to their health and safety arising from exposure at work (Bergkamp and Hunter 1996). 'Biological agents' are defined to include 'micro-organisms, including those which have been genetically modified, cell cultures and human endoparasites, which might be able to provoke any infection, allergy or toxicity' (Directive 90/679, art. 2(a)). A 'micro-organism' means 'a microbiological entity, cellular or non-cellular, capable of replication or of transferring genetic material' (Art. 2(b)). The directive classifies biological agents into four groups based on the magnitude of the risk posed, and requires a prior risk assessment and notification for three of the risk categories (Art. 2(d)).

The directive applies to activities in which workers are or are potentially exposed to biological agents as a result of their work (Art 3(1)). The first use of Group 2, 3 and 4 biological agents must be notified to the national authority. The notification must be made at least 30 days before the intended first use, and include:

- Name and address of the undertaking and/or establishment;
- Name and capabilities of the person responsible for safety and health at work;
- Results of the risk assessment under the directive;
- Species of the biological agent; and
- Protection and preventive measures envisaged (Art, 13).

If an activity is likely to involve a risk of exposure to Group 2, 3 or 4 biological agents, a specific risk assessment must be carried out (Art. 3(2)(a)). If this risk assessment reveals a risk to worker health or safety, worker exposure must be prevented. If that is technically not possible, any exposure must be minimised. The employer intending to use the biological agents must:

- Keep as low as possible the number of workers exposed or likely to be exposed;
- Design work processes and engineering control measures so as to avoid or minimise release of biological agents into the workplace;

- Take collective and individual protection measures;
- Take hygiene measures for prevention or reduction of accidental transfer or release of a biological agent from the workplace;
- Use the biohazard sign depicted in the directive and other relevant warning signs;
- Draw up plans to deal with accidents involving biological agents;
- Conduct testing, where necessary and technically possible, for the presence, outside the primary physical confinement, of biological agents used at work;
- Have means for safe collection, storage and disposal of waste by workers, including the use of secure and identifiable containers, after treatment where appropriate; and
- Make arrangements for safe handling and transport of biological agents within the workplace (Art. 6(2)).

The obligation to take appropriate hygiene measures, as mentioned above, requires that employers ensure that workers do not eat or drink in working areas where there is a risk of contamination by biological agents; that workers are provided with appropriate protective clothing or other appropriate special clothing; that workers are provided with appropriate and adequate washing and toilet facilities (including eye washes and/or skin antiseptics); that there is necessary protective equipment properly stored in a well-defined place, cleaned and checked; and that there are procedures for taking, handling and processing samples of human and animal origin. In addition to these measures, employers must ensure that working clothes and protective equipment that might be contaminated, be removed from the working area and kept separately from other clothing. The employer must also make sure that such clothing and protective equipment is decontaminated and, if necessary, destroyed.

The directive sets forth rules on information disclosure and training and consultation of workers. In addition, the directive requires health surveillance of workers for whom the risk assessment indicates a health and safety risk (Art. 14). The directive thus imposes extensive obligations on employers to protect workers through preventing risks, reducing risks, monitoring and damage control.

Consumer safety regulation – novel foods

Regulation 258/97 on Novel Foods creates a pre-marketing authorisation and labelling scheme for novel foods (See appendix). The regulation applies to foods and food ingredients that (1) have 'not hitherto been used for human consumption to a significant degree within the Community,' and (2) fall in one of the following categories:

a) foods and food ingredients containing or consisting of genetically modified organisms within the meaning of the Deliberate Release Directive/EEC;

b) foods and food ingredients produced from, but not containing, genetically modified organisms;

c) foods and food ingredients with a new or intentionally modified primary molecular structure;

d) foods and food ingredients consisting of or isolated from micro-organisms, fungi or algae;

e) foods and food ingredients consisting of or isolated from plants and food ingredients from animals, except for foods and food ingredients obtained by traditional propagating or breeding practices and having a history of safe food use;

f) foods and food ingredients to which has been applied a production process not currently used, where that process gives rise to significant changes in the composition or structure of the foods and food ingredients which affect their nutritional value, metabolism or level of undesirable substances (Art. 1(2)).

If there is uncertainty as to whether particular products falls within this definition, the Standing Committee for Foodstuffs may be consulted (Arts. 1(3) and 13). This procedure thus provides a mechanism for obtaining legally binding rulings on whether particular foods or types of foods fall within the regulation's scope.

The regulation specifically excludes certain products from its scope, given that those products are covered by other Community legislation. These products include food additives, flavourings and extraction solvents, subject to Directives 89/107, Directive 88/388 and Directive 88/344 (See appendix). However, the regulation provides that these exemptions 'shall only apply for so long as the safety levels laid down in Directive 89/107, 88/388 and 88/344 correspond to the safety level of the Regulation.' (Novel Foods Regulation, art. 2(2)). This peculiar provision, included in the regulation following an amendment of the European Parliament, appears to imply that the exemptions are not automatic. If that interpretation is correct, the regulation does not provide any guidance on the assessment of the exclusion of food additives, flavourings or extraction solvents regulated at EC level, nor does it indicate what procedure would apply in such circumstances. The Commission highlighted this ambiguity in its opinion on the European Parliament's amendments to the Council's Common Position in 1996. (See appendix for Commission Opinion 1996) (Long and Cardonnel 1998).

Vetting

The regulation creates a bifurcated system for the placing on the market of novel foods and food ingredients. Depending on the product's characteristics, the person placing the product on the market must either submit an application for a pre-market authorisation or make a notification. In any case, to be placed on the market, novel foods and novel food ingredients must not: (1) 'present a danger to the consumer'; (2) 'mislead the consumer'; or (3) 'differ from foods or food ingredients which they are intended to replace to such an extent that their normal consumption would be nutritionally disadvantageous for the consumer' (Novel Foods Regulation, art. 3(1)).

Authorisations

One must comply with the regulation's prior authorisation procedure in order to place on the market, among other things, foods and food ingredients containing GMOs, and 'any other product covered by the Novel Foods Regulation (see Appendix) that is not 'substantially equivalent' to existing foods' (Arts 4 and 3(4)). The person responsible for placing the product on the market must submit an authorisation application to the food assessment authority in the member state where the product will first be placed on the market (rapporteur state), and must provide a copy to the Commission (Art. 6(1)). The application must 'contain the necessary information, including a copy of the studies which have been carried out and any other material which is available to demonstrate that the food complies' (Art. 3(4)) the basic requirements cited above – that the product not: (1) 'present a danger to the consumer'; (2) 'mislead the consumer'; (3) 'differ from foods or food ingredients which they are intended to replace to such an extent that their normal consumption would be nutritionally disadvantageous for the consumer' (Art. 3(1)). The Commission published a recommendation in July 1997, concerning the scientific aspects and the presentation of information necessary to support applications for the placing on the market of novel foods and novel food ingredients and the preparation of initial assessment reports under the regulation (O.J. L 253, Sept 16 1997).

- The application must be accompanied by the following documents:
- The consent, if any, to the deliberate release of the GMOs for research and development purposes, together with results of the release(s) with respect to any risks to human health and the environment (Art. 9(1));
- A technical dossier including information equivalent to the information required under the Deliberate Release Directive (Art. 11), the environmental risk assessment based on this information, the results of any studies carried out for purposes of research and

development or, where available, the decision authorising the placing on the market provided for in the Directive (O.J. L 253, Sept 16 1997);

- Any studies and other material demonstrating that the food or food ingredient does not present a danger for the consumer or differs from foods or food ingredients which they are intended to replace to such an extent that their normal consumption would be nutritionally disadvantageous for the consumer (Novel Foods Reg. Arts. 6(1) and 3(1)); and
- A labelling proposal (discussed below).

The rapporteur state proceeds with an initial assessment and then issues an 'initial report' within three months of the date of filing of the application. The rapporteur state forwards the initial report to the Commission, which forwards it to other member states. Member states and the Commission have 60 days to provide comments on or 'reasoned objections' to the initial report, including issues of 'presentation or labelling' (Art. 6). It is not clear from the legislation whether 'reasoned objections' must be limited to scientific aspects or whether they may cover ethical concerns, for example. The Commission informs all member states of comments or objections received.

In the absence of reasoned objections or a request for additional data during the initial assessment, the rapporteur state must inform the applicant that he may place the product on the market (Art. 4(2)). If the rapporteur state decides at the conclusion of its initial assessment that additional data are required, or if a member state or the Commission has issued a 'reasoned objection,' consideration of product authorisation application must follow the technical adaptation committee procedure set forth in the regulation, (arts. 4(2), 7 and 13). Under this procedure (which is essentially the same as the Regulatory Committee procedure under the current Deliberate Release Directive), the matter is referred to the Standing Committee for Foodstuffs. The Commission proposes a draft measure. The Commission proposal must be supported by a qualified majority of member states in order to be adopted. Given the weighting of votes and sensitivity of biotechnology issues, deadlock is the usual result. In that case, the Commission may forward its proposal to the Council of Ministers, which has three months to act on the basis of qualified majority voting. If the Council fails to act in time, the Commission's proposal may (or rather 'shall') become law (Art. 13).

Notifications

The regulation establishes a less burdensome notification procedure for products deemed to pose low risks (Art 5). The notification procedure applies to, inter alia, foods and food ingredients produced from, but not containing GMOs where those food and food ingredients are 'substantially equivalent' to existing foods or food ingredients 'as regards their composition, nutritional value, metabolism, intended use and the level of undesirable substances contained therein' (Art. 3(4)). The determination of substantial equivalence is to be made on the basis of 'the scientific evidence available and generally recognised or on the basis of an opinion delivered by a competent national authority under the regulation'. The Annex of the Commission's Recommendation provides some clarification on the meaning of 'substantial equivalence'. Following OECD terminology, the Annex, point 3.3, states that:

> the concept of substantial equivalence embodies the idea that existing organisms used as foods or food sources can serve as a basis for comparison when assessing the safety for human consumption of a food or food component that has been modified or is new. If a new food is found to be substantially equivalent to an existing food or food component, it can be treated in the same manner with respect to safety, keeping in mind that establishment of substantial equivalence is not a safety or nutritional assessment in itself, but an approach to compare a potential new food with its conventional counterpart.

The establishment of substantial equivalence is an analytical exercise that is to be carried out on a case-by-case basis, in view of a product's particular characteristics. The regulation provides that the technical adaptation committee procedure may be used to determine whether a specific type of novel food or novel food ingredient qualifies for the less burdensome notification procedure (Novel Foods Regulation, arts. 3(4) and 13). In submitting its notification, a company must provide scientific information demonstrating that the novel food or food ingredient is 'substantially equivalent' to an existing food or food ingredient. On receipt of a notification, the Commission must forward copies to all member states within 60 days (Art. 5). This article also provides that (1) any member state may request a copy of the scientific information, and (2) the Commission must publish annually in the Official Journal a summary of notifications. The notification procedure thus does not make provision for possible comments or objections by a member state to overturn a notifier's decision that a product is 'substantially equivalent' and hence subject to notification rather than authorisation procedures. Arguably, however, a member state could use the 'safeguard clause' discussed below to achieve the same result.

Labelling

In brief, the Novel Foods Regulation requires labelling of food products (1) containing or consisting of GMOs, and (2) derived from GMOs where those derived from products are 'no longer equivalent' to conventional foods. Specifically, the Novel Foods Regulation, art. 8(1)(a), requires:

> The final consumer is informed of ... any characteristic or food property such as composition, nutritional value or nutritional effects, and intended use of the food, which renders a novel food or food ingredient no longer equivalent to an existing food or food ingredient. A novel food or food ingredient shall be deemed to be no longer equivalent for the purpose of this Article if scientific assessment, based upon on appropriate analysis of existing data, can demonstrate that the characteristics assessed are different in comparison with a conventional food or food ingredient, having regard to the accepted limits of natural variations for such characteristics. In this case, the labelling must indicate the characteristics or properties modified, together with the method by which that characteristic or property was obtained.

The phrase 'no longer equivalent' is vague, but it appears to be different from the notion of 'substantial equivalence.' The regulation provides that 'no longer equivalent' must be established on the basis of scientific assessment and a comparison of the novel food with a conventional food in terms of its composition, nutritional value or intended use. However, this comparison must consider the 'acceptable limits of natural variations for such characteristics.' Therefore, it would appear that the mere fact that a novel food is not identical to a conventional food does not necessarily mean that it would automatically be considered as 'no longer equivalent.' However, there is considerable scope for diversity of opinion across member states.

Presence of GMOs. The regulation obliges companies to inform consumers of the presence of GMOs in novel foods and novel food ingredients (Art. 8(1)(d)). GMOs, as noted above, are defined as 'an organism in which the genetic material has been altered in a way that does not occur naturally by mating and/or natural recombination.' Hence, this labelling obligation applies only to GMOs that are actually present and viable, and not to products that are derived from, but that no longer contain GMOs.

The regulation's recital nine addresses the issue of bulk shipments that may contain both genetically modified and non-modified products:

> Whereas, in respect of foods and food ingredients which are intended to be placed on the market to be supplied to the final consumer, and which may contain both genetically modified and conventional produce, and without prejudice to the other labelling requirements of this Regulation, information for the consumer on the possibility that genetically modified organisms may be present in the foods and food ingredients concerned is deemed – by way of exception, in particular as regards bulk consignments – to fulfil the requirements of Article 8 the labelling provision.

This recital thus contemplates possibly bulk consignments labelled 'may contain GMOs,' and was intended to address the contentious issue of mandatory segregation of GMO products from conventional products and appears to suggest that segregation is not necessarily required. However, the legal effect of the recital is dubious at best. The labelling of GMOs was the key issue discussed during conciliation negotiations between the European Parliament and Council. The Parliamentary rapporteur explained recital 9 as follows:

> Recital 9 states that, as a matter of principle, bulk consignments must be labelled, but that, by way of an exception, the labelling requirements may be deemed to have been met if 'information for the consumer of the possibility that genetically modified organisms may be present' is provided. This applies, therefore, only to the possible presence of such organisms, as covered by Amendment 51 [Article 8(1)(d) of the regulation]. However, this derogation cannot override the labelling requirement, which continues to apply pursuant to Amendment 55 [Article 8(1)(a) of the regulation]. The new recital merely clarifies Article 8 and does not replace the rules laid down therein.

European Parliament Report on the joint text approved by the Conciliation Committee for a European Parliament and Council Regulation, concerning novel foods and novel food ingredients (Doc En/PR/293/293195, dated Jan. 9, 1997).

Recital 10 suggests as well the possibility of a negative label ('does not contain GMOs'):

> Whereas nothing shall prevent a supplier from informing the consumer on the labelling of a food or food ingredient that the product in question is not a novel food within the meaning of this Regulation or that the techniques used to obtain novel foods indicated in Article 1(2) were not used in the production of that food or food ingredient.

Food Products Derived from GMOs. Novel foods or novel food ingredients derived from GMOs, but not containing GMOs, must be labelled according to the regulation only if (1) they are 'no longer equivalent' to conventional foods, or (2) if they fall within one of two other categories for which labelling is required, specifically if they may (a) have health implications for certain sections of the population, or (b) raise ethical concerns. Novel foods and ingredients must be labelled if they contain material that is not present in an existing equivalent food and which may have 'implications for the health of certain sections of the population' (Novel Foods Regulation, art. 8(1)(b)). This provision is aimed at ensuring proper labelling of, for example, potential allergenic properties in innovative food products. The regulation requires labelling of those containing materials that give rise to 'ethical concerns'. (Art. 8(1)(c)).

This provision is designed to provide sufficient information to individuals who are subject to dietary restrictions on, for example, religious grounds. An example of such a product could be the introduction of a pork gene into a plant.

The Novel Foods Regulation does not provide guidance on how these basic labelling rules are to apply in practice, but authorises the Commission to adopt implementing detailed rules through the technical adaptation committee procedure. The Commission has indicated its intention to prepare guidelines to assist with compliance, but no such guidelines have yet been promulgated.

However, the rules on genetically modified soya and maize do provide some guidance on likely future labelling policies. The Commission adopted a regulation (1813/97) in September 1997 providing for the compulsory labelling of genetically modified soya and maize. This Commission regulation brought genetically modified soya and maize that had been authorised prior to implementation of the Novel Foods Regulation within the scope of the regulation's labelling provisions. The Council, pursuant to the committee procedure, then repealed with its Regulation 1139/98 the Commission regulation, and imposed specific labelling requirements for soya beans and maize.

Under this Council regulation, unless neither protein nor DNA resulting from genetic modification is present, foodstuffs produced from genetically modified soya beans or maize, must be labelled as 'produced from genetically modified' soya or maize (Art. 2(3)). These more detailed labelling policies may well be extended to other novel foods and novel food ingredients. However, the Community has yet to define adequate testing methodologies for determining the presence of modified protein and DNA, leaving companies awash in uncertainty as to how to ensure compliance.

The potential scope of these labelling policies is expansive. British officials, for example, have argued that the labelling obligation should not be limited to manufacturers and distributors of food products containing or derived from GMOs. Rather, they argue that the obligations should extend even to restaurants selling food containing or made from GMOs. Apparently restaurateurs are now expected to indicate on their menus those dishes that contain or are derived from GMOs, at least so long as they are not certain modified protein or DNA is not present.

Safeguard clause

A novel food or food ingredient that has been properly authorised or notified under the directive may be used anywhere in the Community, subject to conditions specified in the consent. A member state may, however, 'temporarily restrict or suspend the trade in and use of' a novel food or food ingredient if, on the basis of either 'new information' or 'a reassessment of existing information,' the state has 'detailed grounds' for considering that the use of the food or ingredient 'endangers human health or the environment' (Art. 11(1)). It is worth noting that this provision does not authorise member states to restrict products merely because of unpopularity with certain political groups, but rather states must be able to provide detailed reasons concerning how the product 'endangers human health or the environment.' If a member state chooses to avail itself of this safeguard provision, it must immediately inform the Commission and other member states of its decision and reasons. The Commission, working with the Standing Committee for Foodstuffs, is to examine the member state's actions and grounds and take 'appropriate measures.' In the meantime, the member state may maintain its restrictions (Art. 11(2)).

Conclusion

As should be manifest from this review of regulation of GMOs, the Community has an elaborate regulatory framework. The regulation imposes extensive environmental risk assessment requirements as a condition for handling GMOs or releasing them into the environment, and requires health-focused risk assessments where workers will be exposed to GMOs, and where GMOs or GMO-derived products will be consumed by humans. Moreover, the prior authorisation procedure for marketing GMO products, and in the food context GMO containing and derived-from products, allows all member states ample opportunity to raise reasoned objections. Further, where a member state can demonstrate that a previously authorised GMO containing or derived-from product presents a risk to human health or the environment, it may legally restrict trade in that product and trigger reconsideration of the authorisation by the Community.

Yet, this comprehensive regulation has not alleviated popular and political concerns. Quite the contrary. And the Commission and member states bear much of the responsibility for this. Given the lack of public understanding of biotechnology, initial popular disquiet over these novel products might be expected. However, European officials have done little to educate the public, and the way Community decision procedures are operated is not calculated to provide confidence. Deliberations of the Commission and national officials in technical committees over whether to authorise the marketing of GMO and GMO-derived products are confidential. Officials typically provide only perfunctory explanations of the grounds for their decisions, and there is no opportunity for interest groups to obtain judicial review of the decisions reached behind closed doors. By dressing the issues up as merely technical and declining to explain their reasoning in detail, officials should not be surprised by public suspicion of the process and anxiety over the products (Hunter 1996).

Further, some national governments have been lawless in their manipulation of the regulation's procedures. It seems clear, for instance, that Austria's resort to the Deliberate Release Directive's safeguard provision to ban Novartis' properly authorised maize is groundless and unlawful. And the Commission's failure to require Austria to remove its ban also appears illegal. Putting aside the international trade law issue this raises, the very fact that the Community tolerates such unlawfulness calls into question the legitimacy of the Community regulation and in turn invites further abuse. With regard to World Trade Organisation law, the EC would likely be vulnerable to a legal challenge under the GATT, articles XI and XX, and the SPS Agreement.

Finally, the Community's overbroad labelling rules and policies seem sure to cause confusion. Perhaps limited mandatory labelling obligations could be justified, though one might have thought that, if consumers really cared, shrewd businessmen seeking profit would segregate and label voluntarily (the Community and member states already have legislation prohibiting misleading advertising). However, the Community's failure to define workable compliance thresholds and testing methodologies for the labelling rules seems destined to result in a wide variety in enforcement practices across member states (this is already happening), and legal uncertainty for producers, distributers and retailers. Prudent businessmen may well decide to attach to virtually everything that could conceivably contain GMOs, or in the case of foods, be derived from GMOs, a label stating that 'this product may contain or be derived from GMOs.' One might wonder how edifying this will be for consumers.

Appendix

Regulations and notes

Directive 90/219 on Contained Uses O.J. L 117/1 (May 8, 1990).

Directive 98/81 amending Directive 90/219 on the contained use of genetically modified micro-organisms, O.J. L 330/13 (Dec. 5, 1998).

Commission Decision 91/448 of July 29, 1991, O.J. L. 239/23 (Aug. 28, 1991) set out more detailed guidelines for the classification criteria. The guidelines introduce additional requirements for the classification of GMMs intended for Type B operations. This Decision refers to guidelines for classification set out in Directive 90/219, as amended by Commission decision 96/134 of January 16 1996, O.J. L. 31/25 (Feb 9 1996).

Directive 90/220 on Deliberate Releases, O.J. L 117/15 (May 8, 1990); O.J. L. 279/42 (Nov. 12 1993).

Commission Directive 97/35 of June 18, 1997 adopting to technical progress for the second time Council Directive 90/220 on the deliberate release into the environment of genetically modified organisms, O.J. L 169/72 (July 27, 1997).

Proposal for a European Parliament and Council Directive amending Directive 90/220/EEC on the deliberate release into the environment of genetically modified organisms, O.J. C 139/1 (May 4, 1998). See also article 1(1) revising article 6(2) of 90/220.

Opinion of the European Parliament on the Proposal for a European Parliament and Council Directive amending Directive 90/220 on the Deliberate Release into the Environment of Genetically Modified Organisms, First Reading, Provisional Edition (Feb. 11, 1999).

Commission Decision 93/584 of October 22 1993 established criteria for simplified procedures concerning the deliberate release into the environment of genetically modified plants pursuant to Art. 6(5) of Directive 90/220.

Commission Decision 94/730 of November 4 1994, pursuant to Decision 93/584, allows certain member states to apply for the simplified notification procedure. These countries are: France, UK, Belgium, Italy, Portugal, Ireland, Spain, Denmark, the Netherlands and Germany.

Commission decisions authorising marketing releases of gmo products:

Commission Decision 93/572 of October 19, 1993 *Raboral V-RG*, O.J. L 276/16 (Nov. 9, 1993).

Commission Decision 94/385 of June 8, 1994 *Herbicide-Resistant Tobacco Variety ITB 1000 OX*, O.J. 176/23 (July 9, 1994).

Commission Decision 94/505 of July 18, 1994 *Vaccine Nobi-Porvac Aujesky live*, O.J. L 203/22 (Aug. 6, 1994).

Commission Decision 96/158 of February 6, 1996 *Hybrid Herbicide-Tolerant Swede-Rape Seeds*, O.J. L 37/.30 (Feb. 15, 1996).

Commission Decision 96/281 of April 3, 1996 *Genetically Modified Soya Beans*, O.J. L 107/10 (Apr. 30, 1996)on herbicide-resistant soya beans stated 'Whereas ... there is no reason to believe that there will be any adverse effects on human health and the environment.'

Commission Decision 96/424 of May 20, 1996 *Genetically Modified Male Sterile Chicory*, O.J. L 175/25 (July 13, 1996) on herbicide-tolerant male sterile chicory states 'Whereas...there is no reason to believe that there will be any adverse effects (from the transfer of the bar gene to wild chicory populations given the fact that such transfer could only confer a competitive or selective advantage to wild populations if the herbicide glufosinate-ammonium were the only means of controlling these populations, which is not the case).'

Commission Decision 97/98 of January 23, 1997 *Genetically Modified Maize*, O.J. L 31/69 (Feb. 1, 1997), states 'The Commission reached the following conclusions: ... there is no reasons to believe that the introduction of these genes into maize will have any adverse effects on human health or the environment.'

Commission Decision 97/392 of June 6, 1997 *Genetically Modified*

Swede-Rape O.J. L 164/38 (June 21, 1997), states 'Whereas... there is no reason to believe that there will be any adverse effects on human health and the environment from the introduction of into swede-rape of the genes coding for phosphinotricin acetyl transferase and for neomycin phosphotransferase II.'

Commission Decision 97/549 of July 14, 1997 *T102-test*, O.J. L 225/34 (Aug. 15, 1997), states 'Whereas ... there is no reason to believe that there will be any adverse effects on human health or the environment from the introduction into Streptococcus thermophilus T102 of the gene coding for chloramphenicol-acetyl-transferase on the plasmid pMJ 763.'

Commission Decision 98/291 of April 22, 1998 *Genetically Modified Spring Swede Rape* O.J. L 131/26 (May 5, 1998),

Commission Decision 98/292 of April 22, 1998 *Genetically Modified Maize* O.J. L 131/28 (May 5, 1998),

Commission Decision 98/293 of April 22, 1998 *Genetically Modified Maize* O.J. L 131/30 (May 5, 1998),

Commission Decision 98/294 of April 22, 1998 *Genetically Modified Maize* O.J. L 131/32 (May 5, 1998).

See also Commission Decisions of April 22, 1998, 98/292/EC (Zea mays L.Bt-11), O.J. L131/28 (May 5, 1998), 98/293/EC (Zea mays L.T25), O.J. L131/30 (May 5, 1998), and 98/294/EC (Zea mays L.MON810), O.J. L131/32 (May 5, 1998).

Directive 90/679 on Biological Agents at Work of November 26, 1990

Concerns the protection of workers from risks related to exposure to biological agents at work (seventh individual Directive within the meaning of art. 16(1) of Directive 89/39/EEC), O.J. L 374/1 (Dec. 31, 1990).

Directive 94/55 on Transport of Dangerous Goods by Road. O.J. L 319/7 (1994).

Regulation 258/97 on Novel Food O.J. L 43/1 (Feb. 14, 1997).

A regulation of the European Parliament and of the Council of January 27, 1997, governing food and food ingredients containing, consisting of or derived from GMOs by specific labelling rules. The Commission has indicated an intention to propose legislation on labelling of genetically modified food additives and flavourings during 1999.

Under the Novel Foods Regulation, art. 3(2), second indent, foods or food ingredients derived from plant varieties subject to Directive 70/457 on the common catalogue of varieties of agricultural plant species, O.J. L 225/1 (Oct. 12, 1970), and Directive 70/458 on the marketing of vegetable seed, O.J. L 225/2 (Oct. 12, 1970), must be authorised in accordance with procedures laid down in these directives. However, the labelling provisions of the Novel Foods Regulation, art. 8, still apply.

Council Regulation 258/97 concerning Novel Foods and Novel Foods

Ingredients, O.J. L 43/1 (Jan. 27, 1997), covers pharmaceutical products derived from biotechnology that are subject to the EC's pharmaceutical legislation. The EU has recently revised its regulation of seeds. See Council Directive 98/95 amending, in respect of the consolidation of the internal market, genetically modified plant varieties and plant genetic resources, Directive 66/400/EEC, 66/401/EEC, 66/402/EEC, 66/403/EEC, 69/208/EEC, 70/457/EEC and 70/458/EEC on the marketing of beet seed, fodder plant seed, cereal seed, seed potatoes, seed of oil and fibre plants and vegetable seed and on the common catalogue of varieties of agricultural plant species, O.J. L 25/1 (Jan. 2, 1999) (setting out environmental risk assessment procedures and requirements for genetically modified seeds).

Directive 89/107 on the approximation of the laws of Member States concerning food additives authorised for use in foodstuffs intended for human consumption, O. J. L 40/27 (Feb.11, 1989).

Directive 88/388 on the approximation of the laws of the Member States relating to flavourings for use in foodstuffs and to source materials for their production, O.J. L 184/61 (July 15, 1988).

Directive 88/344 on the approximation of the laws of the Member States on extraction solvents used in the production of foodstuffs and food ingredients, O.J. L 157/28 (June 24, 1988).

Commission Opinion pursuant to Article 189(2)(d) of the EC Treaty, on the European Parliament's amendments to the Council's common position regarding the proposal for a European Parliament and Council Regulation (EC) on novel food and novel food ingredient, COM (96) 229 Final, at page 3.

European Parliament Resolution (EPR) on genetically modified maize of April 8, 1997, O.J. C132/2 (1997).

Commission Regulation 1813/97 concerning the compulsory indication on the labelling of certain foodstuffs produced from genetically modified organisms of particulars other than those provided for in Directive 79/112, O.J. L 324 (Nov. 27, 1997).

Council Regulation 1139/98 concerning the compulsory indication of the labelling of certain foodstuffs produced from genetically modified organisms of particulars other than those provided for in Directive 79/112/EEC, O.J. L 159/4 (June 3, 1998).

'Failure to Act'

For European Court of Justice jurisprudence on remedies for failure to act under the EC Treaty, art. 175, see Joined Cases C-15/91 and C-108/91 Buckle and Others v Commission (1992) ECR I-6061, and Case T-28/90 Asia Motor France and Other v Commission (1992) ECR II-2285. Under this settled case law, the remedy provided for in the Treaty, art. 175, is based on the premise that unlawful inaction on the part of the Community institutions enables private parties to bring the matter before the Court of Justice in order to obtain a declaration that the failure to act is contrary to the Treaty. Individuals may claim damages from the institutions if there is actual damage and a causal link exists between the unlawful conduct and the harm alleged.

References

Bergkamp, L. (1998) Allocating Unknown Risks: Liability for Environmental Harms from Deliberately Released Genetically Modified Organisms (submitted for publication). Deals with liability rules applying to activities involving GMOs.

Bergkamp, L. and Hunter, R. (1996) Occupational Health and Safety Framework: Current and Proposed European Union Legislation, *World Food Regulation Review (BILLIONA)* Mar, gives an overview of Community occupational health and safety regulation.

Buono, T.P. (1997). Biotechnology-Derived Pharmaceuticals: Harmonising Regional Regulations, 18 *Suffolk Transnational Law Review* 133 (1995).

European Report 2249 (1997) Sept. 10.

European Report 2308 (1998) Apr. 18.

Hunter, R. (1996). Prising Open the Club Doors *European Voice* 14 Jan. 18. (article discussing Community delegated rule-making procedures and suggesting reforms).

Hunter, R. and Muylle, K. (1999) In: *European Community Environmental Law* (Environmental Law Institute) gives a comprehensive survey of EU environmental law as well as a compendium of legislation.

Long, A. and Cardonnel, P. (1998). Practical Implications of the Novel Foods Regulation, 1 *European Food Law Review* 11.

van der Meulen H.E.J. (1997), Biotechnologie en het Besluit genetisch gemodificeerde organismen, Milieu en Recht 1997, pp. 125-128.

Section 3:
So what is the solution?

12 Community markets to control agricultural non-point source pollution*

Bruce Yandle

Summary

Industrial pollution in developed countries has reduced significantly in the last 20 years. Agricultural pollution now makes a significant proportion of total pollution. It is argued that command-and-control (technology specifying) regulation to reduce this pollution is highly inefficient. An alternative approach of allowing a market in tradeable pollution permits will be both more efficient and beneficial to the environment.

Introduction

The regulation of environmental use in the United States is changing. It has to change. Past approaches are just too costly. Aside from cost, the micromanaged technology-based standards of the past cannot readily be applied to pollution that comes from streets, construction sites, and farms.

Instead of hard and fast input regulation, where central authorities mandate cleanup technologies for each and every source of pollution, more flexible performance standards that allow for cost minimization are emerging. The new emphasis is on the result, not the technology.

The quiet revolution now occurring in a few out-of-way places is not the result of scholarly studies or efforts by brilliant bureaucrats to engender more effective control. The change is driven by ordinary people who face the direct cost of meeting stringent and, in some cases,

*This paper originally appeared in 'Taking the Environment Seriously' edited by Roger Meiners and Bruce Yandle, USA: Rowman and Littlefield

impossible pollution control standards. As it turns out, farmers are leading an effort that could bring common sense to American water pollution control strategies.

Until recently, the burden of federal command-and-control regulation was felt heaviest in the industrial sector. That is where pollution control started. Ordinary consumers and investors paid for this in the form of higher prices and reduced dividends, but few individuals faced the full cost head on. The costs were widely dispersed. Agriculture escaped the heavy burden of meeting federally dictated rules for cleaning up farm wastes and limiting the run-off of nutrients, sediments, and animal wastes.

Of course, the agricultural sector bore the burden that came with controls on pesticides and other agricultural chemicals, wetland legislation, and meeting tougher standards for food products. But the burden of direct control continued to be postponed. That has changed. Today, American coastal zone management statutes and state regulations require farmers to implement management practices intended to improve the water quality of streams, rivers, and aquifers in their regions of operation. The costs are high, and the number of farm operators is large.

Unlike industrialists, who generally speak for particular industries, such as steel, computers, and textiles, farmers speak for farming, all of it. Regardless of what farmers produce – corn, soy beans, sugar beets, or wheat – the newly imposed non-point source pollution standards hit them all. Their political voice tends to be heard. And unlike industrialists, farmers usually are part of a well-rooted community. They are wed to the land. Exit costs are high. Political muscle and community orientation combine to yield an interesting possibility. Property rights and community-based trading of rights to water use will likely emerge.

How this chapter is organised

This chapter examines certain features of the newly emerging regulation of pollution from agriculture. Its principal purpose is to examine institutional arrangements that could support market transactions among farmers who seek to reduce discharge from their operations. Related to that are interactions with industrial polluters who continue to face increased pollution control requirements. Other polluters, such as cities that must deal with storm run-offs and construction firms that face erosion controls, may also become a part of a trading community that seeks to minimise the cost of hitting water quality targets.

The next part of the chapter briefly examines the old command-and-control approach that is still the dominant form of pollution control. Turning quickly to market approaches, the section offers a rationale for permit trading, illustrated by a highly simplified example. The section also explains why elements of command-and-control are likely to be observed in the new institutions that emerge. A discussion of point and non-point source pollution is included in the section, along with a summary of potential cost savings from trades that involve the two kinds of sources.

Section three considers some of the major problems assumed away in section two. These have to do with some serious scientific questions about the linkage between farm discharge and run-off and the quality of receiving streams, the tendency for environmental protection, which has to do with improving the quality of water and air, to be transformed to pollution prevention, which may have little effect on environmental quality. The section also discusses other policy conflicts that affect the agricultural sector.

The next major section focuses on permit trading, first discussing the situation where trade occurs between point and non-point sources of pollution. An analysis of trade among non-point sources ends the section.

Section five explores methods that might be used to progress from the current control regime to one that emphasises market instruments. The discussion begins broadly, then narrows. It first focuses on institutions that have emerged for managing natural resources in a market context. Property rights and contracting are crucial to that process. The discussion narrows to the current environment and then provides a summary of traits and characteristics that appear to be necessary for permit trading to emerge.

The last major section is the most applied. It builds a discussion of permit trading in the current environment. The list of ideal characteristics developed in the previous section are folded into a world that compromises the ideal. Finally, the chapter concludes with some brief final thoughts and recommendations.

The old and new control strategies command-and-control versus marketable permits

Command-and-control regulation works. Install enough machinery, hire enough inspectors, and eventually the amount of waste discharged into rivers will be reduced. But the approach is unnecessarily burdensome and basically at odds with American traditions and social norms. While regulation has practically always been a feature of the American enterprise system, the nation's economy is based fundamentally on the operation of free markets where owners of property rights follow price and cost messages received from markets. The messages provide the basis for economising actions and mutually beneficial trade. By contrast, command-and-control is part of an authoritarian tradition, where politicians and their appointees manage economic behaviour.

While the spur of market competition rewards lower cost producers and yields variety, command-and-control pays less attention to the costs borne by the consumers of regulated firms and producers; command-and-control goes for uniformity. Politicians and bureaucrats understandably seek to maintain their positions and minimise their own costs. Widely dispersed costs borne by consumers and investors hardly have a bearing on the politics of control.

Eventually, however, even the bureaucrats receive the message. Pressure to improve the effectiveness of regulation and to accomplish more with fewer resources causes them to consider market alternatives. Indeed, EPA Administrator William K. Reilly recently had this to say about the benefits of harnessing market forces:

> The forces of the marketplace are powerful tools for changing individual and institutional behavior. If set up correctly, they can achieve or surpass environmental objectives at less costs and with less opposition than traditional regulatory approaches.[1]

Marketable pollution permits

Interestingly enough, Mr. Reilly was referring to the possible use of property rights and marketable pollution permits for the control of water pollution. What was he getting at when he pointed to the market alternative? Instead of writing more and more detailed regulations defining technologies to be applied at each and every outfall and then mandating proportional reductions in waste discharge from each and every source, a system of marketable permits simply sets the amount of discharge allowed by the holder of the permit. Market forces take hold from there. But there is more to the story than just saying the market takes hold.

Suppose ten firms are located in the same river basin and for some reason seek collectively to reduce the discharge of a specified pollutant now received in the river. To really simplify things, suppose that the discharge from each of the ten firms is known, and the effects of discharge on the river are identified. After a meeting attended by all the dischargers, each firm agrees to reduce pollution by half. Property rights to the remaining 50 per cent, the allowed discharge, are defined and enforced. The property rights can be exchanged.

If one permit holder can reduce pollution at a lower cost than its neighbour in the same watershed, the higher cost operator can buy pollution discharge rights from the operator with lower cleanup costs. The low-cost producer cleans up far more than stipulated by the community agreement. The higher cost producer cleans up less. The total amount of waste discharged does not exceed the amount allowed. The idea seems so simple that one wonders why it is not been implemented with great frequency.[2] However, the simple description here assumes a lot about knowledge of pollution, its sources, and complex community and legal institutions that will be discussed later.

Whatever the institutional arrangements, the possibilities for trade hinge fundamentally on differences in pollution control costs and the value of pollution control. First off, if all pollution sources have identical cleanup costs, there are no potential gains from trade associated with cost reductions. Each and every producer might as well be told by a government regulator to cleanup specified amounts. Of course, if there are differences in control costs, the possibilities for gains from trade shrink if the methods for cleaning waste are mandated by regulation and all negotiations are managed by a bureaucracy that has little incentive to help other people save money.[3]

The magic of the market relates to the discovery incentive. If ordinary people with common sense can make money by finding cheaper ways to treat waste, they will find and apply new technologies. The discovery incentive is blunted when control techniques are dictated by command-and-control rules. Net gains from trade are reduced even farther when parties to a transaction are required to endure lengthy administrative hearings and engage in costly legal transactions when they seek to engage in trade. Of course, if pollution control costs are subsidized with tax money, polluters tend to accept the bureaucratic costs that bring them the revenues, even if the rules make little sense. Like all other ordinary people, polluters seek to minimize *their* net costs, which are not generally the same as society's net costs.

The value of pollution reduction, which is partly based on potential cost savings, is also related to the market price of goods and services produced by different polluters. Suppose the ten firms in our example include industrial plants, municipal waste treatment plants and farmers who generate waste that consumes dissolved oxygen in a particular river. To simplify things further for the sake of making a point, assume each polluter may be able to apply the same management practice to reduce the amount of waste received in the river. That is, their control costs are identical. There are still potential gains from trade in a polluter permit market.

If the market value of the farmer's product, per unit of waste sent to the river, is larger than that of the industrial plant, the farmer can gain by buying rights to discharge from the industrial plant. The industry will vacate its rights by cleaning up more waste. The farmer will expand discharge. The resulting trade leads to an increase in the total value of goods produced by the two trading parties.[4]

Grappling with the problems

Ignorance and uncertainty

There are immense problems hidden in this simple example. In general, agricultural waste is generated from a wide-ranging set of activities that are not specialized to one particular location. They are 'non-point source' pollution. By contrast, industrial pollution generally emerges from the end of process pipes or other well-identified points. Those are referred to as 'point source' pollution. There is another complicating factor. The linkages between environmental effect and point source discharge are often easier to identify. The gains from control can be estimated with greater accuracy. Identifying the environmental consequences of wide-ranging run-off of sediments and nutrients from hundreds of acres of land presents a more difficult task. Even when controls are implemented, it is difficult and costly to identify specific water quality changes and where they might originate.

Uncertainty is the problem. The direct linkage between agricultural 'discharge' and the water quality of receiving streams is just not there. As one researcher put it: 'The standard solutions that have been successful in controlling point source problems are unworkable for ... non-point pollution partly because it is generally not possible to observe (without excessive costs) the level of abatement or discharge of any individual suspected polluter or to infer those levels from observable ambient pollutant levels.[5]

The same issue surfaces in other research (Milon 1987, 387-95). Referring to non-point source pollution, a USDA report described the problem this way: '[O]ffsite damage associated with water pollution cannot be measured directly and links between farming and affected water uses are not well defined. Many assumptions are made to estimate offsite damage, and both methods and data for estimating damage need to be improved' (Crowder et al. 1988, 2).

Research on non-point source pollution in other countries indicates that phosphorous stream loadings from intense agriculture operations are quite low, in spite of the fact that phosphorous loadings are high in the fields (Loigu 1989, 213-17). The field-generated phosphorous combines with sediment to prevent stream damage. However, phosphorous loadings from industrial and municipal sources are highly interactive. On the other hand, high levels of agricultural nitrogen do reappear as stream nutrients. Taking rather simple steps, such as establishing sod filter strips between fields and streams, offers a viable remedy to this problem.

The uncertain linkage between what happens in a farmer's field and what occurs to the water quality of streams and rivers is a fundamental problem that stands in the way of any effective water quality management approach. Note the choice of words here. Water quality management is different from pollution control. Maintaining some level of water quality is the result, the output. Reducing pollution is an input. We can obviously find ways to reduce sediment and nutrient run-off, that is pollution control. What we cannot do is state emphatically that such actions improve water quality, and that presumably is what environmental protection is about.[6]

The goal shift: water quality improvement to pollution prevention

It is far too easy for regulators to shift environmental goals. Such shifts can be extremely costly. The important goal of improving the quality of streams, rivers, and aquifers can he transformed to pollution prevention, which sounds good but is fundamentally at odds with science and economic logic. Improved water quality enhances life and brings economic benefits. Pollution prevention does not necessarily improve anything but thewallets of machinery manufacturers and the employment of bureaucrats and regulators. Unfortunately, pollution prevention tends to become the goal, and water quality tends to be forgotten.

Instead of monitoring the quality of streams and reporting regularly to concerned people, regulators monitor inputs, whether or not controls are in place, and how production is managed. Almost inevitably, the problem becomes a technical one, and the policy debate gets focused on which technology to use. All along, little attention is paid to environmental protection.

The conflict of policies

A major policy conflict between environmental and agriculture price stability further complicates the problem. U.S. agriculture policy encourages production through a system of loan guarantees and target prices. The farmer is given strong incentives to bring fragile land into production (McSweeny and Krainer 1986, 159-73). In other words, the production effects of improved loan and target prices can completely swamp the effect of soil conservation and pollution control programs, causing the farmer to steer away from pollution control.

Permits, point, and non-point pollution

The US experience

The story of US pollution control has been largely one of command-and-control regulation where technology-based standards are specified for each source of pollution. Permit trading has been the exception to the rule. By and large, the cost of controlling private sector pollution has been borne by consumers who pay more for final goods and services. Special tax treatment of investment in pollution control capital is a relatively common feature, but generally speaking, there has been little in the way of direct subsidies to industrial polluters.

There are exceptions to the command-and-control regime. Marketable pollution permits have been used in limited ways in the United States for at least 15 years.[7] But their use has been confined primarily to air pollution from stationary point sources. Indeed, the 1990 Clean Air Act extends the use of marketable permits to the control of sulphur dioxide emissions from well-identified electrical utilities. Under the statute, electrical utilities are given a fixed number of sulphur dioxide emission allowances, which are fewer in number than their current emission levels. EPA coordinates trades among utilities. The very first trade under the new legislation illustrates the possibilities for gain (Davis 1992, 1217).

In 1992 the Wisconsin Power and Light Company found that it could exceed EPA's cleanup goals and operate well within its allocation of sulphur dioxide allowances. Doing so, the firm sold the annual right to emit an additional 25,000 to 35,000 tons of sulphur dioxide to the Tennessee Valley Authority and the Duquesne Light Company. The transaction was clearly profitable: Wisconsin Power added between $10 million and $20 million to the bottom line.

Tradeable permits have been allowed for the control of water pollution from industrial and municipal sources that discharge into the same river, but trade has been very limited. In all cases thus far, marketable permit systems have related to point source pollution. That is, the source of the pollution was easily identified and controls could be specified.

The fact that it is easy for regulators to identify machines to be controlled in industrial plants or in publicly owned treatment works is just a partial explanation of the dominance of technology-based command-and-control standards. Though cost effectiveness and flexibility are denied to operators of industrial plants, command-and-control makes life easier for regulators and simplifies life for firms that operate plants in multiple locations. Once implemented, command-and-control is assumed to achieve pollution reduction goals. If the control machinery is installed and operating, it is assumed the environment is protected. From industry's standpoint, one set of regulations to be met by all firms in a competitive industry is more desirable than uncertain approaches that might offer a competitive advantage to new and old firms alike.

The point about competitive advantage deserves a little more emphasis. The command-and-control approach generally favors larger more sophisticated operators who have technical expertise and scale of plant to operate effectively in the regulatory system.[8] Smaller operators are frequently forced out of business; output goes down and price rises. In other words, command-and-control regulation can be used for anti-competitive purposes. A system that relies on property rights and market forces cannot.

From a regulator's standpoint, the pollution from a coal-fired boiler at an electrical utility or a petroleum unloading dock can he identified, monitored, and controlled in reasonably well-specified ways. Emission reductions from one source in an airshed can be recognized, and those reductions can be used to offset another well-identified pollution source. Enforceable contracts can be executed. With some degree of workable accuracy, buyers and sellers can tell when the work has been done.

The development of control institutions for point-source pollution is one thing. Implementing controls for waste that washes from streets, construction projects, and agricultural operations is quite another. While the principle involved is the same, the institutions to support trade are quite different. At the same time, it is possible to conceive of arrangements where operators of point and non-point pollution sources in the same control region would he able to trade pollution permits (EPA Office of Water 1992). The gains from trade can be substantial.

Just how large are the potential gains?

It can be far more costly to control the same pollutants by means of command-and-control than by allowing polluters to minimize cost across sources of the same pollutants.[9] An excellent offering of such evidence is found in the work reported by Magat, Krupnick, and Harrington, who analysed a raft of EPA background documents related to effluent limitations for the control of biological oxygen demand (BOD), a basic measure of pollution. The data analyzed covered a large sample of industry and sub-industry groups.[10] But while the coverage is extensive, we must bear in mind that administrative, monitoring, and enforcement costs are not considered in the analysis. In other words, the single focus is on the cost of operating pollution control equipment.

The three analysts identified the incremental cost of removing a unit of BOD across many sources and found the cost varied from 10 cents per kilogram to $3.15 per kilogram, more than a thirty-fold difference in costs (Magat et al. 1986). A survey of capital costs for the same sample of industries found the incremental annual cost of capital for removing a unit of BOD varied from $59.09 per kilogram to one cent. The capital cost data indicate that a unit of BOD reduced at the cost of one penny in one plant can save $59.09 at another plant. The wide-ranging costs offer potential opportunities for significant cost savings if the polluters are allowed to trade permits.

Another glimmer of the possible gains from such an arrangement is reported in an EPA study that focuses on the prospects for permit trading in the control of nutrient run-offs from agricultural operations. The report, which in this case considered the cost of establishing a market framework, described a situation for the Tar-Pamlico estuary in North Carolina, which required estimated expenditures of $2 million to cover administrative personnel, the development of an estuarine and nutrient computer model, and for monitoring costs (EPA Office of Water and Office of Policy 1992). In this case, an industrial point source polluter that faced even higher control costs paid the cost of organizing and managing the institution. At one location in the estuary, further pollution reductions from industrial point sources would cost from $860 to $7,861 per pound eliminated. Non-point source reductions of the same pollutant from farms in the same location would cost between $67 and $119 per pound. The potential for gains from trade is obvious. But the institutional hurdles are still high.

How do we get to permit trading from where we are?

The mandate

According to EPA, there are 18,000 bodies of water that will not attain water quality standards if every point-source polluter meets the letter of the law but non-point sources maintain their current methods of operating (EPA Office of Water and Office of Policy 1992). But a 1987 Conservation Foundation report indicates that the non-point source pollution problem reaches beyond surface water. At the time of the report, 34 states reported nitrates (generally assumed to come from agricultural run-off) as the most common groundwater contaminant. Agriculture was listed as the 'predominant source of the problem for both surface water and groundwater'; state government officials listed agriculture as the most widespread source of water pollution for 60 percent of the states, 11 'causing problems in 64 percent of the river miles and 57 percent of the acre areas assessed'. (Conservation Foundation 1987). But that was the state of knowledge, or ignorance, in the late 1980s.

Whether fully correct or not, something will be done to control agricultural non-point source pollution. Indeed, much is already being done. It is not that agricultural run-off is unchecked. It is the fact that the goals of the Clean Water Act cannot be met unless more is done. To some degree the mandate to clean up is a paper chase. The law requires pollution to be reduced, whether or not our knowledge supports specific actions. Substantial regulation of non-point source pollution is on the way.

In addition to what might be called the legislative necessity of doing so, which is to say that attainment of the goals of other legislation require the action, there are cost savings to consider. The Council on Environmental Quality's mid-1980s estimate of sources of key pollutants tells us about the relative position of various sectors. Some 6.1 per cent of BOD discharge comes from the metals and minerals industry (Council on Environmental Quality 1989, 32-35). Municipal wastes account for 73.2 percent. Agricultural waste adds another 21.6 percent to the total. For suspended particulates, municipal waste accounts for 61.5 percent of the total; industry, 26.6 percent, and agriculture, 13.3 percent. However, since 1970, industry has reduced its level of pollution by some 71 percent. After many years of imposing ever tighter controls on industrial point sources, the incremental cost of additional reductions becomes quite large. That is, larger amounts of pollution may be avoided for lower costs if non-point sources are controlled. Amendments for doing so are being debated in Congress.

Can command and control be avoided?

Granted that the legislation train is headed toward agriculture, is it possible to load the cars with cost-effective rules and allowances for trade that will enhance overall efficiency? Will transaction costs overwhelm the control cost savings? Will the train carry a load of command-and-control regulation?

A naive forecast of any social activity says past behavior is the best predictor of the future, and a naive forecast is hard to beat. That being the case, one can say with confidence that command-and-control, technology-based regulation will dominate the control of non-point source pollution. A review of related documents supports the forecast.[11] There are several reasons for this regulatory dominance. First off, all environmental regulation has evolved in the form of centralized command-and-control; the bureaucracy understands it and is geared to produce it. It is cheaper for the bureaucracy to expand this product line.

Next, command-and-control appeals to planners, and the regulation of natural resources is largely the domain of public sector employees and officials who have little, if any, private market experience. They are a part of the socialized sector of the US economy. When confronted with a problem, they understandably see a challenge to establish committees and commissions, to seek tax money, and then to establish a bureaucracy for dealing with the problem. The notion of leaving a problem in the hands of private citizens and the market is often antithetical to their views of how the world should work and is at odds with their personal experience. Problems are to be solved. Solutions are seen as written documents and reports to be adopted by or imposed on groups who implement the plan. Indeed, if marketlike instruments are used, they will likely be wrapped in bureaucratic clothing that limits their full play.

The tendency to regulate with command-and-control goes even deeper. The management of natural resources in the US is largely a public sector activity. Still organized under the rules of feudalism, the environmental sector is managed by the 'lord of the land' (Yandle 1992, 601-23).

Finally, there are the anticompetitive features of command-and-control discussed earlier. Put simply, command-and-control regulation that requires a reduction in pollution also leads to a reduction in product output.[12] The regulator becomes a cartel manager who enforces the restriction. Prices and profits rise so long as command-and-control rules the day.

The habit of applying command-and-control regulation is well ingrained, but the prospects for doing so in traditional ways for agricultural pollution is not so bright. Non-point source pollution simply does not lend itself well to the system developed for point source pollution. Even the above mentioned rules of best management practices allow for more flexibility than seen in point source regulation. Put differently, there is an opportunity to develop trading. Doing something different will require building institutions that support permit trading.

The necessary institutional characteristics

In a 1991 report based on the experience of the major industrialized nations, the Organization for Economic Cooperation and Development (OECD) focused on how to apply economic instruments in controlling pollution. The report examined a full range of economic incentives, including marketable permits, which are described as being applicable to point and non-point sources.[13] Assuming that pollution is a known quantity that can be measured, preconditions for marketable pollution permits noted in the report include the following:

- Marginal control costs differ across pollution sources.
- Command-and-control standards impose excessive costs.
- The environmental goal is fixed.
- Incentives to discover and implement improved control approaches are needed.
- The number of pollution sources is large enough to support a market where transactions can occur.

All but the last item in the OECD list emerged in the earlier discussion of trading. Even that item was considered less than crucial by OECD. Just one willing buyer and seller will make the market function. Concern about market thickness introduces a focus on transactions and the practical aspects of trading. But that presupposes a willingness to trade that must be based on property rules or the force of law. Legal institutions do not emerge out of thin air.

Property rights: a central issue

Markets can generate cost reductions, but markets are costly to operate. Once in place, they move spontaneously. The definition of property rights is a crucial piece to the puzzle. To bring a focus on property rights and motivation for trade, consider how pollution, including that from agriculture, was controlled under common law prior to the development of national statutes (Yandle 1989, 41-63; Meiners and Yandle 1992).

Under the common law doctrine of nuisance, owners of land downstream had a right to beneficial use of water that was of an undeteriorated quality. In the event of pollution from an upstream user, those damaged downstream could sue the upstream user. Proof of harm was required; injunction, not damages, was the remedy. The polluter was forced to cease polluting. Where more than one polluter was involved, all polluters were held jointly and severally liable, under a rule of strict liability. The common law also allowed for the enforcement of contracts between water users. An upstream discharger could purchase the right to reduce the quality of a receiving stream from downstream users. The party upstream was motivated to do so by rules of property. The burden of control of contract rested with the polluter, who consumed water quality belonging to others.

Common law rules developed from within communities, and they were based on well-specified property rights and measurable damage to those rights; they were not developed by external governments and imposed on communities. Common law courts, organized by communities, enforced the rules. And in ancient times, members of a common law community were ultimately liable, jointly and severally, for one another's actions. If one member imposed costs on the community and skipped town, any other member of the community could be held liable.

Under common law, a nuisance could be ruled private or public. If public, a local or state government official would bring suit on behalf of a group of harmed citizens who could prove damages. The remedy was the same. The Community Markets to Control Non-point Source Pollution polluter would have to stop the harmful action. Again, polluters had the option of contracting with the harmed citizens. Their rights could be purchased.

Sometimes communities moved to collective action and agree to require all property owners to own nontransferable rights to a common-access resource, such as a common pasture or park (the commons). The rights went with the land; acquiring land meant acquiring an undivided proportional share of the commons. Separate courts were operated for the purpose of settling grievances among the land owners. If one land owner abused the commons, the others could bring action, since the value of their interest was affected by abuse of the commons.

Going beyond the common law tradition, communities sometimes developed associations for managing rivers. The modern experience with the Ruhr (Ruhrverband), Wupper, and Emscher river basin associations (genossenschaften) in Germany is a case in point where cities that discharge treated sewage, park operators that impose loadings, and industrial plants that affect a common watercourse, are required to own shares of the association and to be assessed on the cost of maintaining water quality. Systems of fees and charges for discharge and recreational use are a part of this community water quality system. Operators who wish to increase their use of water quality are required to pay the association the cost imposed on other users of the stream or to take steps to reduce discharge, whichever is cheaper. If they pay discharge fees, they effectively pay the cost of maintaining the stream's quality. In effect, the expanding user is buying the right to use the river. No one specifies how a polluter will run a plant or manage a business. The 'owner' of the river is interested in one thing: maintaining the quality of his asset. What might be termed 'best management practice' is determined by market forces.

Consideration of these historical approaches, and recent experience with marketable permits, suggests there are certain characteristics common to the decentralized approaches that rely on established property rights. The following list contains what might be termed an ideal set of such characteristics:

- *Motivation:* Rules of property and liability are developed by members of a community who seek to improve their health and wealth. A surety system is developed that requires each community member to be liable for the costs imposed by any other member. If the system fails, the community must take offsetting action. Gains from trade under a rule of law motivate transactions.
- *Management:* The resulting property rule causes an affected river to be managed with ownerlike concern; its asset value is maintained. Management can be decentralized, under a rule of property law, or centralized, as in a river basin association.
- *Goal Certainty:* Baseline conditions are described for the river and for all uses that affect water quality; agreement is obtained regarding the level of quality to be maintained. Certainty regarding the level of quality to be maintained provides a scarcity parameter that motivates trade and quality controlling activities.
- *Transactions:* The property rule means that all water quality users deal with the owner/manager of the river. A decentralized system, like the land ownership system in the nation, means that individual buyers and sellers seek each other or operate through brokers when developing a permit trade. In a centralized system, the river basin manager is the broker.
- *Monitoring:* The water quality manager monitors water quality; members of the community have an incentive to do the same, since their investments are based on maintaining the asset. Monitoring certainty means that what is bought and sold is delivered. Those who poach on the system will be prosecuted.
- *Trading:* Mutually beneficial trade is allowed among members of the community who own rights to use water quality where the effects of the trade are confined to the trading parties. 'The water quality

manager monitors trade and use of the stream. 'The terms of contracts are enforced. Members of the association have an incentive to the same, since they are liable for any excessive expenditures required to offset the behaviour of cheaters.

- *Accountability:* Water quality managers are required to provide annual reports to each member of the association describing water quality conditions, levels of activity, and steps taken to maintain the asset.
- *Flexibility:* Unless transaction costs dictate it, there are no mandated methods for controlling pollution. Economic incentives and property rules interact to help water quality users find their least cost approach.
- *Entry:* New users of water quality must purchase rights from existing owners or pay the water quality association the cost of offsetting their discharge.

In considering this list, we must recognize that the common law did not work perfectly, nor have modern river basin associations. Indeed, it is unlikely that we will find a water quality management program that is perfect. At the same time, we can take the list of characteristics and observe just how closely they fit currently evolving institutions that might lend themselves to permit trading and the evolution of a market.

Permit trading under current conditions

The current picture

At present, the U.S. system for controlling water pollution motivates industrial sources to seek lower cost alternatives. The system of command-and-control they face is tied to higher water quality goals. Their motivation is dictated by regulation, not by a rule of property. In a sense, the EPA holds the property rights to water quality and is insisting the industrial sources reduce discharge. But the system goes beyond mere insistence. Industrial dischargers are required to reduce their discharge in specified ways. Command-and-control is the baseline.

Agricultural users of environmental quality do not face a binding water quality constraint; they are not bound by EPA technology based standards. In that sense, agricultural users may first be viewed as passive players in a potential permit market. As passive players, they have the property rights to their current use of the environment, but their property rights are threatened by pending rules.

Nonetheless, industrial sources of pollution can reduce their cost of reducing discharge by enticing agricultural interests to transfer some of their discharge rights. How can the transfer take place?

Permit trading in the current regulatory environment

At present, agricultural users of the environment are viewed generally as non-point sources of pollution. Seen that way, they cannot readily install a piece of capital equipment at the end of a pipe and reduce their environmental loadings. Instead, they must change the way they cultivate land; modify their dairy operations; find substitutes for chemical fertilizers, and take other continuing management steps.

However, technology has a way of transforming non-point problems to point sources. A search for sources of agricultural discharge will lead to identification of particular pastures, feedlots, and concentrated dairy operations where waste treatment facilities can be developed in conjunction with sediment basins. Alternatively, an entire farm can be named a point source.[14]

Command-and-control will either dictate methods of operation or installation of treatment facilities, which will likely take on the form of technology-based standards. All of this can be achieved easier if farmers are paid by the operators of industrial point source systems. Still, a middle man is needed. An institutional arrangement has to exist.

Since the burden of control is on industrial sources, they are motivated to organize and fund a community of dischargers so as to minimize control costs, which means finding low cost pollution control. A river basin or watershed association is a natural outcome, where a manager/broker becomes the intermediary between farmers and industrial dischargers. Once the association is formed and gains from trade are witnessed, farmers will be motivated to offer discharge reductions, to sell permits. They will be active participants in the market. Currently, farmers are unlike their industrial counterparts; they will not face penalties for not achieving specified pollution reduction goals. Of course, that condition will change if federal law specifies goals and penalties for agriculture.

The characteristics of a permit trading system, described above, tell us some of the functions of the association manager. The association must develop baseline data, identify the water quality goal to be met for specific pollutants, and document current activities and their impact on achieving the water quality goal. Of course, that is where current uncertainty is largest. The impact of agriculture on water quality is highly uncertain and quite variable. Still, with the baseline identified, the manager must monitor water quality and the behaviour of all users.

If an industrial discharger finds a farmer who can reduce his discharge at lower cost, the industrial discharger will be willing to pay up to his alternative cost for the farmer's reductions. The association manager must monitor the trade and enforce the actions of the farmer, who reduces his discharge. The offsetting increase from the industrial source must also be measured and monitored.

Permit trading among non-point sources

Gains from trade and the binding constraint faced by industrial point sources can motivate trade among industrial point sources and agricultural non-point sources of pollution. A change in the agricultural legal environment could motivate trade among non-point sources. If a binding constraint is imposed on non-point sources of pollution, which means their current rights to discharge are confiscated, then non-point sources will be motivated to minimize cost. The market for permits will expand to include farmers as buyers, not just sellers of permits. The binding constraint will have to specify levels of discharge to be achieved in a stated period of time.

Consider two farmers who face a sediment control constraint. Assume their operations have differing terrain and soil characteristics; their crops are different, as are their tillage and harvest methods. Further assume that a unit of run-off from the two farms has equal impact on the quality of a receiving stream. Both farmers are required by law to reduce their current levels of expected run-off. If farmer A can alter his crops and tillage methods at a lower cost than farmer B, there are potential gains from trade. Now, add to the market an industrial discharger of suspended solids who also seeks to reduce community discharge, his or someone else's. If the industrial discharger faces a higher opportunity cost than either farmer, he may purchase permits from both farmers.

The expanded situation contains binding quality constraints that may differ for all parties but are binding for each of them. The system may include technology-based standards for industrial dischargers and best management practices for farmers. That is, a regulatory baseline sets certain conditions that define alternatives for the trading parties. Even so, there can be gains from trade that result from differences in costs and differences in the value of product.

What do case studies tell us?

At this point, it is safe to say that there have been no permits traded among non-point sources in the United States.[15] However, budding institutions for accommodating trade appear to be emerging. Whether or not they are ready to function is an open question. The EPA has documented the institutional characteristics, potential gains from trade, and the possibilities for point source/non-point source transactions (EPA Office of Water 1992); the regulatory constraints that motivate trade have not become binding, and in some cases there is a subsidized alternative for agriculture that should reduce interest in entering the market for permits.[16] However, a combination of continued economic growth, tighter regulatory constraints, and increased demand for water quality can combine to cause a market to emerge. Scarcity can be contrived by regulation.

As current systems stand, the motivation for trade comes largely from industrial sources. Baseline conditions are defined by management of river basin associations, state regulatory authorities, or the US Geological Survey. Generally speaking, a team of specialists that include faculty from land grant universities and the US Soil Conversation Service participate in baseline identification.

Since there have been no trades, nothing can be said about the form and enforcement of contracts. However, efforts have been made to develop modeling techniques that enable water quality managers to identify the water quality impact of non-point source modifications (EPA 1990). In addition, the EPA has developed proposed guidance for non-point sources that must meet the requirements of the Coastal Zone Act of 1990 (EPA 1991). This regulatory step is supplemented by a memorandum of understanding that addresses a joint effort to reduce agricultural pollution (EPA 1992b). Simply put, the locomotive of change is on the track and moving.

Final thoughts on the evolving system

The evolving system of water quality in the nation has now reached the point where agricultural firms and other non-point sources of discharge are a part of the pollution control scheme. The regulatory path followed to this point predicts that command-and-control, technology-based regulation will continue to play a fundamental role in the overall process. That is particularly the case for industrial sources of pollution and for municipal treatment works. However, the inherent difficulties associated with specifying technology for non-point sources raise the possibilities for a property rights system and enhanced ability to minimize costs.

In some cases, subsidy schemes will reduce the supply of pollution reduction permits from agriculture and work against achieving environmental goals; the potential benefit of gains from trade will be reduced. In other instances, the bureaucratic process involved in permit trading will be so burdensome that firms will be discouraged from attempting to enter the constrained market. However, encouragement can still be mustered when a community of interest develops in a watershed area and the people directly concerned take it upon themselves to organize a water quality management system based on flexible market forces.

At this point, the best possible outcome is for numerous experiments to develop where different approaches and different institutional arrangements emerge. Our best hope lies in the discovery process that always tends to emerge when scarcity beckons.

There is no doubt that scarcity will beckon. The EPA has recently released a draft study of the economic effects on the agricultural sector of meeting the requirements of the Coastal Zone Act Amendments of

1990, which affects 13 percent of all American farms, and 45 percent of all horticultural producers, including 48 percent of vegetable production.[17] Many of the management practices being discussed for all of agriculture are now a part of the requirements of the Coastal Zone Act. The projected impacts are large and significant, especially for smaller agricultural units. The EPA documents the number of vulnerable farms, the expected number of units that will exit the industry and the related price effects. In short, costs are high. The effects on water quality are still speculative.

Some specific recommendations

Given the high costs and disruptions expected from efforts to control non-point sources, it is important that communities, states, and regions be encouraged to find water quality management systems that match their particular soil, water, farming, and other characteristics. Permit trading is one option that will likely emerge in the experiments. It raises community issues that go beyond the notion of experimentation. The following recommendations address some of the issues:

- On a case-by-case basis, a set of scientifically determined baseline data must be obtained. These include determinations of the linkages between agricultural activity, discharge and run-off, and ambient water quality of affected streams. Water quality improvement not pollution prevention, should be the goal for any trading community.
- Once an environmental constraint is set, it should not be changed arbitrarily. The constraint forms a property definition. Those who wish to reduce pollution beyond the total constraint should be required to purchase reductions from members of the affected permit trading community.
- A credible monitor/broker must be identified to maintain a set of baseline data, manage transactions, and monitor environmental conditions and activities across permit traders. The monitor/broker's services can be funded through a system of brokerage and membership fees.
- Farmers who achieve reductions in discharge beyond the requirements of an environmental constraint should be allowed to 'bank' their reductions for future trade. Expanding members of a trading community will be required to purchase allowances from existing farmers.
- Trading communities should be encouraged to establish water courts of law that specialize in settling contract disputes and in handling suits that relate to abuses of the water law.
- Proposed changes in agricultural policy that affect production, commodity prices, and subsidy programs should be evaluated first to

determine their impact on the environmental property rights formed by permit trading communities. If permits are effectively confiscated, the affected parties should be paid the market value of their lost assets.

This chapter has focused on institutions that have evolved for managing water quality and has described some changes that have occurred in the regulatory pageant. Property rights and the possibilities for gains from trade have been the focal points of the discussion.

The development of institutions for managing environmental quality is a study in community action; the environment, whether it be air or water quality, is still a common access resource that is made private through the design of legal arrangements, which is a community activity. If there are to be river basin associations or trade of pollution permits, it will be with the cooperation of members of an affected community.

If trade evolves, we can be certain that the resulting market will be highly regulated. However, given regulation, we can be confident that trading parties will be better off with the right to trade than without it. With property rights protection, we can also be assured that trade will improve the well-being of the larger community.

Notes

1. See Testimony of William K. Reilly. Reilly is not alone in his persuasion regarding the merits of the market. After almost 20 years of history using other approaches, the OECD has concluded that 'economic instruments provide more flexibility, efficiency and cost-effectiveness.' The multinational organization reports that 'declarations at [the] highest political level call for a wider and more consistent use of these instruments.' (See Organization for Economic Cooperation and Development 1991)

2. The story sketched here is certainly not novel; details for such a scheme were described in 1968 by J. H. Dales. The fact that the United States has not chosen to follow the scheme reflects a deliberate choice, not a lack of knowledge.

3. A much documented and analyzed situation comes to mind here. The Fox River of Wisconsin is a case where a permit trading system was developed. Because of bureaucratic inertia and heavy regulation of technologies, the scope for trade finally vanished. For the optimistic story on the possibilities offered, see Joeres and David. For a description of what has not happened, see Yandle 1991a. Also see Maloney and Yandle.

4. The trading analysis here assumes that discharge from the industrial source is a perfect substitute for discharge reduction from the agricultural source. In a geographic sense, one can imagine a farm and industrial plant operating in close proximity. Trading ratios may be established to cause discharge from different locations to be equivalent. That is, one unit of waste reduction from a farm may be equal to two units from an industrial source, if the effect of industrial waste on water quality is half that of farm-generated waste.

5. See Segerson; her analysis recommends a reward/penalty scheme that deals with the problem on a collective basis.

6. EPA's document on the rural clean water program effectively describes the non-point source problem and management practices that can reduce discharge. However, in the extensive discussion of specific cases, the report generally fails to give solid, unambiguous evidence that the reductions in agricultural discharge translate into improvements in water quality. This scientific weakness is obviously a comment on the complex nature of the problem being addressed. See U.S. Department of Agriculture and U.S. Environmental Protection Agency.
7. For relevant discussion, see Yandle 1978; Hahn; and Hahn and Hester. For a useful recent survey of literature, see American Petroleum Institute.
8. See Pashigian 1984, Pashigian 1985, and Maloney and McCormick 1982.
9. The research on air pollution control is rather extensive. For a survey of findings, see Yandle 1991.
10. See Magat et al. A survey of studies of air pollution control costs by Yandle (1991a) indicates similar opportunities for reducing costs. He calculated the ratio of command-and-control to least cost controls (which assumes marketable permits) and found the ratio to be as high as 22. The typical ratio was in the order of 4 to 5 to one.
11. For example, see U.S. EPA, *Managing* Non-point *Source Pollution.* For a survey of state programs, see National Water Quality Evaluation Project. For discussion of activity underway in North Carolina, see Division of Soil and Water Conservation. In general, these and other reports speak of and rely on 'best management practices,' which are technology based, as the basis for regulation.
12. For the theoretical insight, see Buchanan and Tullock.
13. Ibid., pp. 30-31. The report suggests that point source polluters can obtain the right to increase pollution by financing the expansion of 'best available agricultural practices.' (Ibid., 30.)
14. Both these ideas have been suggested by Alm.
15. This statement is based on the EPA 1992b report on the topic.
16. The North Carolina Department of Natural Resources and Community Development administers a state-funded program that provides 75 percent of the cost of adopting improved management practices in agriculture. The remaining 25 percent can be paid in cash or in kind.
17. The related EPA documents are: EPA 1992b, EPA 1992c, EPA 1992d, and EPA 1992e. The latter two items are a part of the draft reports. All are from EPA's Non-point Source Control Branch, Washington, D.C.

References

Alm, Alvin L. 1991. Non-point Source Pollution. *Environmental Science and Technology* 25: 1369.

American Petroleum Institute. 1990. *The Use of Economic Incentive Mechanisms in Environmental Management*. Research Paper #051. Washington, DC: American Petroleum Institute.

Buchanan, James M., and Gordon Tullock. 1975. Polluter's 'Profit' and Political Response. *American Economic Review* 65 (March): 139-47.

Conservation Foundation. 1987. *State of the Environment: A View Toward the Nineties.* Washington, DC: Conservation Foundation.

Council on Environmental Quality. 1989. *Environmental Trends.* Washington, DC: Council on Environmental Quality.

Crowder, Bradley M., Marc Ribaudo, and C. Edwin Young. 1988. *Agriculture and Water Quality*. Washington, DC: U.S. Department of Agriculture.

Dales, J. H. 1968. *Pollution, Property and Prices.* Toronto: University of Toronto Press.

Davis, Erroll B. 1992. Cleaning Up Pollution. *New York Times* (May 17): 12F.

Division of Soil and Water Conservation, North Carolina Department of Natural Resources and Community Development. 1987. *North Carolina Agriculture Cost Share Program for* Non-point *Source Pollution Control.* Raleigh, NC: Division of Soil and Water Conservation (May).

Environmental Protection Agency. 1990. *Biological Criteria*. Washington, DC: EPA (April).

______1991. *Proposed Guidance Specifying Management Measures for Sources of* Non-point *Pollution in Coastal Waters*. Washington, DC: EPA (May).

______. Office of Water. 1992. *Managing* Non-point *Source Pollution*. EPA 506/19-90. Washington, DC: EPA (January).

______.Office of Water and Office of Policy. 1992. *Incentive Analysis for Clean Water Act Reauthorization: Point Sourcel*Non-point *Source Discharge Reductions*. Washington, DC: EPA, April.

______.1992a. *Fact Sheet: Agriculture and Pollution Control*. Washington, DC: EPA, April.

______. 1992b. *Draft: Economic Impact Analysis of Coastal Zone Management Measures Affecting Confined Animal Facilities*. Washington, DC: EPA (June 11).

______. 1992c. *Preliminary Economic Achievability Analysis: Agricultural Management Measures*. Washington, DC: EPA, (June 12).

______. 1992d. *Agricultural Impacts of Erosion Management Measures in Coastal Zone Drainage Basins*. Washington, DC: EPA.

______. 1992e. *Agricultural Impacts of Requiring Alternative Conservation Systems in Coastal Zone Drainage Basins*. Washington, DC: EPA.

Hahn, Robert W. 1989. Economic Prescriptions for Environmental Problems: How the Patient Followed the Doctor's Orders. *The Journal of Economic Perspectives* (Spring): 95-114.

Hahn, Robert W., and Gordon L. Hester. 1989. Marketable Permits: Lessons from Theory and Practice. *Ecology Law Quarterly* 16:361-406.

Joeres, Erhard R, and Martin H. David. 1983. *Buying a Better Environment.* Land Economics Monograph No. 6. Madison: University of Wisconsin Press.

Loigu, E. 1989. Evaluation of the Impact of Non-Point Source Pollution on the Chemical Composition of Water in Small Streams and Measures for the Enhancement of Water Quality. *In Advances in Water Pollution Control*, cd. by H. Laikari-

MeSweeny, Williain T., and Randall A. Kramer. 1986. The Integration of Farm Programs for Achieving Soil Conservation and Non-point Pollution Control Objectives. *Land Economics* 62 (May): 159-73.

McSweeny, Williain T., and Jaines S. Shortle. 1990. Probablistic Cost Effectiveness in Agricultural Non-point Pollution Control. *Southern Journal of Agricultural Economics* 22 (July): 95-104.

Magat, Wesley A., Alan Krupnick, and Winston Harrington. 1986. Rules in the Making: *A Statistical Analysis of Regulatory Agency Behavior*. Washington, DC: Resources for the Future.

Maloney, Michael T., and Robert E. McConnick. 1982. A Positive Theory of Environihental Quality Regulation. *Journal of Law and Economics*. 25: 99124.

Maloney, Michael T., and Bruce Yandle. 1983. Building Markets for Tradable Pollution Permits. *In Water Rights*, ed. by Terry Anderson. San Francisco: Pacific Institute for Public Policy Research.

Meiners, Roger E., and Bruce Yandle. 1992. Constitutional Choice in the Control of Water Pollution. *Constitutional Political Economy* 3 (Fall): 359-80.

Milon, J. Walter. 1987. Optimizing Non-point Source Controls in Water Quality

Regulation. *Water Resources Bulletin* 23 (June): 387-95.

National Water Quality Evaluation Project. 1989. *NWQEP 1988 Annual Report: Status of Agricultural* Non-point *Source Projects*. Raleigh, NC: Biological and Agricultural Engineering Department, North Carolina State University, May.

Organization for Economic Cooperation and Development. 1991. *Environmental Policy: How to Apply Economic Instruments*. Paris: OECD.

Pashigian, Peter. 1984. The Effects of Environmental Regulation on Optimal Plant Size and Factor Shares. *Journal of Law and Economics* 27 (April): 1-28.

______.1985. Environmental Regulation: Whose Interests Are Being Protected? *Economic Inquiry* 23 (October): 551-84. -

Sergerson, Kathleen. 1988. Uncertainty and Incentives for Non-point Pollution Control. *Journal of Environmental Economics and Management* 15: 87-98.

Testimony of Williain K. Reilly, U.S. Environmental Protection Agency. 1991. Presented before the U.S. House of Representatives, March 20, Washington, DC, 32-33.

U.S. Department of Agriculture and U.S. Environmental Protection Agency. 1991. *The Rural Clean Water Program: A Report*. Washington, DC.

Yandle, Bruce. 1978. The Emerging Market in Air Pollution Rights. Regulation (July/August): 21-29.

______.1989. *Political Limits of Environmental Regulation*. Westport, CT: Quorum Books, Inc.

______.1991a. A Primer on Marketable Permits. *Journal of Regulation and Social Costs* 1: 25-41.

______.199lb. *Why Environmentalists Should Be Efficiency Lovers*. St. Louis: Washington University Center for the Study of American Business (April).

______. 1992. Escaping Environmental Feudalism. *Harvard Journal of Law and Public Policy* 15 (March): 601-23.

13 Meeting global food needs: the environmental trade-offs between increasing land conversion and land productivity

Indur M. Goklany

Summary

Despite this century's dramatic population increase, the global food situation, possibly excepting sub-Saharan Africa, has improved remarkably – mainly due to the interdependent forces of economic growth, technology and trade. However, low purchasing power and strife keep certain populations vulnerable. Due to increased land conversion and inputs (fertilisers, pesticides and water), the improvements have exacted an environmental price. That price would have been higher without technological change which, since 1961, forestalled additional conversion of 3,550 million hectares (Mha) of habitat globally to agricultural uses, including 970 Mha to cropland. Increasing – and richer – populations may raise food demand 120 per cent between 1993 and 2050, which can only be met by increasing cropland, productivity, or both. The precise combination is critical for global biodiversity. Increasing average productivity 1 per cent a year implies losing 368 Mha of habitat to cropland by 2050; while a 1.5 per cent a year increase would reduce cropland by 77 Mha. Either is plausible, given productivity-enhancing opportunities. To the extent productivity increases result from additional inputs, the environmental benefits of reducing habitat loss may be partially offset. Given the severity of habitat conversion, prudence suggests increasing productivity while using inputs efficiently and mitigating their impacts. That requires sustained commitment to economic growth, technological change and freer trade. Otherwise, technologies – whether to maintain or increase productivity or mitigate impacts – can neither be developed nor afforded and access to food will diminish. That would – by reducing food supplies, increasing vulnerability of the poorest to hunger, escalating habitat loss and increasing environmental degradation – deprive humanity and despoil the rest of nature.

Introduction

At least since Malthus (1926) wrote his first essay on population in 1798, raising the spectre of geometric population growth outstripping humanity's ability to expand food supply – which he presumed could only increase arithmetically – scholars and policy makers have debated whether future food needs can keep pace with population growth. To this debate, a new issue should now be added, namely, whether these needs can be met without displacing the rest of nature (Goklany and Sprague 1991; Goklany 1992, 1993; Waggoner 1994). In the past two centuries global population increased from under one billion to 5.7 billion and ever larger shares of the world's land, water and other resources have been expropriated for human needs (Vitousek et al. 1986; Turner et al. 1990; Cohen 1995; Goklany 1995a). In particular, conversion of land to agriculture is the single greatest agent of habitat conversion and associated displacement of species and increasing stress on biological diversity (Goklany and Sprague 1991; Goklany et al. 1992; Goklany 1995a, 1998a). These pressures will only increase if, as many project, the world's population doubles over the next 100 to 150 years (World Bank 1994) and becomes wealthier (Goklany and Sprague 1991; Goklany 1996, 1998b).

This Chapter first addresses whether Malthus' prognostications regarding an increasing imbalance between population growth and food supplies have been vindicated and, if not, why not? In this context, it will examine the correspondence, or lack of it, between trends in global population and cropland over the last two centuries and, over a more limited period, trends in agricultural land. The Chapter will then address whether past progress in meeting global food demand can be continued into the future and, equally importantly, whether and to what extent the quest to meet that demand could diminish habitat for the rest of nature. Finally, based on the lessons of the past, it will recommend a set of self-consistent, practical policies that would help meet the increasing food demand of a larger and wealthier population into the middle of the next century, while reducing environmental harm and minimising land conversion and associated loss of biodiversity.

A historical perspective

During the last two centuries, global population has sextupled; yet the average person is fed better, more cheaply and spends less time and effort getting food on the table. Thus, Malthus' countryman is about three inches taller, healthier, less prone to disease and lives twice as long today as his contemporary of 1798 (Fogel 1994; Wrigley and Schonfeld 1981; UNFPA 1995). The situation elsewhere around the world has also improved markedly, particularly since World War II. Since their post war lows, average daily food supplies per capita in China and the Indian subcontinent – 42 per cent of humanity – increased about 30 per cent and 45 per cent, respectively (Table 13.1). The number of people in areas with reported famine dropped from over 700 million annually (1950-56 average) to 35 million in 1992, a drop of 95 per cent (Chen and Kates 1994). Chronic under-nourishment decreased from 941 million in 1969-71 to 781 million in 1988-90, declining from 36 per cent of the population of developing nations to 20 per cent (Alexandratos 1995b). In fact, it is a measure of humanity's success that some suggest the issue is no longer one of food quantity or security, but quality, or nutritional security (Swaminathan 1989).

Despite increased demand due to both increasing numbers and more affluent populations, the real price of food and produce has dropped. Between 1950 and 1992, international food commodity prices dropped 78 per cent in constant 1990 prices (Mitchell and Ingco 1993). Following that, prices increased, taking back 8 per cent of the drop by the first half of 1996 (World Bank 1996; WRI 1996). This was mainly due to a combination of several singular and temporary factors including reductions of agricultural subsidies in both Eastern and Western European nations; poor winter wheat prospects in the US; and delayed spring wheat plantings in the US, Canada and Western Europe in 1996, coupled with historically low levels of reserve stocks (FAO 1996b). Despite these recent increases, food prices are much lower today than for much of history. Some have ascribed these increases to a fundamental turnaround, presaging increasing future difficulties (Brown and Kane 1994), but there seems to be little evidence of that (Alexandratos 1995a; Harris 1996). In fact, while one should be cognisant of them, it would be imprudent to base long-term policies on just short-term trends – hence the attention in this paper to both long- and short-term trends.

For the US, from the turn of this century (1897-1902) to 1992-94, retail prices of flour, bacon and potatoes relative to per capita personal income dropped 92 per cent, 87 per cent and 80 per cent, respectively, even as the length of the average work week declined 40 per cent (Maddison 1989); bread dropped 78 per cent between 1919-21 and 1992-94 (Figure 13.1). As relative prices of food and agricultural products have declined, so has agriculture's share of the Gross Domestic Product (GDP) and labour force. It now contributes about 2 per cent to the US GDP and 3 per cent to its civilian labour force as opposed to about 75 per cent in 1800 (Bureau of the Census 1975, 1993). Ironically, the very success of the agricultural sector, by decreasing its economic and demographic importance for the rest of society could, as will be discussed later, adversely affect the ability to meet future food needs. Nevertheless Churchill's immortal words, 'never ... has so much been owed by so many to so few,' may be just as apt to today's American farmer as it was to the victors of the Battle of Britain.

Still problems linger: as noted, famine and malnourishment while reduced, have not been eliminated. However, today's problems are due as much to failures of institutions, policies and political systems causing greater poverty, civil strife and disincentives to food production, as to the inability to physically produce food (Sen 1981, 1993; Drèze and Sen 1990; Kumar 1990; Tapsoba 1990). In Africa, despite a 60 per cent increase in food production between 1969-71 and 1992 (FAO 1994b), available food supplies have barely kept pace with population growth rates (Table 13.1) due, in part, to economic systems which inadequately reward farmers for their investments and risks, or subsidise some consumers at their expense (Tapsoba 1990). Civil strife in Ethiopia, Somalia and Sudan converted what may have been more manageable droughts into full-scale famines. Similarly, the problems of Peru can be traced, in part, to civil strife. The regression indicated by the 1992 figures for the formerly centrally planned economies (ex-CPEs or 'transition' nations) is due to the transition from an unsustainable socio-economic system to a market

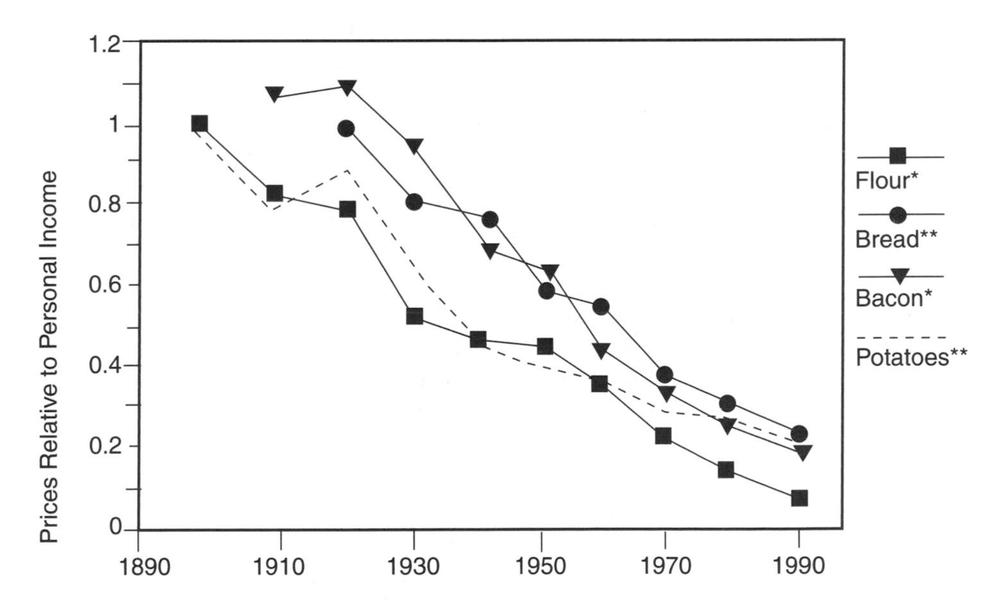

Note: Data are 3-year averages centred on the year for which data are shown, except for 1899 and 1909, which are 5- year averages centred atround those years.
*price relative to income = 1 in 1887-1901; **price relative to income = 1 in 1919–1921.

Sources: Historical Statistics of the United States, various issues of CPI Detailed Report and Statistical Abstracts.

Figure 13.1 Retail prices of various food items relative to per capita personal income (1899–1993)

economy, aggravated in the former Yugoslavia by an uncivil war. With the current peace in Bosnia, food supplies are recovering in that area (FAO 1996b).

In summary, the last two centuries' experience suggests that so far at the global scale, at least, Malthus' fears regarding the ability of food supply to keep pace with population growth have not yet been borne out. However, at local to regional scales, there have been several instances where food supplies have effectively diminished, at least temporarily, but not necessarily for the reasons he anticipated. Is it possible that, as the world's population doubles over the next century he may yet be vindicated at the global scale? The following attempts to shed light on this question by first examining the reasons for past successes and failures before evaluating future prospects into the next century.

Factors contributing to improved food security: past lessons

Land, water and technology

Between 1800 and 1993, while global population increased about 500 per cent, cropland increased 250 per cent from about 408 million hectares (Mha) to 1,448 Mha (Figure 13.2). The difference between these increases hints at the contribution of technological change to increasing global food supplies. In fact, technology reversed the gradual, centuries-long global trend of increasing cropland per capita: from 0.43 ha per capita in 1700, it peaked at 0.48 in the first few decades of this century; today it is at 0.26 (Figure 13.2).

Technologies responsible for this turnaround affect each link of the entire food chain, from farm to consumer. Specific technologies – many of which rely on increased use of inanimate rather than human or animal energy – include scientifically-bred high yielding varieties (Hives) of crops which mature faster, making multiple crops more possible; improved livestock management practices; fertiliser and pesticide usage; mechanisation; refrigeration; canning and other means of preserving food and produce; and meteorological forecasts (Goklany and Sprague 1995). Irrigation – an ancient technology, freshly applied – today has been extended to a sixth of the world's cropland (compared to about 2 per cent in 1800 and 6 per cent in 1950) and provides a third of the crop production, on average increasing yields 2.5-fold (Crosson 1995a; Rozanov et al. 1990). Finally, an elaborate transportation and distribution infrastructure makes possible local, regional and international trade in agricultural products and inputs (fertilisers, pesticides, fuels) with greater rapidity – cheaper costs and lower losses than ever before.

Table 13.1. Per capita food supplies (Kcal/Day), selected countries (1934/38 to 1992).

Country	*1934-38*	*1950-51*	*1961-63*	*1970*	*1980*	*1990*	*1992*
France	2,830	2,827	3,299	3,349	3,455	3,650	3,633
Germany	3,070[a,h]	2,807[a]	2,945[a]	3,217	3,382	3,455	3,344
Portugal		2,234[b]	2,570	2,966	2,925	3,590	3,634
Czechoslovakia			3,367	3,357	3,330	3,638	3,156
Yugoslavia		2,400	3,119	3,327	3,594	3,551	
EUROPE			3,098	3,255	3,390	3,470	3,410
US	3,150[i]	3,085[c]	3,195	3,192	3,333	3,680	3,732
Cuba	2,609[i]	2,682	2,286	2,638	2,998	3,104	2,833
Brazil	2,150[i]	2,353[c]	2,315	2,448	2,705	2,731	2,824
Peru		2,077[d]	2,225	2,294	2,042	1,825	1,882
S. AMERICA			2,395	2,500	2,662	2,621	2,689
China	2,226[e,j]	2,115[e,f]	1,666	2,032	2,332	2,679	2,727
India	1,970[g]	1,635	1,997	2,082	1,959	2,297	2,395
Pakistan	[k]	1,624	1,705	2,200	2,114	2,431	2,315
Bangladesh	[k]	[l]	1,953	2,196	1,902	1,994	2,019
Japan	2,180	2,100	2,591	2,691	2,758	2,906	2,903
ASIA			1,887	2,144	2,276	2,544	2,585
Egypt	2,450	2,342	2,296	2,515	3,119	3,336	3,335
Zimbabwe		2,224	2,053	2,226	2,292	2,173	1,985
Somalia			1,704	1,819	1,788	1,769	1,499
AFRICA			2,086	2,244	2,285	2,294	2,282
WORLD			2,289	2,464	2,559	2,718	2,718

[a] West German portion only
[b] 1949/50
[c] 1951
[d] 1952
[e] 22 provinces only
[f] 1947/48
[g] Includes what is currently Pakistan and Bangladesh
[h] 1935-1938
[i] 1935-39
[j] specified as 'pre-war'
[k] see India
[l] see Pakistan
(FAO 1952, 1954, 1991, 1994)

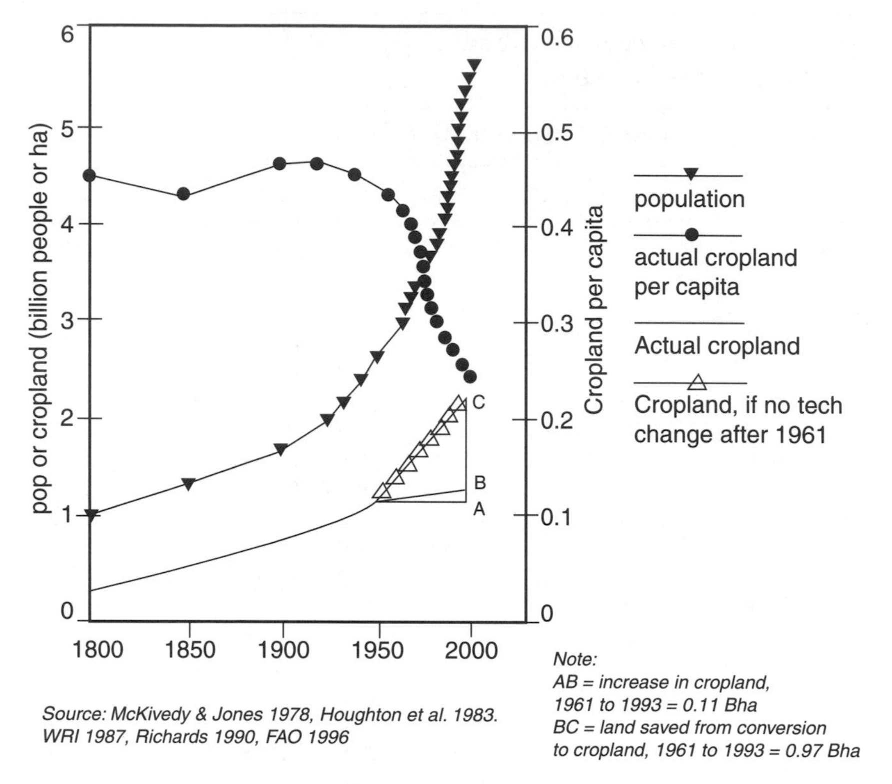

Figure 13.2 Global cropland area, 1800–1993.

Of course, these technologies also create their own problems. The adverse environmental effects of diverting water for irrigation and inefficient use of fertilisers and pesticides are well known (Ehrlich et al. 1993; Pimentel et al. 1994a, 1994b, 1995). Less appreciated is that technology, in addition to feeding a much larger population better, has also prevented the conversion of enormous amounts of the world's habitat of forests and grasslands into cropland or other agricultural uses (Goklany and Sprague 1991; Goklany 1993).

Between 1961 and 1993, population increased from 3.08 to 5.54 billion (FAO 1996a). If technology had been frozen at 1961 levels, then in 1993 the world would have needed to convert at least 1,073 Mha of other habitat to cropland (indicated by line AC on Figure 13.2), rather than 107 Mha (line AB). This (AC) is probably an underestimate. This assumes that after 1961, cropland would expand proportionally to population and the world's 1993 population would be fed no better than the inadequate levels of 1961 (particularly considering the unfortunate realities of unequal distribution). In other words, daily per capita calorific and protein intakes would be at 2,259 Kcal and 59.5 gm compared to current (1992) levels of 2,718 Kcal and 70.8 gm, respectively (FAO 1991, 1994a).

Similarly, Figure 13.3, which compares the 1961-1993 trends in total global agricultural area (cropland and permanent pasture), population and agricultural area per capita, also shows that under a technology freeze at 1961 levels, at least 3,546 Mha (BC) – in addition to the 375 Mha actually converted (AB) – would need to be converted to agricultural uses. That would have increased agriculture's share of all land (excluding Antarctica) from the current 34 per cent to 61 per cent.

To gain another perspective on the amount of habitat 'saved' from conversion since 1961, consider that: 1) globally, forest and woodlands (including natural and planted tree stands and logged areas slated to be reforested) declined 144 Mha between 1961 and 1993 (FAO 1996a); and 2) all 'protected' areas in the world, excluding Antarctica, total 960 Mha (WRI 1996), of which 510 Mha are 'totally protected'.

In the US, the amount of cropland harvested in 1995 was 7 per cent less than in 1910, despite a 186 per cent increase in the population and an increase in exports relative to production (Bureau of the Census 1975, 1995; USDA 1996). Goklany and Sprague (1991) calculated that if technology had been frozen at the 1910 levels, in order to produce the same quantity of food as in 1988, an additional 370 Mha of cropland would have had

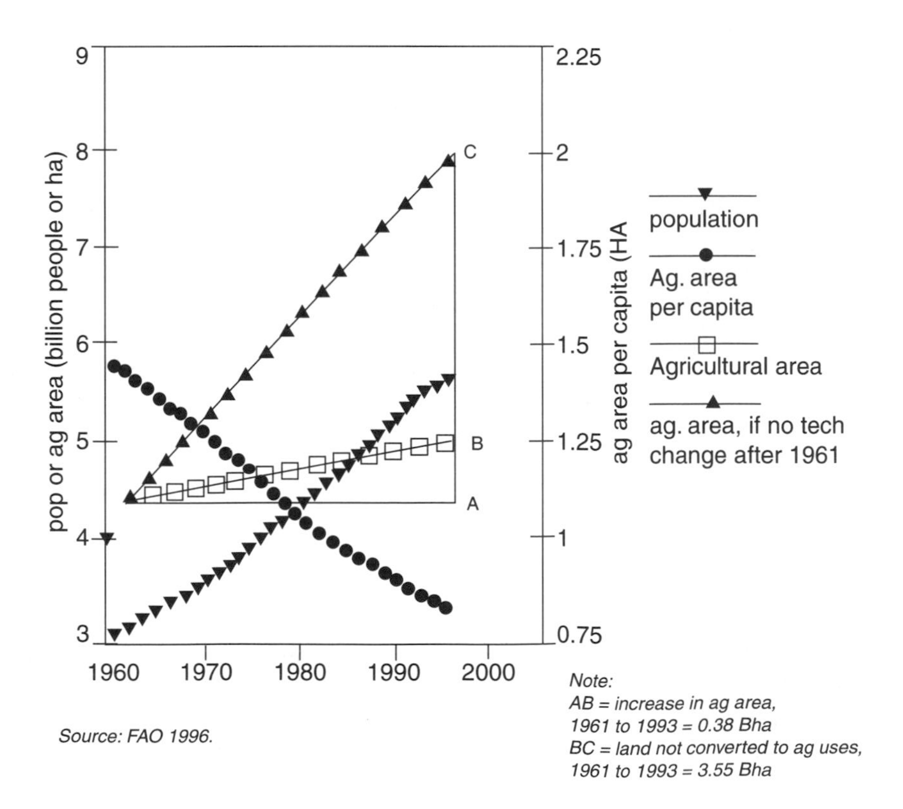

Figure 13.3 Global agricultural area, 1961–93.

be harvested in 1988 – more than all the remaining arable and forested ınds, combined, in the US. The resulting land conversion, equivalent to ›3 per cent of the US excluding Alaska (or more than three times all federally owned lands including all National Parks, Forests, Grasslands, Wildlife Refuges, Recreation Areas and multiple-use lands), would have devastated many species and rendered irrelevant any discussions in the US about preserving biodiversity *in situ*. Unwittingly and virtually unheralded, technological change leading to increased agricultural productivity, despite its numerous adverse environmental affects, seems to have been the single most important agent for conserving habitat and species (Goklany et al. 1992).

The development and adoption of new technology in the US was aided by a high level of general and technical education; strong commitment to research and development; and elaborate systems for technology transfer. However, the US experience, confirmed by the decades-long experiments of the CPEs, including China, for example, indicate that these conditions are insufficient, by themselves, to sustain high-yield agriculture and a modern food supply system. The legal, economic, and political framework also has to encourage innovation and adoption of new technologies by economically rewarding individuals for their endeavours and risk taking (Goklany and Sprague 1991).

Economic growth/reduction of poverty

Economic growth affects food security and agricultural production in several ways. First, as the recent food-related problems in Africa and the transition nations illustrate, greater wealth translates into greater purchasing power which, in turn, increases food security for society, families and individuals (Goklany 1998a). Table 13.1 hints at this; the poorer nations seem to have lower food supplies. Second, it takes more than the availability of technology to have technological progress: technology, first, has to be created and, then, it has to be adopted (Goklany 1996). That product of economic growth – capital, both human and financial – is critical to both stages. Wealthier nations have more highly-educated populations and greater resources for research and development (R&D) and poverty directly reduces the affordability of new and existing technologies (new seeds, fertiliser, irrigation and transportation systems). Thus, technological progress and affluence co-evolve (Goklany 1995a). In fact, the legal and economic institutions that nurture one, sustain the other. Third, affluence increases incentives for lowering fertility rates which, in turn, would reduce demands on the food system – all else being equal (World Bank 1984; Livi-Bacci 1992).

Finally, consistent with the theory of an environmental transition, economic growth enables societies, sooner or later, to reverse water pollution problems, such as phosphorus, nitrate, or dissolved oxygen levels caused or aggravated by agricultural practices (Goklany 1994, 1995a, 1995b). This theory holds that a society's environmental problems will, in general, first be aggravated by economic growth, then go through a transition after which further growth (and greater affluence) will reduce those problems. The point at which the transition occurs depends upon the magnitude and urgency of the specific problem and the cost to society of action (or inaction) (Goklany 1998a). Many economists refer to such behaviour which results in an inverted U-shape curve if environmental degradation is plotted against affluence (Grossman and Krueger 1991; Shafik and Bandyopadhyay 1992), as indicative of a 'Kuznets curve' – after Simon Kuznets (1955) who showed that economic inequality between the richest and poorest segments of society first increased and then decreased as a function of affluence. Thus, in the long run, wealth eventually increases the overall sustainability of agriculture.

Trade

Trade – both within and between nations – is the mechanism for allowing non- or low-producers to obtain food. It translates purchasing power (or economic growth) into food security. Trade globalises sustainability by allowing one area's shortage to be willingly made up by another's surplus, that is, it converts a locally unsustainable system to one sustainable in a wider context (Goklany 1995a). As a gauge of its importance consider that in 1993-96, developing nations' cereal imports amounted to 15 per cent of their production (FAO 1996b); and in 1991-93, of the 153 countries for which the World Resources Institute (1996) provides data on trade in cereals, 127 were net importers. Of the 26 net exporters, only one was in Africa; six were in Asia; and fourteen, who provided the bulk of the exports, were from the wealthy Organisation for Economic Co-operation and Development (OECD) block of nations. Similarly, virtually all food aid is donated by the richer nations. Thus, patterns of trade and aid also confirm the significance of affluence in increasing the world's food security.

Trade also helps reduce the exploitation of marginal lands for growing crops – provided neither trade nor production is subsidised. Thus, because of internal trade in the US, land used for crops and pastures in the Northeastern US declined 59 per cent between 1949 and 1987, despite increases in: 1) regional and national populations of 31 per cent and 70 per cent, respectively; 2) agricultural exports from 13 per cent of production to 35 per cent and 3) agricultural subsidies (Goklany and Sprague 1991). Most of the land so freed up was returned to nature and many species, previously under stress, have rebounded.

Free trade is also an argument against a rationale often advanced successfully for subsidies, that food security demands national self-sufficiency. Free trade will also increase economic growth and purchasing power in those developing nations whose economies depend heavily on agricultural products or, increasingly, on exports (Goklany 1995a).

In summary, past and present success in humanity's quest for food security is founded on three mutually-reinforcing pillars: economic growth, technological change and trade.

Future population growth and food supply scenarios

Future food consumption depends, among other things, on population size and the economic circumstances of that population. Annual global population growth rates peaked in the 1960s and are expected to decline further, from about 2.0 per cent and 1.8 per cent in the 1970s and 1980s to 1.3 per cent (1990-2010), 1.1 per cent (2010-2025) and 0.7 per cent (2025-2050) under the World Bank's (1994) 'standard fertility' projection. This projection suggests that global population will increase from the current 5.7 billion to 9.6 billion in 2050 and 11.4 billion in 2150. The low and high fertility estimates are 8.6 and 10.1 billion for 2050 and 9.7 and 12.9 billion for 2150. Most of this growth is expected to occur in developing nations. Population may not double until the last quarter of the next century even under the high fertility assumption. Projections for the US are less robust because of the added uncertainty of immigration rates. The Bureau of the Census (1995) estimates that US population will grow from 263 million today to 300 million (+/-19 million) in 2010 and 392 million (+130/-107 million) in 2050.

If, as is hoped, the world enters an era of sustained economic growth, per capita food demand should increase more rapidly than population – particularly in developing nations (Poleman and Thomas 1995). The increased demand for meat, fish and dairy products would increase cereal demand even faster because it takes about two to seven pounds of cereal to produce one pound of such products (Brown and Kane 1994). Assuming that between 1993 and 2050, global per capita food supplies (using cereals as a surrogate) increase at the same rate as they did between 1969-71 to 1989-91, calculated using data from the Food and Agricultural Organisation (FAO) (1996a), then aggregate supplies in 2050 should increase 121 per cent over 1993 levels to meet food demand for 9.6 billion, the standard World Bank projection. Per capita food supplies, in terms of cereal production, would have to increase 28 per cent. (Note: if the past pattern of growth in average daily calorie intake per capita between 1969-71 to 1989-91 is also replicated, then average per capita calorie intake would increase 35 per cent, to over 3,600 kcal/capita. However, given that much of the additional cereal supplies would probably be used for producing animal protein, such replication is unlikely.)

A 121 per cent increase in aggregate supplies by 2050 could also be viewed either as an increase in per capita cereal supplies of 21 per cent for 10.1 billion people (the World Bank's high projection), a 42 per cent increase for 8.6 billion (its low projection), or a 11 per cent increase for a doubling over 1993 levels.

For the US, between 1970 and 1990 the effective cereal consumption per capita (considering all animal products) dropped, mainly because of decreased demand for beef and eggs. Therefore, it is reasonable to assume US food supplies need only keep pace with population growth and increase about 50 per cent by 2050.

The neo-Malthusian case

The projected increases in food demand led Neo-Malthusians to argue that the earth's carrying capacity will soon be exceeded, if it has not already been. In general, they contend that the factors responsible for the phenomenal growth in the world's food supply since 1950 are not sustainable into the future (Ehrlich et al. 1993; Brown and Kane 1994; Pimentel et al. 1994a). First, they argue, technological change can no longer be relied upon to bail out the population. The backlog of unrealised yield increases due to existing technologies is small. Past improvements were due, in large measure, to greater use of fertilisers and pesticides, but these inputs are at a point of diminishing returns. Also, pesticides are a treadmill; ever more powerful ones are needed as pests adapt to existing chemicals. Similarly, the biological limits for improving yields are being reached. Moreover, there are no dramatic breakthroughs on the horizon; if anything, loss of biodiversity compromises the ability to devise such breakthroughs. Second, past yield increases have been at the expense of future productivity. On land, current practices have increased desertification and soil erosion to rates exceeding natural regeneration; on the seas, they have resulted in over-harvesting and collapsing wild fisheries. Moreover, future productivity will decline further because of increased air pollution (including acidic deposition), ultraviolet radiation and, possibly, climate change. Third, the ability to bring new land under cultivation is limited. Sea level rise due to climatic change and increased urbanisation will further reduce land available for agriculture. Fourth, irrigation is unlikely to expand much: the best irrigation sites are already in use and new ones will be economically and environmentally costly. Moreover, silting, salinisation and water logging will reduce the effectiveness of existing irrigation projects. Fifth, over-reliance on monocultures makes the entire food system susceptible to catastrophic diseases and pests. Finally, many areas of the world lack sufficient fresh water to meet the future needs of agriculture and other competing demands, which could seriously limit future productivity increases (Postel et al. 1996; Falkenmark and Biswas 1995; Engelman and LeRoy 1993).

Thus, Ehrlich claims the world's carrying capacity has already been exceeded, which he currently estimates at between 2.5 and six billion people depending upon whether it adopts a North American-style meat-eating, or a vegetarian, diet (Ehrlich 1995). Similarly, Brown and Kane (1994) project that in 2030, 2.5 billion could be fed at the current US level or 10 billion at the Indian level, assuming no major breakthroughs in technology.

Non-Malthusian projections

There are, of course, alternative visions of the future. Several theoretical analyses suggest that the earth could feed between 10 to 40 billion people. Cohen (1995) summarises several of these analyses. Most of these studies do not explicitly address whether environmental constraints could (or would) be surmounted. However, they implicitly or explicitly assume that with due diligence and additional R&D, it is possible to continue to increase yields and expand cropland to meet food demand. Waggoner (1994), for instance, estimates that it should be possible to feed 10 billion on 2.8 Bha of cropland (compared to 1.4 Bha today) with an average daily per capita food supply of 3,000-6,000 Kcal. Smil (1994) estimates that 10 to 11 billion could be fed without relying on biotechnological advances by increasing cropland by about 300 Mha – using and optimising currently available technologies and farming practices and reducing beef consumption and post-harvest losses.

An elaborate FAO study (Alexandratos 1995a, 1995b) concludes that it should be possible to meet global food demand to the year 2010 – and possibly to at least 2025 – assuming evolutionary rather than revolutionary changes in technology (continuing emphasis on R&D and policy reforms to encourage production and economic growth) through a combination of increased yields and modest increases in cropland (about 93 Mha in developing countries), irrigated areas and cropping intensities. Global crop production would increase 1.8 per cent per year between 1990 and 2010, lower than historic annual rate increases (3.0 per cent, 2.3 per cent and 2.0 per cent in the 1960s, 1970s and 1980-92, respectively) but higher than the expected population growth rates noted above. In 2010, compared to 1988-89 levels in developing nations, chronic under-nourishment would drop from 20 per cent to 11 per cent afflicting 140 million fewer people (Table 13.2). The improvement in Sub-Saharan Africa would be small until 2010, but could accelerate thereafter. The ex-CPEs would maintain calorific input at the 1988-90 level but meat consumption would remain lower than their pre-transition levels due to loss of subsidies. Food security would, as now, not pose a significant challenge to the richer nations. Despite increased production, developing nations would become net importers of agricultural products as their populations expand and become richer. Real prices may continue to decline unless population or economic growth accelerates or yield growth rates decline (Mitchell and Ingco 1995; Alexandratos 1995a).

Without judging which camp's projections are more plausible, the most significant neo-Malthusian arguments will be examined, i.e., that land, water and technology are running out.

Land availability

Two facets to the issue of future land for agriculture are whether existing cropland's long-term productive potential can be conserved and whether

Table 13.2 FAO projections, 2010 and 2025

Region	*Population (Billions)*			*Per capita Food Supplies (Kcal/Day)*			*Chronic under-nutrition (millions*	
	1990	2010	2025	1988-90	2010	2025	1988-90	2010
WORLD	5.30	7.07	8.38	2 700	2 880	3,000		
DEVELOPED COUNTRIES	0.86	0.97	1.48	3,400	3,470	3,470		
FORMER CPEs	0.39	0.43	included above	3,380	3,380	included above		
DEVELOPING COUNTRIES	3.90	5.67	6.90	2,470	2,740	2,900	781 (20%)	637 (11%)
Sub-Saharan Africa	0.49	0.87	1.28	2,100	2,170	2,700	175 (37%)	296 (32%)
Near East/N. Africa	0.31	0.51	0.67	3,010	3,130	3,180	24 (8%)	29 (6%)
South Asia	1.16	1.68	1.94	2,215	2,450	2,700	265 (24%)	195 (12%)
East Asia	1.60	2.01	2.31	2,600	3,045	3,060	258 (16%)	77 (4%)
Latin America Caribbean Islands	0.44	0.59	0.69	2,690	2,950	3,030	59 (13%)	40 (6%)

(Alexandratos 1995a, 1995b)

additional suitable lands are available for conversion to cropland.

Land conservation technologies

Several technologies are currently available for mitigating soil erosion, salinisation and water logging (Pimentel et al. 1995; McLaughlin 1993). Controlling erosion involves minimising soil disturbance, maintaining vegetative cover or managing the slope of the land. Existing technologies include conservation tillage (no-till cultivation efficiency, 80 per cent to 99 per cent), crop rotations with legumes (50 per cent), intercropping with grass (65 per cent), contour planting (80 per cent), crop residue management (stubble mulching), level terracing (95 per cent) and windbreaks. Many techniques provide multiple benefits, for example, mulching conserves soil moisture and organic matter; crop rotation with legumes reduces run-off and fixes nitrogen. Erosion rates on US croplands dropped about 25 per cent between 1982 and 1992 due to greater penetration of such technologies (NRCS 1995; Kellogg et al. 1994). Similarly, salinisation can be mitigated by selecting appropriate salt-tolerant crops (such as wheat instead of corn or rice) or managing soil moisture through crop rotations or surface drainage and water logging through well-maintained drainage systems.

Some techniques entail relatively large initial costs, for example, foregoing crop residue as a source of scarce fuel or fodder, or manpower for terracing which requires 750-900 man-days of hand labour per ha, or 15 man-days using a 60 hp tractor, which may preclude or delay their wider adoption – particularly in developing nations (Dazhong 1993).

Uncultivated and potential cropland in the US

An early 1980s estimate placed the amount of arable land in the US at about 218 Mha (Batie and Healy 1983). More recently, the 1992 National Resources Inventory of non-federal rural lands estimated that there were 228 Mha in land capability classes I to III which could be cultivated using, at most, 'special conservation practices' and/or choice of crops (NRCS 1995; Kellogg et al. 1994). This estimate excludes areas classified as 'other rural land', or Conservation Reserve Programme (CRP) areas, areas under structures, windbreaks, barren land such as exposed rock or salt flats and marshland. Another 73 Mha of Class IV could be used as cropland with 'very careful management' and/or a reduction in the choice of crops. An additional 21 Mha of cropland were in Class V to VIII lands or enrolled in the CRP.

These estimates compare with 132 Mha that were cultivated or harvested in 1995 (USDA 1996). In 1990, based on Bureau of the Census (1993: Table 1128) data, domestic needs would have been met with about 102 Mha. Without agricultural subsidies, these figures may have been even lower. Clearly, the US has ample potential cropland to meet its domestic food needs even if population and amount of developed land (estimated at 37 Mha in 1992) were to double, provided at least current technology is deployed.

Potential cropland worldwide

Excluding unexploited potential cropland in the developed nations (including transition nations) and China, there are at least 3,335 Mha of potential rainfed cropland worldwide (Alexandratos 1995a), about 2.3-times the 1,448 Mha that were arable or in permanent crops in 1993. However, the 1,887 Mha surplus in developing nations (excluding China) is unevenly distributed (relative to where population growth may occur). Much of it is in Latin America and Sub-Saharan Africa. Second, between 45 per cent and 57 per cent of the surplus is in forests or protected areas. Third, it may be inherently less productive than existing cropland. About two-thirds of this surplus cropland is constrained by low natural fertility, poor soil drainage, steep slopes, or sandy and stony soils. Thus, substantial investment will be required to: 1) bring these lands into production; 2) develop the infrastructure to sustain agriculture, including transportation and distribution networks for inputs and produce; and 3) ensure that these lands receive adequate inputs. Compared to these constraints, Alexandratos estimates that the loss of productive potential due to urbanisation will be modest (a projected 70 Mha by 2010, an increase of 20 Mha since 1990).

Water

As noted previously, irrigation increases yields 2.5-fold on average. Nevertheless, competing demands (direct human use, industry and instream uses) are increasingly winning out over agriculture for water. This is virtually inevitable because as countries get richer and more urbanised, the relative economic and political clout of the agriculture sector diminishes. Moreover, as noted earlier, agriculture has been a victim of its own success in that it now commands a smaller share of GDP and employment than ever before.

Because of its importance, societies have generally been loath to treat water as just another economic commodity. In fact, most countries subsidise its usage; when it is in short supply, it is rationed explicitly or implicitly by, for example, limiting its availability by time of day or week or banning certain uses. Often, even where private entities pay for extracting water from the subsurface, the pumps and tubewells are indirectly subsidised through, for example in India, low electricity rates. By the law of perverse consequences, these very sensibilities – by providing direct or indirect subsidies – effectively encourage wastage, discourage investments in conservation and result in less-than-socially-optimal use of water. The reverse side of this coin is that there are numerous unexploited opportunities for conserving water. In the US, for instance, the efficiency of use of water diverted for irrigation, which accounts for about 80 per cent of consumptive water use nationally, was estimated at 47 per cent in 1987 – suggesting the high potential for water conservation (USDA 1989). In fact, a small increase in irrigation water use efficiency goes a long way toward meeting water needs for other sectors. Thus, increasing that efficiency from 41 per cent in 1975 to 47 per cent in 1982 saved water equivalent to 80 per cent of domestic consumption. It is encouraging to note that both water withdrawal and consumptive use per capita seemed to have peaked around 1975 in the US. In 1990, they were down 17 per cent and 18 per cent from 1975 levels, respectively; and despite a 13 per cent population increase, total withdrawal and consumption in 1990 were below 1975 levels (Bureau of the Census 1995).

Treating water as an economic commodity, by establishing realistic prices and allowing trading so that unused water could be sold to other users, would give farmers and other users the necessary incentive for conserving water and lead to greater adoption of available conservation technologies. For instance, it would stimulate research into, and adoption of, drought tolerant cultivars. Other available conservation technologies include methods of managing cultivation and soil moisture (centre pivot irrigation; drip irrigation; maintenance of existing canals, ditches and on-farm storage facilities; optimising the timing and amount of water used; and land levelling). Alternatively, water could be diverted to higher value crops. Both the amount of irrigation water applied per hectare and the economic efficiency of water use vary with the crop, the latter by more than fifty-fold (Cervinka 1989). For example, rice requires an order of magnitude more water per dollar of crop than potatoes and about twice as much as wheat.

Fundamental to water trading is the vesting of legally-enforceable property rights in the water allocated to the various users. Successful cases of formal and informal water trading have been recorded in socio-economic milieus as diverse as the US, Chile, Jordan and India (Rosegrant et al. 1995). In Chile, trading increased efficiency of water use by 22-26 per cent between 1976 and 1992, which effectively expanded the irrigated area by that much. Moreover, water can be and is often, reused, which itself offers technological opportunities for reducing overall water consumption. In fact, much of the water used for human consumption and industrial use – being treated prior to consumption – is second or third (or more) hand.

Finally, in the ultimate analysis, if the price is right, one can resort to desalination. 97.5 per cent of the world's water is in the oceans; 69 per cent of the freshwater is locked up in glaciers and permanent snow cover (Engelman and LeRoy 1993). Today, desalination, because it relies on fossil fuels, is economically and environmentally costly. Its use is restricted mainly to relatively rich Middle Eastern nations with access to cheap natural gas or oil. However, new technology, in the form of solar powered plants, could reduce these barriers (*The Economist* 1995).

Technologies to increase net yields

Increasing yields with existing technologies

The high potential for using existing technologies to expand yields is suggested by Tables 13.3 and 13.4 which are based upon FAO (1996a) data. For the ten most important types of crops, which cumulatively account for about 60 per cent of the world's cropland, Table 13.3 shows the yield ceiling [Y©] in 1991-93, defined as the average yield for the nation with the highest yield and the average yields in 1992-94 for developing, transition and other developed nations [Y(DING), Y(TR) and Y(DPED-TR), respectively]. This Table also shows the area planted in each of the crops worldwide and the portion of that in developing and transition nations. Table 13.4, calculated from Table 13.3, shows the increases in 1992-94 production if various yield gaps were eliminated. For corn, for instance, raising both Y(DING) and Y(TR) to Y(DPED-TR) would have increased production by 84 per cent, while increasing all yields to Y© would have resulted in an increase of 154 per cent. Of course, Y© is not, by itself, limiting. Yields achieved by research stations or individual farmers are often higher; at 21 T per hectare, the

Table 13.3. Average yields in developing, transition, other developed and highest yielding nations for major crops (1992-1994 averages)

Crop	Global Crop Area (million ha)	Average Yields (t/ha) Transition Nations	Developing Nations	Yield Ceiling (t/ha) Other Developed Nations	Highest Yield for a Single Country, Y©, Average for 1991-1993*		Percent of Global Crop Area in Developing & Transition Nations
		Y(TR)	Y(DING)	Y(DPED-TR)	Y©	Country	
Rice	146.1	2.93	3.56	6.15	9.23	Australia	97
Wheat	220.5	1.98	2.44	3.07	8.15	Netherlands	71
Corn	135.1	2.96	2.62	7.15	9.87	Greece	72
Sorghum	45.1	0.94	1.10	4.01	5.76	Italy	89
Millet	37.9	0.70	0.75	1.28	1.81	China	~100
Barley	73.5	1.96	1.42	3.36	6.05	Switzerland	70
Cereals, Total	*700.0*	*1.96*	*2.52*	*4.26*	*7.48*	*Netherlands*	*80*
Pulses	*67.1*	*1.64*	*0.67*	*1.94*	*4.79*	*France*	*92*
Roots and tubers	*48.6*	*12.22*	*11.40*	*31.25*	*45.17*	*Blgm/Luxmbg*	*95*
Soybean	*60.3*	*0.90*	*1.70*	*2.51*	*3.32*	*Italy*	*58*
Peanuts	*21.4*	*1.00*	*1.20*	*2.31*	*3.82*	*Greece*	*96*

* Area harvested annually must be at least 5,000 acres on average. (FAO 1996. Database available on World Wide Web).

1992 champion US corn grower's yield was 136 per cent greater than Y© for corn, that is, the average yield for the highest nation (Waggoner 1994). Similar logic applies to all the other crops listed in Tables 13.3 and 13.4.

These yield gaps can be bridged by using more inputs, particularly fertiliser, in developing and transition countries; and modifying hives so they are better adapted to specific locations around the world which would increase their adoption rates. Furthermore, developing location-specific integrated nutrient management systems; making water use more efficient; and improving extension services would be useful. There is also a need to help optimise the timing and quantities of inputs, develop more effective pest control and improve feed supplies for livestock in pastures and farms using (singly or in combination) the various techniques mentioned previously for reducing erosion, water logging and salinisation, maintaining soil organic matter, liming acidic soils and adding micronutrients to utilise the full potential of hives.

Raising the yield ceiling

Maximum theoretical yields [Y(MAX)] exceed what can be achieved by even the best farmers. So long as Y(MAX) is not reached, it should be possible to

Table 13.4. Increases in global crop production if yield gaps were erased (calculated from Table 13.3)

Crop	*Crop Area in Developing and Transition Nations (percent)*	*Increase in Global Production (percent) if Y(DING) and Y(TR) are Raised to Levels of:*	
		Other Developed Nations Y(DPED-TR)	*Yield Ceilings Y©*
Rice	97	70	155
Wheat	71	22	225
Corn	72	84	154
Sorghum	89	184	308
Millet	–100	71	142
Barley	70	50	170
Cereals, Total	80	54	170
Pulses	92	131	472
Roots and tubers	95	148	258
Soybean	58	24	64
Peanuts	96	86	208

NOTE. Y© = average yield (1991-93) for the nation with the highest average yield; nation must harvest at least 5,000 ha in 1991-93.

Y(TR) = average yield for nations whose economies are in transition.

Y(DING) = average yield for developing nations.

Y(DPED-TR) = average yield for developed nations excluding nations whose economies are in transition.

increase the yield ceilings (Plucknett 1995; Oram and Hojjati 1995). One estimate of the theoretical maximum global cereal yield for rain-fed agriculture placed it at 13.3 T (grain equivalents) per hectare (Plucknett 1995). By comparison, the average global cereal yield in 1992-94 was 2.77 T per hectare ranging from 0.75 T per hectare for millet to 3.89 T per hectare for corn (FAO 1996a). Yield ceilings may be raised through continued emphasis on plant breeding – both locally and professionally, more intensive management of soils and development of location-specific integrated nutrient and pest management systems.

Biotechnology will increase ceilings by, for instance, allowing breeders to confer resistance to stresses, such as droughts, freezes, salinity, pests and herbicides, and by regulating flowering (Coupland 1995). The livestock sector can be made more efficient by producing growth promoters which can improve feed efficiencies, increasing reproductive success and reducing losses from diseases (OTA 1994). It can also be used to develop plant and livestock breeds with favoured characteristics, for example, plants that produce more oil or starch, ripen later, or look better; meats with less fat. Moreover, by conferring fruits and vegetables the ability to ripen later, as in the Flavr Savr tomato, biotechnology can help reduce post-harvest and end-use losses, estimated at about 47 per cent of global calorie consumption (Bender 1994).

Future prospects: the trade-off between land conversion and land productivity

The above discussion suggests that it should be possible for the US to meet domestic food needs relatively easily in 2050. Global food security, however, is more problematic. In theory, it should also be possible – at an economic and environmental price and through a combination of measures increasing cropland and overall productivity – to augment food supplies by 121 per cent by 2050 (relative to 1993 levels) which, as noted, would expand per capita food supplies between 10 per cent and 42 per cent depending on whether population doubles or hews to the World Bank's low fertility projection. Increases in productivity, as used here, includes anything that augments the amount of food used by eventual consumers without putting additional land into agriculture. Productivity may be boosted by increasing yields and cropping intensities, using feed more efficiently, increasing pre- and post-use shelf life, or otherwise decreasing post-harvest and end-use losses. Several different scenarios can be constructed to achieve this. For instance, cropland could be increased from the 1993 level of 1,448 Mha by 121 per cent to 3,201 Mha, assuming average productivity is unchanged. Alternatively, in lieu of the massive habitat loss implicit in that scenario (1,753 Mha of new cropland), while cropland is kept constant, productivity could increase 121 per cent, or an average of 1.4 per cent per year for 57 years from 1993 to 2050. Thus, there is a trade-off between increased productivity and increased habitat loss. Figure 13.4 illustrates this trade-off for average annual increases in productivity between 1993 and 2050 ranging from zero to 2 per cent per year and the corresponding cumulative increases in productivity.

This trade-off is sensitive to even small changes in the rate of productivity increase. Thus, accelerating the average annual productivity increase from 1.0 per cent to 1.1 per cent (or 76 per cent over 57 years to 87 per cent) would, in 2050, reduce net habitat conversion by 100 Mha.

Productivity increases at an annual rate of 1.0 per cent to 1.5 per cent for 1993-2050 are plausible. First, they are within the range of historical experience. Between 1969-71 and 1979-81, yields alone, calculated from changes in the crop production indices and cropland areas (FAO 1995), increased at an annual rate of 2.1 per cent; between 1979-81 and 1991-93, they increased 2.0 per cent. They are also below Alexandratos' (1995a) projection of an annual increase of 1.8 per cent in crop production for 1990-2010, preponderantly due to increases in productivity rather than cropland. Second and more importantly, the corresponding cumulative improvements from 1993-2050 would be between 76 per cent and 134 per cent (Figure 13.4), which is realisable with relatively modest technological changes, considering the increases possible if current yield gaps were closed (Table 13.4) and considering the potential for raising yield ceilings and reducing downstream losses, provided current constraints to increasing production can be reduced (see below).

The annual 1.0 per cent to 1.5 per cent increase also translates into a change in total cropland by 2050 ranging from an increase of 368 Mha to a decline of 77 Mha. Clearly, the precise combination of productivity increases and new cropland employed to meet future food needs will be critical for the globe's biological health.

To the extent productivity improvements come from increased inputs, like fertilisers, pesticides and water, that will, unless mitigated, increase one set of environmental and public health-related impacts. For fertilisers and pesticides, these impacts include surface and ground water contamination, eutrophication, oxygen depletion, build-up of pesticide residues in fish and avian species and in human tissue. Where water is the input, diverting it could have drastic consequences for other existing in-stream uses and for species dependent on that water. These environmental impacts must be weighed against the effects of habitat loss that will inevitably result if those inputs are foregone (Goklany 1998a).

Knutson et al. (1990) estimated that foregoing use of pesticides and inorganic nitrogen fertiliser in the US would reduce US yields for soybean, wheat, corn, cotton, rice and peanuts by 37, 38, 53, 62, 63 and 78 per cent, respectively. Thus, to make up all this lost production, the amount of land cultivated would have to be increased at least 170 per cent for soybean to 350 per cent for peanuts. However, it would be uneconomic to do so. Knutson et al. estimated that because of higher prices, cultivated area would, in fact, increase only 10 per cent. Nevertheless, because of the production shortfall, prices for wheat would still increase 24 per cent, while those for other crops would be doubled or more. They estimated the lower 20 per cent of the population's outlays on food would increase from 38 per cent to 44 per cent. US grain exports would decline 50 per cent by volume and stock carry-overs would be reduced between 42 per cent for wheat to 83 per cent for corn. While Knutson et al. did not estimate impacts on global food security, clearly the ability of poorer nations and peoples to import and purchase food would be significantly – and adversely – affected.

Similarly, if pesticide use were eliminated, cropland devoted to fruits and vegetables would have to be increased by 33 per cent to 150 per cent to compensate for any lost production, based upon Taylor's (1995) estimates of declines in yields of 25 per cent to 60 per cent. The resulting increases in costs would also reduce fruit and vegetable consumption which could itself have negative impacts on nutrition and public health (NRC 1996).

Thus, the trade-off between increasing productivity and increasing habitat loss indicated in Figure 13.4 also involves a trade-off between the environmental and public health impacts of increased inputs against those of habitat loss. Either could lead to reductions in biological diversity. However, it would seem that the two sets of impacts are not quite equivalent. Habitat loss is much more final compared to the impacts of fertilisers

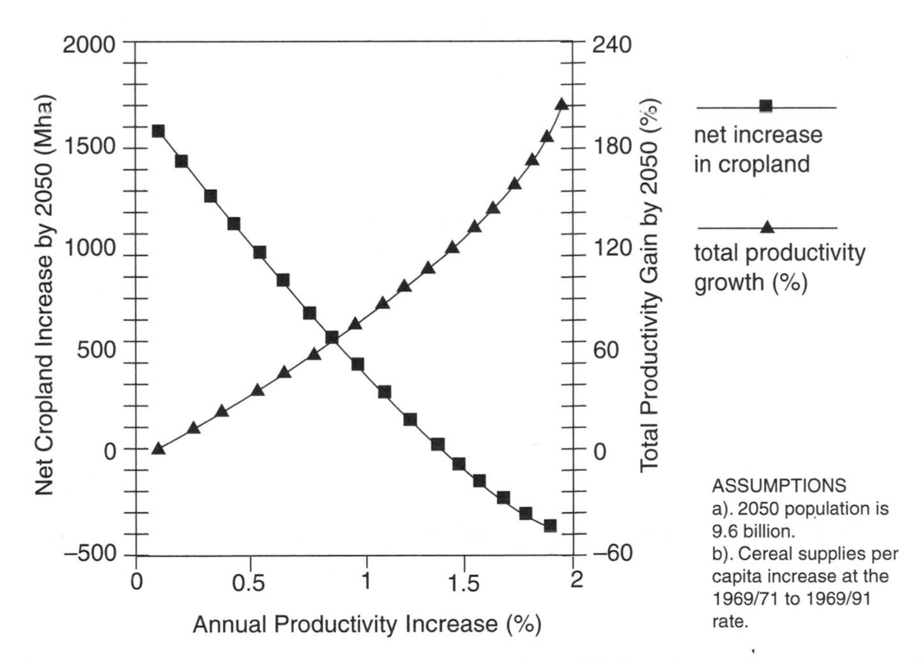

Figure 13.4 Trade-off between productivity growth and habitat loss net conversation of land to cropland from 1993 to 2050

and pesticides, particularly if they are used efficiently and cautiously. Second, given time, the effects of fertilisers and pesticides seem to be reversible, albeit at some cost and with great difficulty. For instance, the levels of DDT and other organochlorine residues have declined in fish, human and animal tissue in the US and other OECD nations – in some cases by more than an order of magnitude, for example, in Canada and the Netherlands for some forms of DDT in human adipose tissue (Gunnarson et al. 1995; Goklany 1994, 1998a). Consequently, populations of several avian species, such as bald eagles and peregrine falcons in the US which were affected by these residues are now recovering (Goklany et al. 1992). Similarly, nitrogen loading in the streams and water have been reduced and, at long last, dissolved oxygen levels are now increasing. The number of species in some of the major rivers in the developed world are also recovering. For instance, the number of species in the Rhine River which had declined to 27 in 1971 had increased to 97 in 1987 (CEC 1992).

On the other hand, because of diminishing returns to yield at higher levels of pesticide and fertiliser usage, it may well be environmentally unsound to attempt to squeeze out as much productivity as possible for every hectare of cropland.

The situation is at least as complex if the input in question is water: total loss of water can be just as drastic as loss of habitat. However, in making choices regarding such a trade-off it is important to recall that the yield of one irrigated hectare is, on average, equivalent to 2.5 unirrigated hectares. Moreover, diversion of water for agriculture does not have to be an all-or-nothing enterprise. Also, the adverse effects of water diversions can be mitigated, if not eliminated, by returning or reusing some of the water or adjusting timings of flows.

Finally, because the choice among the numerous scenarios illustrated in Figure 13.4 involves trade-offs, there is an optimisation problem which, because of the uncertainties surrounding it, is not accurately solvable today, though, *prima facie*, habitat loss would seem to be overall worse than greater use of inputs – particularly if the latter are used efficiently, with due caution and below the point of diminishing returns. These uncertainties suggest a research agenda to help increase efficiency of input use and make the educated trade-offs necessary to meet future global food needs while limiting net environmental impacts.

Realising the world's food potential while limiting the environmental price: policy recommendations

The above discussion suggests that it ought to be possible to meet the world's food needs in 2050 even if the population doubles, but at an environmental price. Whichever route is taken to meet future global food needs, whether or not it is represented by one of the several scenarios of Figure 13.4, underlying each is a basic set of assumptions, namely: 1) there will be sufficient capital to obtain the inputs, implement technologies needed to maintain or increase productivity and to bring any needed new cropland into production; 2) R&D, particularly for the scenarios with the higher productivity increases, will continue to bring some new technologies on line and adapt technologies – current and new – to local conditions around the world; 3) knowledge of these adaptations will continue to be transferred to farm enterprises, and; 4) infrastructure and the trading system will expand, as necessary, to ensure that surpluses are moved to areas of shortage. Therefore, at a minimum, policies are needed that will convert these underlying assumptions into reality, while simultaneously reducing or mitigating any resulting environmental impact and minimising disruption to habitat, forests and biodiversity.

First, complacency because of past successes should be avoided; and the national, regional and international systems for agricultural research and the extension services that transfer technology to farm enterprises should continue to be vigorously supported.

Second, R&D and extension services should focus on optimising location-specific methods of maintaining and increasing productivity (including closing yield gaps, reducing post-harvest and end-use losses) and reducing or mitigating environmental consequences of agricultural technologies, per previous discussions. Such programmes should include developing methods to reduce the amount of inputs that are applied or mobilised into the environment by, for example, controlling the timing and ratios of inputs so as not to over-apply one input when another may be the limiting factor. In addition to increasing net production, they would also help make agricultural practices more economically efficient. Greater effort should also be made to develop drought- and salinity-resistant crops and cultivars. Such research, by increasing agricultural options, will also help the world adapt to global change, whether or not the agents of change are human-induced (Goklany 1992, 1995a).

A research programme is also needed to evaluate the relative public health and environmental risks and social and economic consequences of the different methods for increasing net production. Such information should be incorporated into policy-related exercises, such as cost-benefit or environmental impact assessments, related to fertiliser and pesticide use and water and other projects affecting agriculture. Current assessments tend to ignore the effects on land conversion and broad impacts on nutrition and public health on various segments of society (Pimentel et al. 1994b, 1995).

Third, the public, environmentalists and policy makers need to be educated about the trade-offs involved in feeding the world's billions. The awareness of agriculture's role in polluting and diverting waters and their environmental impacts, particularly in richer, developed nations – while necessary to generate pressure for much-needed improvements in fertiliser and pesticide use, water projects and environmental cleanup – unfortunately is not matched by an appreciation of technology's role, discussed above, in reducing habitat loss while improving food supplies, nutrition and public health for today's billions. Without this, more habitat loss, increased pollution, less but costlier food and increased vulnerability of the poor to malnourishment and hunger could result, short-changing both humanity and the rest of nature. Education is also needed to help address 'new source bias' evident in the reactions of various segments of the public to biotechnology-derived products, like the Flavr Savr tomato or milk produced from BST-treated cows, despite their potential role in reducing other inputs and habitat loss (Goklany and Sprague 1991).

Education may also help ensure that there is support in the richer nations for continued production of crop surpluses, as dictated by the market place, as opposed to subsidising overproduction. Otherwise, richer nations – untroubled by food shortages; safe in the knowledge that they can always purchase, if not produce, food; motivated by the desire for a cleaner environment and a balanced budget and, possibly, some new source bias – may discourage the production of surpluses, reduce investments in agricultural R&D at home and abroad and decrease food aid. These are essential for global food security today and into the foreseeable future.

Fourth, the institutions responsible for the co-evolution of technological change and economic growth should be bolstered. Such support cannot automatically be assumed even in richer nations. Some transition and developing nations may well waver from, if not abandon, the path to long-term economic growth. These institutions include market economies, freer trade, conferring and enforcing property rights for land, water and innovations and requirements for risk analyses to address trade-offs and 'new source bias' which support entrenched interests or old technologies at the expense of competitors. These will help create new technologies, as well as the necessary capital and other incentives for adopting new or unused technologies to raise or maintain yields, conserve soil and water and control pollution resulting from yield-enhancing technologies. As noted previously, the richest nations, in general, have the best environmental quality.

Equally importantly, these institutions would ensure that farm enterprises are fairly compensated for their investments in capital, labour and technology, eliminating a major cause of the relatively poor performance in the agricultural sector in many developing nations. Implicit in this is also the reduction, if not total elimination, of agricultural subsidies. This should be accomplished gradually, over five to 10 years, so that producers and non-producers can adjust to a new, subsidy-free environment, without abandoning insurance and other safeguards against catastrophic crop failures. Several strides have been taken in this regard in OECD – and, out of necessity, transition – nations. The passage in the US of the Freedom to Farm Act was a giant step in this direction.

Moreover, as noted, economic growth and trade increase food security of both producers and non-producers. Barriers to free, unsubsidised trade should be further reduced. Also, removal of agricultural subsidies in the richer nations, in particular, would benefit their economies and environment as well as accelerate economic growth in developing nations, in part, because that would make the latter's agricultural sectors more competitive (Goklany 1995a). Moreover, trading systems allowing the free movement and repatriation of capital (and profits) across borders may help generate the capital developing nations will need to expand cropland and obtain and employ both new and underused existing technologies (Goklany 1998a). Despite their imperfections, the conclusions of the Uruguay Round of GATT (General Agreement on Trade and Tariffs) and NAFTA (the North American Free Trade Agreement) negotiations bode well on these scores.

Also, trade – both external and internal – militates against the strife and disruption which have been responsible for much of the hunger and malnutrition over the past few decades. It is particularly difficult for groups involved in trading the most basic of commodities, food, to be engaged in hostilities. For the seller, it makes little sense to attack and impoverish one's customer; while for the buyer, that would be tantamount to biting the hand that feeds it. The recommendation on trade implies the abandonment of the notion of national self-sufficiency in food production. Post-war statistics on food trade and aid indicate that this goal is unrealistic. Moreover, as noted, too often this goal has been used to justify economically and environmentally unsound agricultural subsidies. Social and political institutions should strive to reconcile themselves to the reality that trade has globalised sustainability (Goklany 1995a).

Economic growth will, as noted, increase the incentives for smaller families, thereby helping secure the population growth trajectory implied in the World Bank's standard projection of virtual population stabilisation in the 22nd century. Finally, economic growth may reduce pro-natalist sentiments among policy makers as the world's population becomes older which will increase pressures on social safety nets (Goklany 1995a).

Fifth, the agricultural sector needs to strengthen its ability to compete for land and water. Increasingly, as economies grow, the shrinking of agriculture's share of GDP and total employment will reduce its economic and political clout. Thus, agricultural interests will be well served if they establish property rights to water and land sooner rather than later. Care should be taken that as subsidies for food crops are reduced, they are not replaced by subsidies for non-food uses of crop and agricultural land. One can well imagine future constituencies for subsidising carbon sequestration and fuel farms similar to the one for ethanol today.

Sixth, whether or not food prices decline and despite economic growth, there will be some people unable to afford an adequate diet. This can and ought to be addressed without creating disincentives for producers by providing transfer payments, food stamps and/or food banks for the truly needy. Again, a wealthier society will be more able to afford – and less likely to begrudge – such programmes.

Finally, to relieve the pressure on the land, the oceans may have to be utilised more. Most of the world's food needs are currently supplied by the photosynthetic product of 30 per cent of the world's area, which is land. It would be sensible to obtain more from the rest of the globe – but in a manner quite different from what has been used to date. By and large, human beings have treated the ocean's bounty as our hunter-gatherer forebears treated the land's bounty [Serageldin, quoted in CGIAR (1995)]. This hunter-gatherer mentality creates havoc, particularly when technology is pressed into service for harvesting but not for increasing the population of the harvested species. The demise or near-demise of species such as the beaver, American buffalo and various whale species are cases in point. In fact, only after conscious and more intensive management of the land, namely agriculture, started to replace hunting and gathering was it possible for the earth's human-carrying capacity to go from a few to hundreds of millions (Livi-Bacci 1992). In order to allow the billions of humans to co-exist with the rest of nature, the realm of conscious management may have to be expanded to a portion of the seas. On this score, it is encouraging to note that aquaculture (though some of it is land-based or inshore) is one of the fastest expanding sections within the food and agriculture sector, growing for example, 12 per cent just between 1993 and 1994 with the growth being led by the most populous nation, China and other developing nations (FAO 1996c). But to bring this about, once again, additional inputs – capital and R&D – will be required and environmental impacts on seas may have to be traded-off for additional habitat loss and environmental degradation on land. Perhaps if more time is spent in the water, fewer footprints will be left on the land.

Conclusion

Despite a sextupling of population, food supply around the world has improved markedly over the last two centuries, due primarily to the forces of science-based and market-driven economic growth, technological change and trade. Compared to the dismal situation immediately following World War II, today famine and malnourishment are down both in absolute and relative terms; people are generally better fed and they spend much less time and effort getting food to the table. Nevertheless, hunger and malnutrition, while reduced, has not been banished. Most vulnerable to shortfalls are populations or areas with low purchasing power; civil strife or dysfunctional relations with the external world, such as Iraq and North Korea; or some combination thereof. While today's improvements have come at an environmental price, mercifully the wholesale destruction of forests, habitats and biodiversity, which would inevitably have occurred without technological change, has been avoided. Since 1961 alone, technology has forestalled the conversion of at least 966 Mha to cropland worldwide (370 Mha in the US since 1910). Another 2,580 Mha would have been converted to other agricultural uses worldwide. Indirectly and despite its poor reputation among many environmentalists, technological change is responsible for conserving more habitat than any other conservation measure.

With respect to the future to 2050, the richest nations should have few problems. As a group, they are currently running large surpluses. Their populations have, more or less, stabilised, with the notable exception of the US, whose population continues to expand fuelled, in part, by immigration. Nevertheless, it should be able to meet its own food needs – even if its population doubles. It has sufficient land, water and the capital to deploy new and existing technologies and if it did not, it could purchase food security with its wealth – as can other rich nations.

Future food security is more uncertain for the least developed nations. Despite recent declines, they still have the highest population growth rates. Moreover, because of poverty and low incomes, they have less ability to adopt productivity-enhancing technologies or to purchase food security in the market place.

This chapter explored whether global food supplies in 2050 could be augmented by 120 per cent over 1993 levels, an amount sufficient to increase available grain supplies per capita even if population were to double which, under the World Bank's high-end estimate, would occur in the last quarter of the next century. It is, indeed, mathematically possible to more than double global food supplies by 2050 through various combinations of increased cropland and diverse methods of raising productivity. First, the amount of potential rainfed cropland is more than 2.3 times what was used in 1993. Second, several techniques are available to maintain and increase productivity on existing croplands. If current gaps between average yields and yield ceilings are erased, global production of virtually all major crops would double without adding a single hectare to net cropland (Table 13.4). Third, productivity can also be raised by increasing cropping intensity, raising yield ceilings and decreasing post-harvest and end-use losses. Fourth, supplies of water – perhaps the most critical of all inputs – can, effectively, be stretched through institutional changes which would treat it as an economic commodity and allow trading. That would provide everyone the incentives needed to apply existing and stimulate R&D of new, technologies to conserve, reuse or increase water supplies.

However, it is not enough to produce sufficient food. Food security also means ensuring that non- or low-producers have economic and ready access to food. Moreover, the world's food potential should be realised without undue environmental impacts, with the least disruption to habitat, forests and biodiversity.

The precise combination of increased cropland and increased productivity – and the mix of productivity-enhancement measures – used to meet future food needs will be crucial for the earth's future biological diversity. For instance, average annual increases in productivity of 1 per cent versus 1.5 per cent between 1993 and 2050 – both plausible given the opportunities to raise productivity – means the difference between converting 368 Mha of habitat (globally) to new cropland or reducing cropland by 77 Mha. However, to the extent productivity increases are due to additional fertiliser and pesticide use or water diversions, their environmental impacts may partially offset the benefits of reducing habitat loss. Given the rather drastic nature of habitat conversion, perhaps the most prudent approach would be to focus on increasing productivity while using inputs efficiently and cautiously and simultaneously working to mitigate their impacts.

To help assure that the world's food security needs are met without displacing the rest of nature or inflicting irreparable environmental harm, a long-term commitment is needed toward strengthening the very forces responsible for the current victory, however incomplete or tarnished, over hunger and malnutrition. Specifically:

1. Economic growth is essential. It is needed to generate the fiscal resources needed for the adoption and application of new or unused existing technologies to maintain or enhance productivity; to reduce or mitigate environmental effects due to agricultural technology; to bring any new cropland into production; and to develop the infrastructure for integrating that cropland into the existing food and agricultural system. Economic growth also helps ensure political support and funding for social safety nets, as well as R&D, extension services and food aid. It will augment the purchasing power for non- or low-producers so they can better afford food. Finally, it also helps create the conditions for families to voluntarily limit their sizes, possibly leading to eventual population stabilisation.
2. Technological change is needed. Without it, productivity of the land cannot be maintained or increased, while simultaneously keeping environmental impacts in check. This means support for R&D and extension services for agriculture, food and the environment, embedded within a larger effort to bolster the institutions that stimulate such change. This includes greater emphasis on education in general and science education in particular, protection of intellectual property rights, elimination of new source bias and favourable tax treatment for R&D expenditures. The larger effort will help develop a culture conducive to technological change. It is also necessary because the course of scientific and technological progress is unpredictable; advances in one field often spread to others - consider the importance of computer chips or synthetic materials in and out of the food and agricultural sector.
3. Freer and unsubsidised, trade is critical to ensure efficient movement of food surpluses and capital across borders. Developing nations will need to import more food, as well as capital to open up new cropland or afford modern technologies. Moreover, to pay for food imports, they will need exports and growth in other economic sectors. Nations should eschew the economically and environmentally counterproductive notion of food self-sufficiency. Trade, particularly in foodstuff, also gives nations another incentive to live in harmony.

Strengthening these interdependent forces will also enhance the world's adaptability to global environmental change, regardless of its cause (Goklany 1995a). In addition, to relieve humanity's burden on the land – and the effects of technological harvesting – a concerted effort may be needed to develop sustainable methods of enhancing the oceans' productivity. Moreover, it is crucial that developed nations continue to produce crop surpluses. Conceivably, their desire to maintain and enhance their own environmental quality may lead to policies effectively reducing such surpluses. Finally, there has to be acceptance of and the ability to make, trade-offs between imperfect outcomes.

Because they would harness human nature rather than seek to change it, the above measures, while no panacea, are more likely to be successful than fervent and well-meaning calls, often unaccompanied by any practical programme, to reduce populations, change diets or lifestyles, or embrace asceticism. Heroes and saints may be able to transcend human nature, but few ordinary mortals can. Sadly, vanity and the personal desire for a longer and healthier life have changed many more diets than urgings to save the globe. Holding out for perfection and rejecting the admittedly 'second-best' solutions outlined above would reduce quantities of and access to, food supplies while increasing habitat loss and environmental degradation – short-changing both humanity and the rest of nature.

References

Alexandratos, N., ed. (1995a). *World agriculture: Towards 2010*. Chichester: John Wiley: 163.

Alexandratos, N. (1995b). The outlook for world food and agriculture to year 2010. In: Islam, N., ed. Population and Food in the Early Twenty-First Century: Meeting Future Food Demand of an Increasing Population. Washington, DC: Int. Food Policy Res. Inst.: 25-48.

Batie, S. S., Healy, R.M. (1983). The future of American agriculture. *Sci. Am.* **248**, 45-53.

Bender, W.H. (1994). An end use analysis of global food requirements. *Food Policy* **19**, 381-395.

Brown, L.R.; Kane, H. (1994). *Full House: Reassessing the World's Population Carrying Capacity*. New York: W.W. Norton.

Bureau of the Census. (1995). Statistical abstract of the United States 1995. Washington, DC: US Bureau of the Census.

Bureau of the Census. (1993). Statistical abstract of the United States 1993. Washington, DC: US Bureau of the Census.

Bureau of the Census. (1975). Historical statistics of the United States, Colonial times to 1970. Washington, DC: US Bureau of the Census.

BLS (Bureau of Labour Statistics). CPI Detailed Reports. Various January issues. Washington, DC: Bureau of Labour Statistics.

CEC (Commission of the European Communities) (1992) The state of the environment in the European community: Overview, volume III. Brussels: Comm. Eur. Communities.

Cervinka, V. (1989). Water use in agriculture. In: Pimentel, D.; Hall, C.W., eds. *Food and Natural Resources*. San Diego: Acad. Press;141-162.

Chen, R.S.; Kates, R.W. (1994). World food security: Prospect and trends. Food Policy 19: 193-208.

Cohen, J.E. (1995). *How Many People Can the Earth Support?* New York: W.W. Norton;532.

CGIAR (Consultative Group on International Agricultural Research). (1995). From hunting to farming fish – rapid production increases are possible. Press release.

Washington, DC: Consult. Group Int. Agric. Res.; May 14.

Coupland, G. (1995). LEAFY blooms in Aspen. *Nature* **377**, 482-483.

Crosson, P. (1995a). Future supplies of land and water for world agriculture. In: Islam, N., ed. *Population and Food in the Early Twenty-First Century*: Meeting Future Food Demand of an Increasing Population. Washington, DC: Int. Food Policy Res. Inst.

Dazhong, W. (1993). Soil erosion and conservation in China. In: Pimentel, D., ed. *World Soil Erosion and Conservation*. Cambridge: Cambridge Univ. Press.

Drèze, J.; Sen, A.K. (1990). The political economy of hunger, volume 1. Oxford, U.K.: Clarendon Press, Oxford Univ. Press.

Ehrlich, P.R. (1995). Interview on Diane Rehm show. WAMU. Washington, DC: October 9. Cassette available from WAMU 88.5 FM, The American University, Washington, DC 20016-8082.

Ehrlich, P.R.; Ehrlich, A.H.; Daily, G.C. (1993). Food security, population and environment. *Popul. Dev. Rev*. 19: 1-32.

Engelman, R.; LeRoy, P. (1993). Sustaining water: Population and the future of renewable water supplies. Washington, DC: Popul. Action Int. 9-11.

Falkenmark, M.; Biswas, A.K. (1995). Further momentum to water issues: Comprehensive water problem assessment in the being. Ambio. 24 (6): 380-382.

FAO (Food and Agricultural Organisation of the United Nations) (1996). FAOSTAT database available at http://apps.fao.org/lim500/Agri_db.pl.

FAO (Food and Agricultural Organisation of the United Nations) (1996). Food outlook. June. Available at
http://www.fao.org/waicent/faoinfo/economic/
giews/english/fo/fo9606/tables /fot96069.htm.; 1996b.

FAO (Food and Agricultural Organisation of the United Nations) (1996c) Major trends in global aquaculture production: 1984-1994. Available at
http://www.fao.orgWAICENT/faoinfo/ fishery/aqtrend/aqtrend.htm.;

FAO (Food and Agricultural Organisation of the United Nations) (1995). Country tables. Rome: Food Agric. Org: 332.

FAO (Food and Agricultural Organisation of the United Nations). 1994a Country tables. Rome: Food Agric. Organ: 332.

FAO (Food and Agricultural Organisation of the United Nations). (1994b). FAO quarterly bulletin of statistics 7 (2/3/4): 71-72.

FAO (Food and Agricultural Organisation of the United Nations) (1991). Country tables. Rome: Food Agric. Organ.

FAO (Food and Agricultural Organisation of the United Nations) (1954). Yearbook of food and agricultural statistics: Production 1954 vol. VIII, part 1. Rome: Food Agric. Organ: 205-207.

FAO (Food and Agricultural Organisation of the United Nations). (1952). Statistical yearbook 1952. Rome: Food Agric. Organ:264-266.

Fogel, R.W. (1994). The relevance of Malthus for the study of mortality today: Long-run influences on health, mortality, labour force participation and population growth. In: Lindahl-Kiessling, K.; Landberg, H., eds. *Population, Economic Development and the Environment*. Oxford, U.K.: Oxford Univ. Press; 231-284.

Goklany, I.M. (1998a). Saving habitat and conserving biodiversity on a crowded planet. *BioScience* **48**, 941-953.

Goklany, I.M. (1998b). Conserving habitat, feeding humanity. *Forum for Applied Research and Public Policy* **13**, 51-56.

Goklany, I.M. (1996). Factors affecting environmental impacts: The effects of technology on long term trends in cropland, air pollution and water-related diseases. *Ambio*. 25:

297-503.

Goklany, I.M. (1995a). Strategies to enhance adaptability: Technological change, sustainable growth and free trade. *Climatic Change* **30**, 427-449.

Goklany, I.M. (1995b.) Richer is cleaner: Long term trends in global air quality. In: Bailey, R., ed. *The True State of the Planet*. New York: Free Press.

Goklany, I.M. (1994). Air and inland surface water quality: Long term trends and relationship to affluence. Washington, DC: Off. Policy Anal., Dep. of the Interior.

Goklany, I.M. (1993). Climate change and natural resources: Is it too soon to start adapting? *Climate Change Newsletter* **5** (4): 3-6.

Goklany, I.M. (1992). Adaptation and climate change. Prepared for 1992 Annual Meeting of the American Association for the Advancement of Science, Chicago, February 6-11. Available from author, Off. Policy Anal., Dep. of the Interior, 1849 C. St., NW, Washington, DC 20240.

Goklany, I.M. et al. (1992). America's biodiversity strategy: Actions to conserve species and habitats. US Department of Agriculture and Department of the Interior. Washington, DC: Off. Policy Anal., Dep. of the Interior.

Goklany, I.M.; Sprague, M.W. (1995). Technological progress increases food production. In: Barbour, S., ed. *Hunger: Current Controversies*. San Diego, CA: Greenhaven Press: 118-125.

Goklany, I.M.; Sprague, M.W. (1991). An alternative approach to sustainable development. Conserving Forests, Habitat and Biological Diversity by Increasing the Efficiency and Productivity of Land Utilization. Washington, DC: Off. Programme Anal., US Dep. of the Interior.

Grossman, G., Krueger, A. (1991). Environmental impacts of a North American free trade agreement, discussion paper no. 158. Princeton, NJ: Woodrow Wilson School, Princeton Univ.

Gunnarson, J. et al. (1995). Interactions between eutrophication and contaminants: Toward a new research concept for the European aquatic environment. Ambio. 24 (6): 382-385.

Harris, J.M. (1996). World agricultural futures: regional sustainability and ecological limits. *Ecol. Econ.* **17** (2), 95-115.

Houghton, R.A. et al. (1983). Changes in the carbon content of terrestrial biota and soils between 1860 and 1980: A net release of CO_2 to the atmosphere. *Ecol. Monogr.* **53**, 235-262.

Kellogg, R.L., TeSelle, G.W., Goebel, J.J. (1994). Highlights from the 1992 national resources inventory. *J. Soil Water Conserv.* **49** (6), 521-527 .

Knutson, R.D.; Taylor, C.R.; Penson; J.B.; Smith, E.G. Economic impacts of reduced chemical use. College Station, TX:

K & Associates; 1990.

Kumar, B.G. (1990) Ethiopian famines 1973-1985, a case study. In: Drèze, J.; Sen, A., eds. *The Political Economy of Hunger*, Volume II. Oxford, U.K.: Clarendon Press, Oxford Univ. Press; 1990: 173-216.

Kuznets, S. Economic growth and income inequality. Am. Econ. Rev. 45: 1-28; 1955.

Livi-Bacci, M. (1992) *A Concise History of World Population*. English edition translated by C. Ipsen. Cambridge, MA: Blackwell; 220.

Maddison, A. (1989) The world economy in the 20th century. Paris: Organ. Econ. Coop. Dev.

Malthus, T.R. (1926) First essay on population, 1798. In: Bonar, J., ed. Reprinted for the Royal Economic Society. London: McMillan

McEvedy, C.; Jones, R. Atlas of world population history. New York: Penguin; 1978: 342.

McLaughlin, L. (1993) A case study in Dingxi County, Gansu Province, China. In: Pimentel, D., ed. World Soil Erosion and Conservation. Cambridge: Univ. Press; 87-107.

Mitchell, D.O.; Ingco, M.D. (1995) Global and regional food demand and supply prospects. Chapter 4. In: Islam, N., ed. Population and Food in the Early Twenty-First Century: Meeting Future Food Demand of an Increasing Population. Washington, DC: Int. Food Policy Res. Inst.

Mitchell, D.O.; Ingco, M.D. (1996) The world food outlook. Hunger Notes 19: 20-25; 1993.

NRC (National Research Council). (1996) Excess calories pose more of a cancer threat than natural or synthetic carcinogens on food. News from the NRC. Washington, DC: Natl. Res. Counc.; February 15.

NRCS (National Resources Conservation Service). (1995) Summary report: 1992 national resources inventory. Natural Resources Conservation Service. Ames, IA: Iowa State Univ. Statistical Lab.; 1995: 3, 19-21.

OTA (Office of Technology Assessment). (1996) A new technological era for American agriculture. Washington, DC: Off. Technol. Assess.; 1994: 452.

Oram, P.A.; Hojjati, B. (1995) The growth potential of existing agricultural technology. In: Islam, N., ed. Population and Food in the Early Twenty-First Century: Meeting Future Food Demand of an Increasing Population. Washington, DC: IFPRI; 1995: 167-89.

Pimentel, D. et al. (1995) Environmental and economic costs of soil erosion and conservation benefits. *Science* **267**, 1117-1123.

Pimentel, D. et al. (1994) Natural resources and an optimum human population. *Popul. Environ.* 15: 347-369.

Pimentel, D. et al. (1993) Environmental and economic impacts of reducing US agricultural pesticide use. In: Pimental, D.; Lehman, H., eds. *The Pesticide Question: Environment*, Economics and Ethics. New York: Chapman and Hall.

Plucknett, D.L. (1995) Prospects for meeting future food needs through new technology. In: Islam, N., ed. Population and Food in the Early Twenty-First Century: Meeting Future Food Demand of an Increasing Population. Washington, DC: Int. Food Policy Res. Inst.: 207-219.

Poleman, T.T.; Thomas, L.T. (1995) Report: Income and dietary change. *Food Policy* **20**, 149-159.

Postel, S.; Daily, G.C.; Ehrlich, P.R. (1996) Human appropriation of renewable fresh water. Science 271 (5250): 785-788.

Richards, J.F. (1990) Land transformation. In: Turner, B.L., II et al., eds. *The Earth as Transformed by Human Action: Global and Regional Changes in the Biosphere Over the Past 300 Years*. Cambridge: Cambridge Univ. Press; 164.

Rosegrant, M.W.; Schleyer, R.G.; Yadav, S.N. (1995) Water policy for efficient agricultural diversification: Market-based approaches. *Food Policy* **20**, 203-223.

Rozanov, B.; Targulian, V.; Orlov, D.S. Soils. (1990) In: Turner, B.L.,II et al., eds. *The Earth as Transformed by Human Action: Global and Regional Changes in the Biosphere Over the Past 300 Years*. Cambridge: Cambridge Univ. Press; 203-214.

Sen, A.K. (1981) Poverty and famine. An essay on entitlement and deprivation. Oxford, U.K.: Oxford Univ. Press.

Sen, A.K. (1993) The economics of life and death. *Sci. Am.* **268** (5), 40-47.

Shafik, N. and Bandyopadhyay, S. (1992) Economic growth and environmental quality: Time series and cross-country evidence. Policy Research Working Papers. Washington, DC: World Bank.

Smil, V. (1994) How many people can the earth feed? Popul. Dev. Rev. 20: 255-292.

Swaminathan, M.S. (1989) Agriculture and food industry in the 21st century. Presented at the Resource Use and Management Subgroup, Intergovernmental Panel on Climate Change, Geneva; October 30, 1989. Available from Intergovernmental Panel on Climate Change, Geneva, Switzerland.

Tapsoba, E.K. (1990) Food security policy issues in West Africa: Past lessons and future prospects. A critical review. FAO Economic and Social Development Paper 93. Rome: Food Agric. Organ.

Taylor, C.R. (1995) Economic impacts and environmental and food safety trade-offs of pesticide use reduction on fruit and vegetables. ES 95-1. Auburn, AL: Auburn Univ.

The Economist. Sunshine and showers. October 21, 1995: 84.

Turner, B.L., II et al., (1990) eds. The earth as transformed by human action: Global and regional changes in the biosphere over the past 300 years. Cambridge: Cambridge Univ. Press.

UNFPA (United Nations Population Fund) (1995) The state of the world population 1995. New York, NY: United Nations Popul. Fund 65.

USDA (US Department of Agriculture) (1996) Agricultural statistics 1995-96. US Department of Agriculture, National Agricultural Statistics Service. Washington, DC: US Gov. Print. Off.; IX-16.

USDA (US Department of Agriculture) (1989) The second RCA appraisal: Soil, water and related resources on non-federal lands in the United States, analysis of conditions and trends. Washington, DC: US Dep. Agric.; 7.

Vitousek, P.M.; Ehrlich, P.R.; Ehrlich, A.H.; Matson, P.A. (1986) Human appropriation of the products of photosynthesis. BioScience 36: 368-73.

Waggoner, P.E. (1994) How much land can ten billion people spare for nature? Ames, IA: Counc. Agric. Sci. Technol.; 26-27.

World Bank. (1996) The pink sheet. Commodity Price Data. June issue, available at http://www.worldbank.org/html/ieccp/pink.html.

World Bank. (1994) World population projections 1994-1995. Washington, D.C.: World Bank; 13.

World Bank. (1984) World development report 1984. New York, NY: Oxford Univ. Press.

WRI (World Resources Institute) (1996) World resources 1996-97. New York, NY: Oxford Univ. Press.

WRI (World Resources Institute) (1987) World resources 1987. New York, NY: Basic Books.

Wrigley, E.A.; Schonfeld, R.S. (1981) The population history of England 1541-1871: A reconstruction. Cambridge, MA: Harvard Univ. Press; 529.

Index